S0-DSE-427

Basics of AIR CONDITIONING

V. PAUL LANG

Third Edition

VNR
VAN NOSTRAND REINHOLD COMPANY
NEW YORK CINCINNATI TORONTO LONDON MELBOURNE

Library of Congress Catalog Card Number 78-24591
ISBN 0-442-23153-9

Printed in the United States of America.

Published in 1979 by Van Nostrand Reinhold Company
A division of Litton Educational Publishing, Inc.
135 West 50th Street, New York, NY 10020, U.S.A.

Van Nostrand Reinhold Limited
1410 Birchmount Road
Scarborough, Ontario M1P 2E7, Canada

Van Nostrand Reinhold Australia Pty. Ltd.
17 Queen Street
Mitcham, Victoria 3132, Australia

Van Nostrand Reinhold Company Limited
Molly Millars Lane
Wokingham, Berkshire, England

16 15 14 13 12 11 10 9 8 7 6 5 4 3

Library of Congress Cataloging in Publication Data

Lang, V Paul.
Basics of air conditioning.

Published in 1961 and 1972 under title: Principles of air conditioning.
Includes index.
1. Air conditioning. I. Title.
TH7687.L29 1979 697.9'3 78-24591
ISBN 0-442-23153-9

DEDICATION

This text is dedicated to my loving wife who examined the first several units of manuscript and blithely announced that the style of writing and technique of presentation were palatable; and to my children who suffered countless hours of "self-imposed" sound level restrictions.

PREFACE

The accelerated demand for air-conditioning applications has brought with it the need for greater numbers of practical, technical, and sales personnel who have a sound training in the basic principles of modern air conditioning.

The technical information presented in this book is intended to satisfy the immediate and fundamental needs of the reader in the field of air conditioning. This information will help prepare the reader to make a real and immediate contribution in entering this field.

It is not the intent of the author that this book be complete in all phases of air-conditioning practice and theory; rather, the material presented is intended to create a practical sense of good judgment and a level of air-conditioning knowledge that will permit the user to enter the more specialized areas of study in the field of air conditioning.

BASICS OF AIR CONDITIONING serves three main purposes:

- The content provides a broad background of the principles of science, refrigeration, and air conditioning which must be understood by apprentices, designers, and technicians. This technical information is fundamental to all types of domestic, commercial, and industrial systems. The planning, designing, construction, operation, and servicing of such systems depend on a complete understanding of this material.
- The instructional units can be considered as terminal for those who apply immediately the basic skills developed in the laboratory and shop with this background technical information.
- The instructional units can be considered as preparatory for those who plan to enter more advanced phases of refrigeration and air-conditioning work, including design and engineering.

The text contains seven sections representing major blocks, divisions, or areas of instruction. The important area of refrigeration was omitted (except for a brief review) because it is covered in detail in other books. Each section contains a number of instructional units. These units are arranged in the natural order of dependence of one principle, law, or set of conditions upon another. The material within a unit is organized in a pattern which is designed to help the reader see relationships.

A summary of the contents appears at the end of each unit. These statements emphasize important items for the trainee.

The appendix of the text contains tables to be used for reference. These tables include U-factor data for masonry and frame exterior walls, interior walls and partitions, ceilings and floors, and roofs. Also included in the Appendix are troubleshooting charts for mechanical and electrical service problems.

The back cover envelope contains materials which are designed to be studied with specified units. These materials and the instructional units to which they apply are as follows:

- Psychrometric Chart (Units 6, 7, 8, 9, 16, and 17)
- ARI Commercial Cooling Load Estimate Form (Unit 12)
- ARI Residential Cooling Load Estimate Form (Unit 12)
- Plans of a Seven-room Residence (Unit 16)
- Completed Commercial Cooling Load Estimate for a Restaurant (Unit 17)

V. Paul Lang has been associated with the air-conditioning industry for many years. Formerly he was a member of the Carrier Air Conditioning Company Application Engineering Department, and prior to that, was employed by the Minneapolis Honeywell Company. Presently, Mr. Lang is manager of the technical writing services at Carrier. Mr. Lang received his undergraduate degree at St. John's University, Collegeville, Minnesota and completed graduate master's credits at the University of Minnesota.

ACKNOWLEDGMENTS

It should be recognized that this book could not have been written without the generous contributions of training literature and product literature from numerous corporations and companies in the air-conditioning and related industries. The standards' manuals and guides provided by the various societies and associations concerned with air conditioning were of invaluable assistance. A portion of the information submitted by these sources was used in several of the units.

Appreciation is expressed to the following for permission to use illustrations and data specified:

- to Carrier Corporation for use of the halftone illustrations in Units 1 through 11, and for use of their psychrometric chart in the back cover envelope.
- to the Trane Company for permission to use their psychrometric chart in the back cover envelope.
- to the Air Conditioning and Refrigeration Institute for permission to reproduce the standard tables in Unit 12, and the estimating forms in the back cover envelope.
- to the National Environmental Systems Contractors Association for permission to adapt their technical data and illustrations in Unit 18.
- to the American Society of Heating, Refrigeration and Air-Conditioning Engineers for permission to use excerpted tables from the ASHRAE Guide (see the Appendix), and friction loss charts in Unit 15.
- to Minneapolis Honeywell Regulator Company for permission to use the data and circuits on electrical controls in Unit 23.
- to McGraw-Hill Book Company for permission to use line drawings from their publication, *Residential and Commercial Air Conditioning,* by C.H. Burkhardt.
- to Windsor Press for permission to use line drawings from their publication, *Summer Air Conditioning,* by Konzo, Carrol, Bareither.
- to Lennox Industries, Inc. for the troubleshooting charts in the Appendix.

The following companies contributed generously by supplying both illustrations and technical data:

Admiral Corporation
Chicago, Illinois

Arkla Air Conditioning Corporation
Little Rock, Arkansas

B-I-F Industries
Providence, Rhode Island

Bryant, DayNite, Payne Co.
Indianapolis, Indiana

Carrier Air Conditioning Company
Division of Carrier Corporation
Syracuse, New York

Chrysler Corporation
Airtemp Division
Detroit, Michigan

Dunham Bush, Inc.
West Hartford, Connecticut

General Controls Company
Glendale, California

General Electric Company
Air Conditioning Division
Tyler, Texas

General Motors Corporation
Frigidaire Division
Dayton, Ohio

Johnson Service Company
Milwaukee, Wisconsin

Lennox Industries, Inc.
Marshalltown, Iowa

Lima Register Company
Lima, Ohio

The Marley Company
Kansas City, Missouri

Minneapolis Honeywell Regulator Co.
Minneapolis, Minnesota

Powers Regulator Company
Skokie, Illinois

Reynolds Metals Company
Richmond, Virginia

Sporlan Valve Company
St. Louis, Missouri

The Trane Company
LaCrosse, Wisconsin

Tuttle and Bailey Division
Allied Thermal Corporation
New Britain, Connecticut

York Division of Borg-Warner Corporation
York, Pennsylvania

Westinghouse Electric Corporation
Air Conditioning Division
Staunton, Virginia

CONTENTS

SECTION 5 APPLIED LOAD ESTIMATING

SECTION 6 RESIDENTIAL AND COMMERCIAL EQUIPMENT

SECTION 7 RESIDENTIAL AND COMMERCIAL CONTROLS

SECTION 8 BALANCING THE SYSTEM

SECTION 9 THE METRIC SYSTEM IN AIR CONDITIONING

SECTION 1
INTRODUCTION

UNIT 1
INTRODUCTION TO AIR CONDITIONING

OBJECTIVES

After completing the study of this unit, the student will be able to

- recall early and present day uses of air conditioning.
- define air conditioning,
- identify the contributions air conditioning has made to progress.

EARLY USES

As early as 1500 A.D., Leonardo da Vinci built a water-driven fan to ventilate a suite of rooms for the wife of his patron. This was possibly the first attempt to provide an automatic way of changing the condition of the air in an enclosed space. Another such device was the punka which originated in India many years ago. The punka was a large fan which extended from the ceiling. It was operated manually by pulling a rope. Some of the later models were machine operated.

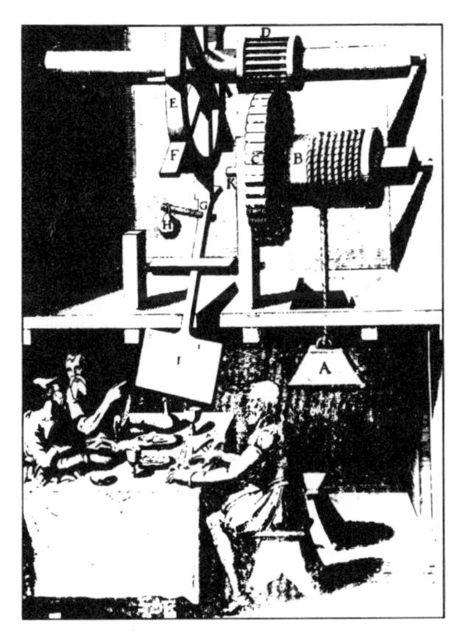

Fig. 1-1 Fan, wheel, gear combination

Fig. 1-2 Arm and leg movement operates series of fins

Figures 1-1 through 1-4 show early attempts at automatic ventilation. Although these may appear ludicrous today, they represent the progress in our attempt to control the surrounding air.

Fig. 1-3 Clock mechanism moves fan device

PRESENT DAY USES

Before 1922 conditioned air was used to produce items such as candy, gum, cheese, and matches. During 1922 the first comfort installation was made in a theater. This installation consisted of a central station spray-type, downdraft, bypass system. Since that time, almost every major type of building has been air conditioned, from giant skyscrapers to small homes.

DEFINITION OF AIR CONDITIONING

The basic concepts of air conditioning are not understood or even thought about by countless millions who enjoy the comfort produced by it. Yet, air conditioning is a readily accepted part of modern life. Because of this, air conditioning requires a definition:

> Air conditioning is defined as a process which heats, cools, cleans, and circulates air, as well as controlling the moisture content of air. Ideally, air conditioning does all of the tasks at the same time and on a year-round basis.

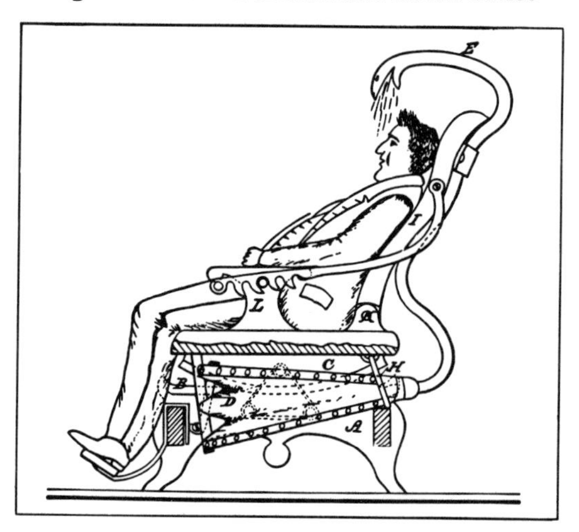

Fig. 1-4 Rocking motion operates bellows

Thus, air conditioning makes it possible to change the condition of the air in an enclosed area. Since most people spend a large portion of their lives in enclosed areas, air conditioning is actually more important and produces more benefit than is realized by a majority of people.

CONTRIBUTION TO 20TH-CENTURY PROGRESS

The discovery of the principles of air conditioning is one of the most important events in the twentieth century. Human beings work harder and more efficiently, play longer, and enjoy more leisure in comfort because of air conditioning. Since the first scientific air-conditioning system was used in a printing house over half a century ago, the scientific achievement and uses of air-conditioning principles have been outstanding. For example,

- Military centers which can track and intercept hostile missiles are able to operate around-the-clock only because the air is at a controlled temperature. Without air conditioning, the computers quickly cease to operate because of the intense self-generated heat.
- Atomic submarines can remain submerged almost indefinitely due, in part, to air conditioning.
- Modern medicines such as the Salk and Sabin vaccines are prepared in a scientifically controlled atmosphere.
- The exploration of space is aided by air conditioning.

The uses of air conditioning show that when new products are made, or when new discoveries take place, and certainly in the exploration of space, air conditioning plays a key role.

UNIT 2
BODY COMFORT

OBJECTIVES

After completing the study of this unit, the student will be able to

- describe the processes by which the body produces and rejects heat.
- discuss the conditions – temperature, humidity, and air movement – which affect body heat.

The normal temperature of the human body is 98.6° Fahrenheit (or 37° Celsius in the metric system). This temperature is sometimes called *subsurface* or *deep tissue temperature* rather than skin or surface temperature. An understanding of the way the body maintains this temperature helps in understanding the way the air-conditioning process helps keep the body comfortable.

THE BODY PRODUCES HEAT

All food taken into the body contains heat in the form of calories. The large or great calorie is used to express the heat value of food. The *large calorie* is the amount of heat required to raise the temperature of one kilogram of water one degree Celsius (in the metric system). As the calories (food) are taken into the body, they are converted into energy. This energy is then stored for future use. The conversion process generates heat. All body movements also add to the heat generated by the conversion process and use up the stored energy.

For body comfort, all of the heat produced must be given off by the body. Since the body produces more heat than it needs, heat must be constantly given off or removed.

THE BODY REJECTS HEAT

The constant removal of body heat takes place through three natural processes which usually occur at the same time. These processes are:

- convection
- radiation
- evaporation

Convection

The convection process of removing heat is based on two phenomena:

- Heat flows from a hot surface to a cold surface. For example, heat flows from the body to the air surrounding the body. This surrounding air is below the body skin temperature.
- Heat rises. This action can be seen by watching the smoke from a burning cigarette.

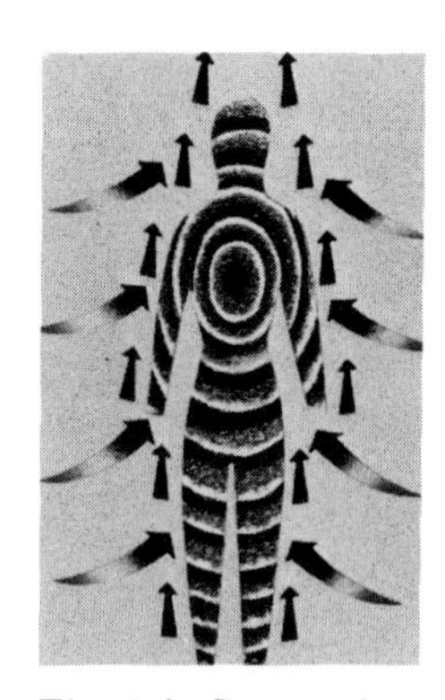

Fig. 2-1 Convection

When these two phenomena are applied to the body process of removing heat, the following happens:

- The body gives off heat to the cool surrounding air.
- The surrounding air becomes warm and moves upward.
- As the warm air moves upward, more cool air takes its place, and the convection cycle is completed.

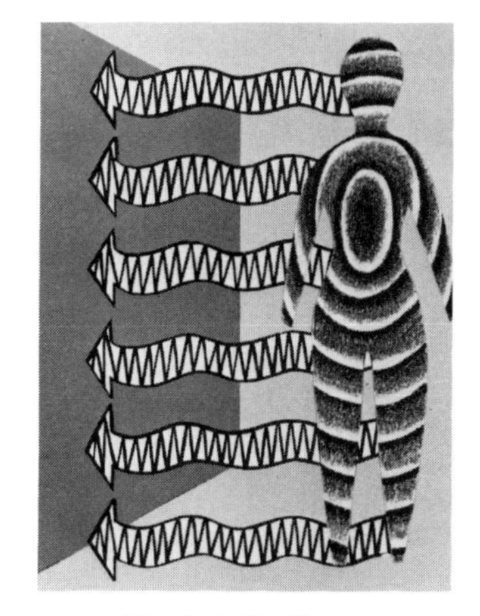

Fig. 2-2 Radiation

Radiation

Radiation is the process by which heat moves from a heat source (such as the sun or fire) to an object by means of heat rays. This principle is based on the phenomena that heat moves from a hot surface to a cold surface. Radiation takes place independently from convection, and does not require air movement to complete the heat transfer. Radiation also is not affected by air temperature; but, it is affected by the temperature of surrounding surfaces.

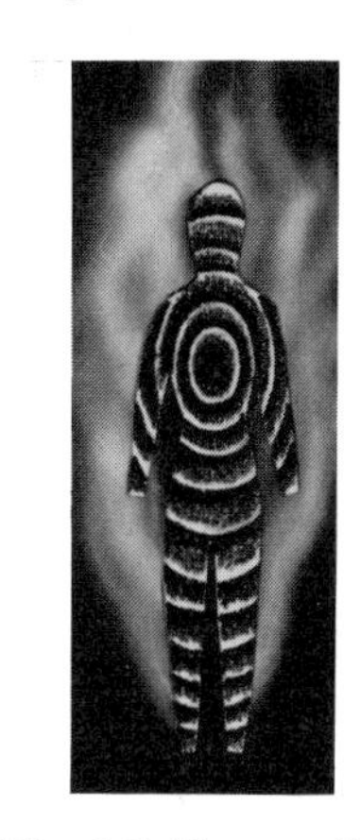

Fig. 2-3 Evaporation

The body quickly experiences the effects of sun radiation when one moves from a shady area to a sunny area. Radiation effects are also experienced when the body surface closest to a fire becomes warm while the opposite body surface remains cool. Just as the heat from the sun and the fire moves by radiation to a colder surface, the heat from the body moves to a colder surface.

Evaporation

Evaporation is the process by which moisture becomes vapor. As moisture evaporates from a warm surface, heat is removed and the surface is cooled. This process takes place constantly on the surface of the body. Moisture is given off through the pores of the skin. As the moisture evaporates from the skin, it removes heat from the body.

The body can produce more heat than can be removed by these processes. This excess of heat is indicated by perspiration that appears as drops of moisture on the body.

CONDITIONS THAT AFFECT BODY HEAT

Temperature

- Cool air increases the rate of convection; warm air slows it down.
- Cool air lowers the temperature of surrounding surfaces; therefore, cool air increases the rate of radiation. Warm air increases the surrounding surface temperature; therefore, the radiation rate is decreased.

- Cool air increases the rate of evaporation and warm air slows it down. This process depends on the amount of moisture already in the air and the amount of air movement.

Humidity

Moisture in the air is measured in terms of humidity. For example, 50 percent relative humidity means that the air contains one-half the amount of moisture that it is capable of holding. To simplify the measurement of humidity, a unit called *a grain of water vapor* is used. A grain is so small that there are approximately 2,800 grains in one cup of water and 7,000 grains in one pound of water.

As an example of how humidity is measured, consider the following situation. Assume that the room shown in figure 2-4 has a temperature of 70°F. Also assume that there are four grains of water vapor for each cubic foot of space.

If the room temperature remains at 70°F and water vapor is added to the air, a point is eventually reached where the air in the room can absorb no more water. At this point, the air is saturated and one cubic foot of room space now holds eight grains of water vapor. A concentration of eight grains per cubic foot, at 70°F, represents 100% relative humidity. The original room condition of four grains per cubic foot at 70°F represents 50% relative humidity (4 grains ÷ 8 grains = 0.50 or 50%).

> Relative humidity is obtained by dividing the actual number of grains of moisture in a cubic foot of room air, at a given temperature, by the maximum number of grains that the cubic foot of air can hold when it is saturated.

The relative humidity changes as the air temperature changes. For example, at 80°F, a cubic foot of air can hold a maximum of 11 grains of water vapor instead of eight. If a cubic foot of air holds four grains of moisture and the temperature is 80°F, then the relative humidity is: 4 ÷ 11 = 0.37 or 37%, figure 2-5.

However, if the air temperature is 70°F and the actual moisture content of the air is decreased from 4 grains to 3 grains per cubic foot, then the relative humidity is: 3 ÷ 8 or 37%, figure 2-5.

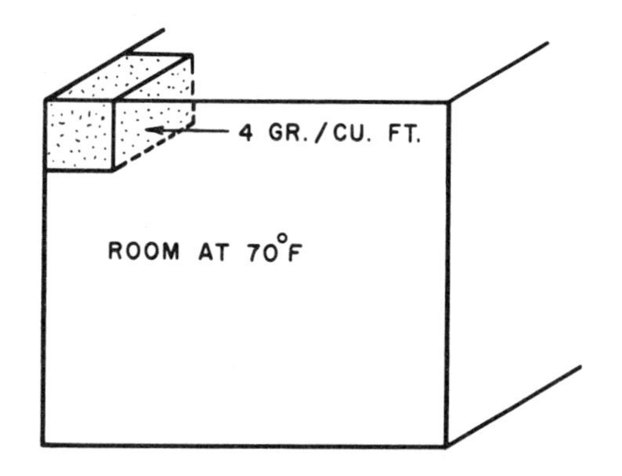

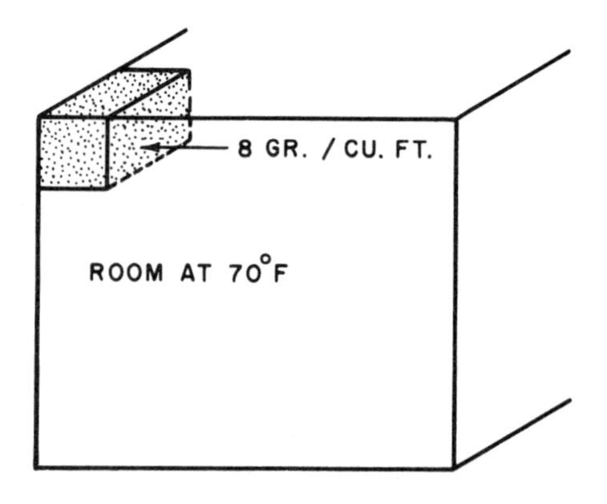

Fig. 2-4

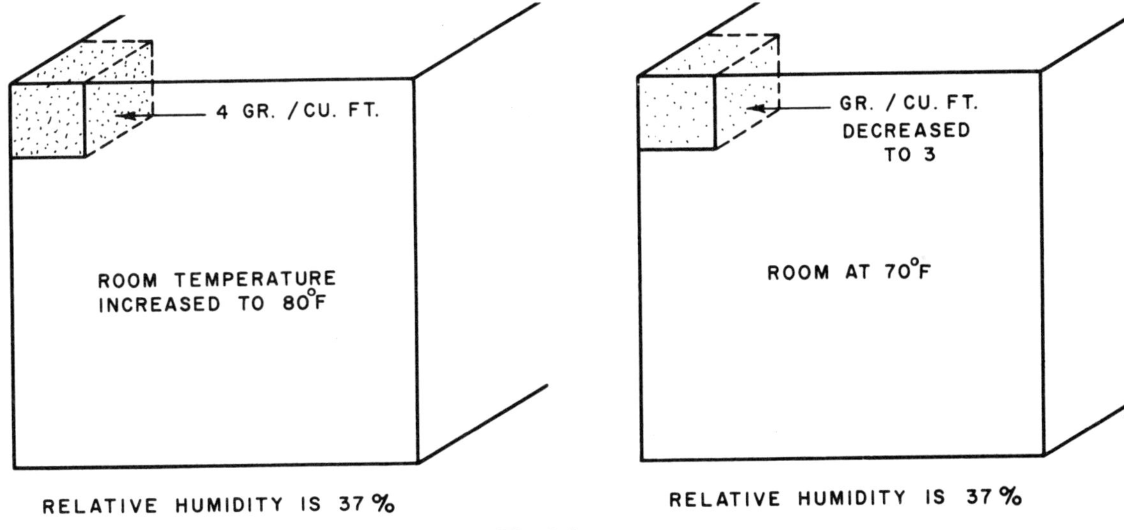

Fig. 2-5

The ways in which the relative humidity can be changed are summarized as follows:

- To increase relative humidity, either the actual moisture content of the air is increased or the air temperature is decreased.
- To decrease relative humidity, either the actual moisture content of the air is decreased or the air temperature is increased.

A low relative humidity permits heat to be given off from the body by evaporation. In other words, because the air at low humidity is relatively dry, it can readily absorb moisture. A high relative humidity has the opposite effect: the evaporation process is retarded. Thus, the speed at which heat can be removed from the body by evaporation is decreased. An acceptable comfort range for the human body is 72° to 80°F at 45% to 50% relative humidity.

Air Movement

Another factor which affects the ability of the body to give off heat is the movement of air around the body. As air movement increases:

- The rate of the evaporation process of removing body heat speeds up since moisture in the air near the body is carried away at a faster rate.
- The convection process increases since the layer of warm air surrounding the body is carried away more rapidly.
- The radiation process tends to speed up because the heat on the surrounding surfaces is removed at a faster rate; thus, heat radiates from the body at a faster rate.

As air movement decreases, the evaporation, convection, and radiation processes decrease.

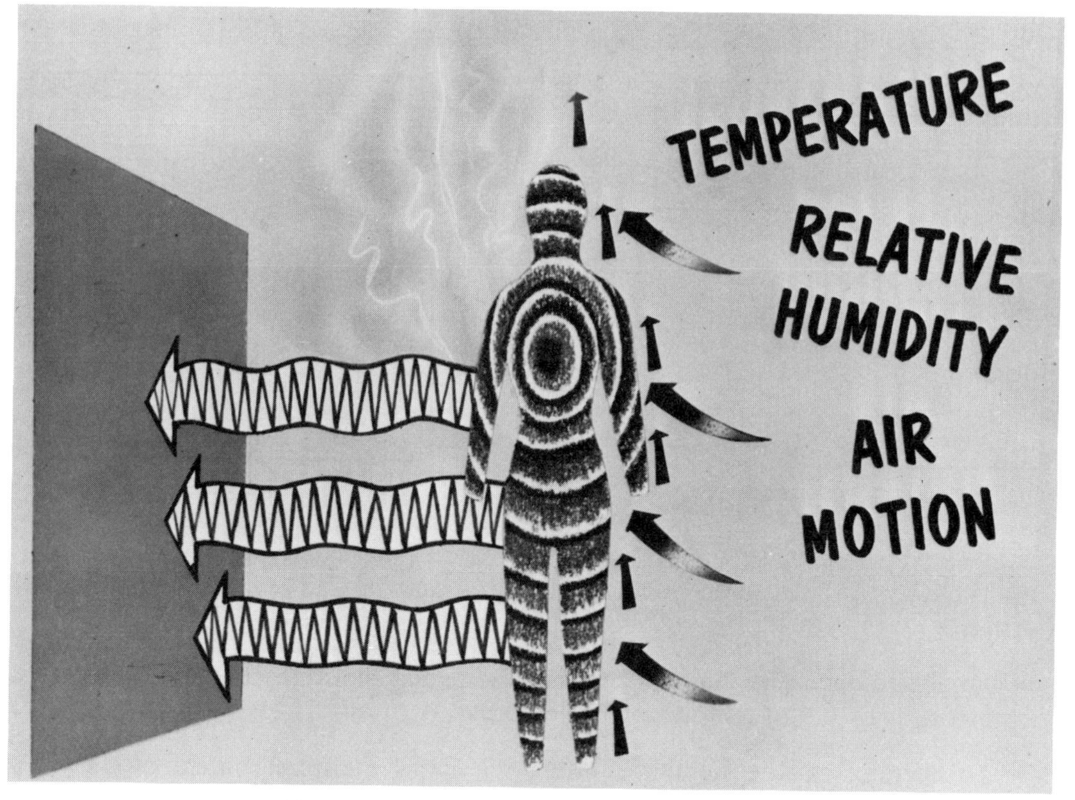

Fig. 2-6 Conditions which affect body comfort

SUMMARY

- Normal body temperature is 98.6°F (37°C).
- The body constantly generates heat by converting food to energy and by body movement.
- The body produces more heat than it needs.
- The body continuously gives off heat by convection, radiation, and evaporation.
- Convection is based on the phenomena that heat flows to colder surfaces and that heat rises. An increase in air movement speeds up the convection process on the body surface.
- Radiation is the process by which heat flows from a hot surface to a cold surface by means of rays, such as the sun's rays. The heat flow is not affected by temperature.
- Evaporation is the process by which moisture evaporates from a warm surface. As the moisture evaporates, the surface is cooled. Moisture is continuously evaporating from the surface of the body. An increase in temperature slows down the evaporation process. An increase in air movement speeds up the evaporation process. A high relative humidity decreases the evaporation process.
- Relative humidity is the actual amount of moisture in the air as compared with the maximum amount of moisture the air can hold.

- A grain of water vapor is the unit of measurement used to determine the relative humidity in percent.
- One pound of water contains 7,000 grains of water vapor.
- The comfort range of the human body is 72° to 80°F at 45% to 50% relative humidity.
- Body comfort is affected by temperature, relative humidity, and air movement.

REVIEW

Select the one best answer for each of the following questions.

1. Under normal conditions, the heat which the body produces

   a. equals the heat the body requires.
   b. is more than the body needs.
   c. is less than the body needs.
   d. is equal to 2000 Btuh.

2. The direction of heat flow is

   a. from a cold to a warm surface.
   b. from a warm to a cold surface.
   c. from a warm to a warm surface.
   d. always to the north.

3. The body cannot lose heat by convection if the blood temperature is

   a. colder than the surrounding air temperature.
   b. warmer than the surrounding air temperature.
   c. 96 F.
   d. 102 F.

4. Radiation depends upon one of the following for heat transfer:

   a. temperature.
   b. humidity.
   c. heat flow from a cold to a hot surface.
   d. heat rays.

5. The evaporation process cools because

   a. moisture, in becoming a vapor, absorbs heat.
   b. air moves.
   c. it creates reverse airflow.
   d. moisture, in becoming a vapor, gives up heat.

6. Humidity is a measure of air

   a. movement.
   b. speed.
   c. moisture content.
   d. pollution level.

7. Humidity is measured in

   a. grains of water vapor.
   b. grains of sand per hour.
   c. tons of moisture.
   d. grains of ice.

8. If 8 gr./cu. ft. at 78 F represents saturated air, then 2 gr./cu. ft. indicates

   a. 10% relative humidity.
   b. 25% relative humidity.
   c. 40% relative humidity.
   d. 60% relative humidity.

9. At 78 F, 6 gr./cu. ft. represents

   a. wind speed.
   b. 40% relative humidity.
   c. 90% relative humidity.
   d. 100% relative humidity.

10. Air temperature, humidity, and air movement directly affect

    a. wind speed.
    b. body comfort.
    c. rainfall.
    d. air direction.

Select the word or phrase which best completes each statement.

11. Cool air (increases) (decreases) (has no effect on) the rate at which the body loses heat.

12. Warm air (increases) (decreases) (has no effect on) the rate of radiation of heat from the body.

13. Cool air (increases) (decreases) (has no effect on) evaporation of heat from the body.

14. An increase in air movement (increases) (decreases) (has no effect on) the evaporation, convection, and radiation processes.

15. High relative humidity (increases) (decreases) (has no effect on) the evaporative process.

UNIT 3
AIR CYCLE

OBJECTIVES

After completing the study of this unit, the student will be able to

- list the methods of conditioning air.
- describe the air-conditioning air cycle.
- identify the purpose and function of each piece of equipment in the air-conditioning cycle.

Unit 2 showed how the body is able to remain comfortable if the air temperature, relative humidity (moisture), and air movement are within favorable limits. Since there are few days in the year during which all three conditions are ideal, it is necessary for human beings to adjust to maintain even a minimum of comfort:

- When conditions are uncomfortably warm, less clothing is worn.
- When conditions are uncomfortably cool, more clothing is worn.

For most people, these solutions are, at times, unsatisfactory. Although a *practical* method of maintaining air comfort conditions out-of-doors has not yet been achieved, the problem of conditioning indoor air has been solved. This unit describes how air is conditioned indoors and how it is then delivered where it is needed.

TYPICAL AIR CYCLE

Indoor air can be too cold, too hot, too wet, too dry, too drafty, and too still. These conditions are changed by treating the air. Cold air is heated, hot air is cooled, moisture is added to dry air, moisture is removed from damp air, and fans are used to create and maintain an adequate air movement. Each of these air treatments is accomplished in the air-conditioning air cycle.

AIR-CONDITIONING AIR CYCLE

The cycle described in this section is typical in the sense that it contains the basic air-conditioning system parts that are required to supply and remove air from a room. During this process, the air is conditioned as needed, figure 3-1, page 12.

The cycle begins when the fan forces air into ductwork leading to openings in the room. These openings are called *outlets* or *terminals.* The air is directed from the ductwork through these outlets and into the room where the air either heats or cools the room as needed. Dust particles in the room enter the air stream and are carried along with it.

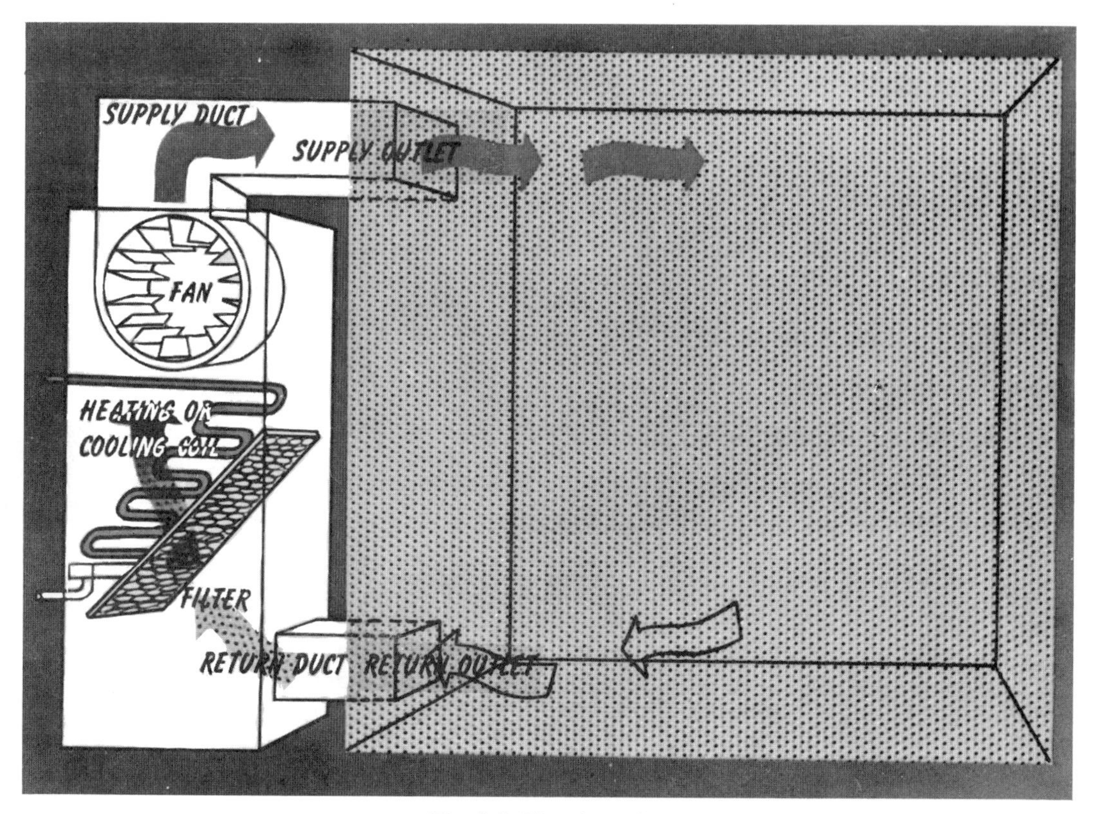

Fig. 3-1 The air cycle

The air then flows from the room through a second outlet or return outlet. Dust particles are removed from the air by a filter installed in the return ductwork. After the air is cleaned, it is either heated or cooled depending on the requirements of the room. If cooling is required, the air is passed over the surface of a cooling coil. If heat is required, the air is passed through a combustion chamber or over the surface of a heating coil. Finally, the air is returned to the fan and the cycle is completed.

AIR-CONDITIONING CYCLE EQUIPMENT

The major pieces of equipment required to complete the air-conditioning cycle are listed as follows:

- fan
- supply ducts
- supply outlets
- space to be conditioned
- return outlets
- return ducts
- filter
- heating coil (combustion chamber) or cooling coil

The purpose and function of each of these components are covered in the following sections.

The Fan

The fan moves air to and from an enclosed space. In an air-conditioning system, the fan moves air that consists of:

- all outdoor air
- all indoor or room air (this is also called recirculated air)
- a combination of outdoor and indoor air

The fan pulls air from the outdoors and from the room at the same time.

Since drafts in the room cause discomfort, and poor air movement slows the body heat rejection process, it is necessary to regulate the amount of air supplied by the fan. To accomplish this regulation, a fan is selected that can deliver the correct amount of air. By controlling the speed of the fan, the air stream in the room can be regulated to provide good circulation without drafts.

The Supply Duct

The supply duct directs the air from the fan to the room. A typical duct arrangement is shown in figure 3-2. The supply duct should be as short as possible and have a minimum number of turns to insure that the air can flow freely.

Supply Outlets

Supply outlets help to distribute the air evenly in a room. Some outlets fan the air and other outlets direct the air in a jetstream. Still other outlets combine these actions. As a result of these actions, the outlets are able to exert some control on the direction of the air delivered by the fan. This directional control plus the location and the number of outlets in the room contribute greatly to the comfort or discomfort resulting from the air pattern.

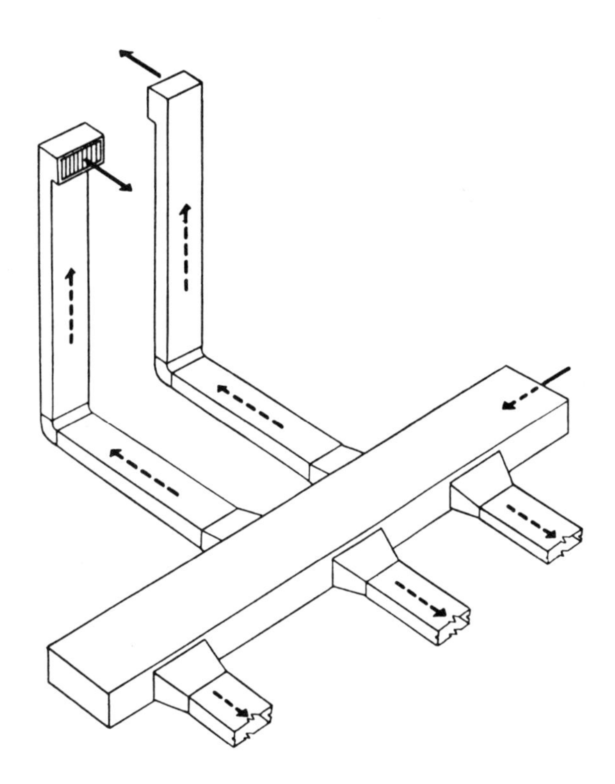

Fig. 3-2 Duct system

Room Space

The room or the space to be conditioned is one of the most important parts of the air cycle. The dictionary states that a room is an enclosed space set apart by partitions. If an enclosed space does not exist, then it is impossible to complete the air cycle. This is due to the fact that the conditioned air from the supply outlets simply flows into the atmosphere. In fact, the material and the quality of workmanship used to enclose the space are also important since these factors help to control the loss of heat or cold that is confined in the enclosed space. Note: the materials used to enclose rooms and the effect of these materials are described in detail in units 10 and 11 in the sections "Reducing the Load."

Return Outlets

As stated previously, return outlets allow room air to enter the return duct. The main function of the return outlet is to allow air to pass from the room. These outlets are usually located on the opposite wall from the supply outlet. For example, if the supply duct is on the ceiling, or on the wall near the ceiling, then the return duct may be located on the floor or on the wall near the floor. This situation is not true in all installations, however. Some systems are provided with both the supply and the return near the floor or near the ceiling. (See unit 14).

Filters

Filters clean the air by removing dust and dirt particles. Filters are located within the return air duct. These devices are made of many materials including spun glass and composition plastic. Other filter materials maintain an electrostatic charge, and attract and capture dust and dirt particles from the air flowing through them.

The Cooling Coil and Heating Coil or Combustion Chamber

The cooling coil and the heating coil, or combustion chamber, are located either ahead of or after the fan. In all installations, however, these devices are placed after the filter. Such an arrangement prevents excessive dust and dirt particles from covering the coil surface.

Winter Operation (Heating Coil or Combustion Chamber). During winter operation, the air-conditioning cycle adds heat to the air. The return air from the room is passed over the surface of a heating coil or over the surface of a combustion chamber. The air is heated to the required temperature. It is then delivered to the room through the supply duct. The air loses its heat to the room and passes through the return duct to the coil or chamber. The cycle is repeated as long as the heated air is required. If the room air is too dry, moisture can be added by installing shallow pans in the bonnet above the combustion chamber or in the ductwork after the coil. The pans are automatically filled with water to a preset level. Thus, moisture is added to the air by the process of evaporation as the air passes over the pans.

Summer Operation (Cooling Coil). For summer operation, the air-conditioning cycle cools the air. The return air from the room passes over the surface of the cooling coil. The air is cooled to the required temperature. If there is too much moisture in the room air (high humidity), it is removed automatically as the air is cooled by the coil.

The following example illustrates the cooling and moisture removal process.

EXAMPLE: COOLING AND MOISTURE REMOVAL

If air leaves a room at 78°F and holds approximately 5 1/2 grains of moisture per cubic foot, its relative humidity is 50%. (Air at 78°F can hold about 11 grains per cubic foot. Therefore, $5.5 \div 11 = 50\%$ RH). As the air passes over the cooling coil, it is cooled to 50°F. At this temperature, the air can hold only 4 grains of moisture per cubic foot. Since the air enters the coil with 5 1/2 grains and leaves the coil with only 4 grains, 1 1/2 grains of moisture are removed in the cooling process.

The temperature of the supply air entering the room is now 50°F. The moisture content is 4 grains per cubic foot. Although 4 grains is equivalent to 100% relative humidity at 50°F, the supply air is actually dryer by 1 1/2 grains than the 78°F air leaving the room. As the supply air mixes with the room air, it removes heat and moisture, because it is colder and dryer. As a result, comfort conditions are maintained. If the cooling coil is sized properly, it provides enough cool air and removes enough moisture to offset (or balance) the heat and moisture that is constantly entering the room from a variety of sources such as the body, lights, motors, cooking, and outside air.

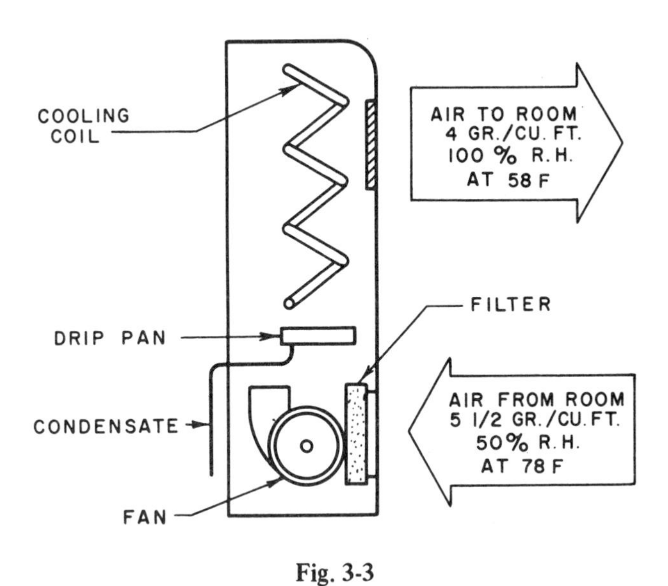

Fig. 3-3

SUMMARY

- In the absence of air conditioning, humans must adjust their manner of dress in an attempt to balance the extremes of weather conditions.
- As the air passes through the air-conditioning cycle, it can be heated, cooled, dried, moistened, and directed into a specific air pattern.
- The main pieces of equipment in the air-conditioning cycle are: the fan, supply duct, supply outlet, rcom, return duct, filter, and heating coil or combustion chamber, or cooling coil.
  1. The fan moves the air.
  2. The supply duct guides the air to the room outlet.
  3. The room outlet directs the air flow in the room.
  4. The room provides an enclosed space in which the air can be confined.
  5. The return outlets allow air to pass from the room.
  6. The return duct guides the air back to the fan and usually contains the filter and the heating or cooling device.

7. The filter cleans the air by removing dirt, dust, and dirt particles.
8. The heating coil or combustion chamber heats the air (for winter operation).
9. The cooling coil cools and dries the air (for summer operation).

REVIEW

Select the one best answer for each of the following questions.

1. Indoor air conditions can be changed
   a. by treating the air supply.
   b. by changing filters.
   c. by changing clothes.
   d. by putting on more clothes.
2. The design of the air supply duct
   a. affects the distribution of the air.
   b. does not affect air distribution.
   c. adds moisture to the air.
   d. lowers the temperature of the air.
3. The filter should always be located:
   a. in the attic.
   b. in the supply duct.
   c. in the return duct.
   d. in the garage.
4. Duct outlets
   a. may be located near the top or the bottom of the room to be air conditioned.
   b. should never be round.
   c. should never be rectangular.
   d. are never located in the room.
5. Moisture in the conditioned air is controlled during the cooling process by
   a. a humidifier in the system or in the room.
   b. opening the windows.
   c. closing the windows.
   d. setting out pans of water.

Match the items in Column II with the items in Column I by inserting the correct letter in the space provided.

Column I	Column II
______ 6. Supply ducts	a. Source of air movement
______ 7. Return outlets	b. Treats air by adding heat
______ 8. Filter	c. Pathway through which air flows to the room
______ 9. Cooling coil	d. Allows room air to be removed
______ 10. Combustion chamber	e. Cleans the air
______ 11. Fan	f. Treats air by removing heat
______ 12. Return ducts	g. Pathway through which air flows from the room
______ 13. Supply outlets	h. Distributes air evenly within the room

14. Beginning with the fan, arrange the following items in the proper sequence to represent the winter air-conditioning cycle: supply ducts, return outlets, filter, cooling coil, combustion chamber, fan, return ducts, supply outlets.

Supply the word or phrase which best completes each of the following statements.

15. The fan should supply adequate air circulation without causing ____________ .
16. Supply ducts should be __________ as possible and contain __________ turns.
17. By their design, supply outlets may either ______ or ______ the air stream.
18. The space to be conditioned must be ______________________________ .
19. The location generally used for return outlets is __________ that used for supply outlets.
20. Filters are located at some point in the __________ air duct.
21. Filters which trap dust and dirt particles by the attraction of unlike electrical charges are called __________ filters.
22. The cooling coil should always be located after the ____________ .
23. The cooling coil __________ and __________ the air.
24. Moisture is added when required by the process of ____________ .

UNIT 4
REFRIGERATION CYCLE

OBJECTIVES

After completing the study of this unit, the student will be able to

- list the components of the refrigeration cycle.
- describe the purpose and function of each of the components in the refrigeration cycle.

Unit 3 described the basic parts of the air cycle of an air-conditioning system. At one point in the system, the air is cooled as it passes over the surface of a cooling coil. If heating is required, the air is passed over the surface of a heating coil or combustion chamber.

This unit covers the refrigeration cycle in detail. It will be seen that the cooling coil is a basic part of both the air cycle and the refrigeration cycle. In fact, the cooling coil is the key part of both cycles because it is at this point that heat is removed from the air and transferred through the coil walls to the refrigerant flowing inside the coil.

The refrigeration cycle is concerned with the heat after it is removed from the air by the refrigerant in the coil. This cycle is based on the following two principles:

- As a liquid changes to a vapor, it absorbs large quantities of heat.
- The boiling point of a liquid can be changed by changing the pressure exerted on the liquid. In other words, the boiling point of a liquid can be raised by increasing its pressure, and it can be lowered by reducing its pressure.

The most commonly used refrigerants boil at very low temperatures, some as low as -21°F.

TYPICAL REFRIGERATION CYCLE

The principle parts of a refrigeration cycle are the cooling coil or evaporator, the compressor, the condenser, and the expansion valve, figure 4-1.

As the cold refrigerant liquid moves through the cooling coil, it picks up heat from the air passing over the surface of the coil. When the liquid refrigerant picks up enough heat, it changes to a vapor. The heated refrigerant vapor is then drawn into the compressor where it is subjected to a higher pressure. As a result of being pressurized, the temperature of the vapor increases.

The refrigerant vapor is now at a high pressure and a high temperature. In this state, vapor passes to the condenser where heat is removed and the vapor changes back to a liquid; however, this liquid is still under pressure.

The liquid then flows to the expansion valve. As the liquid passes through the valve, its pressure is reduced immediately. The pressure decrease lowers the temperature of the liquid even more and it is now ready to pick up more heat.

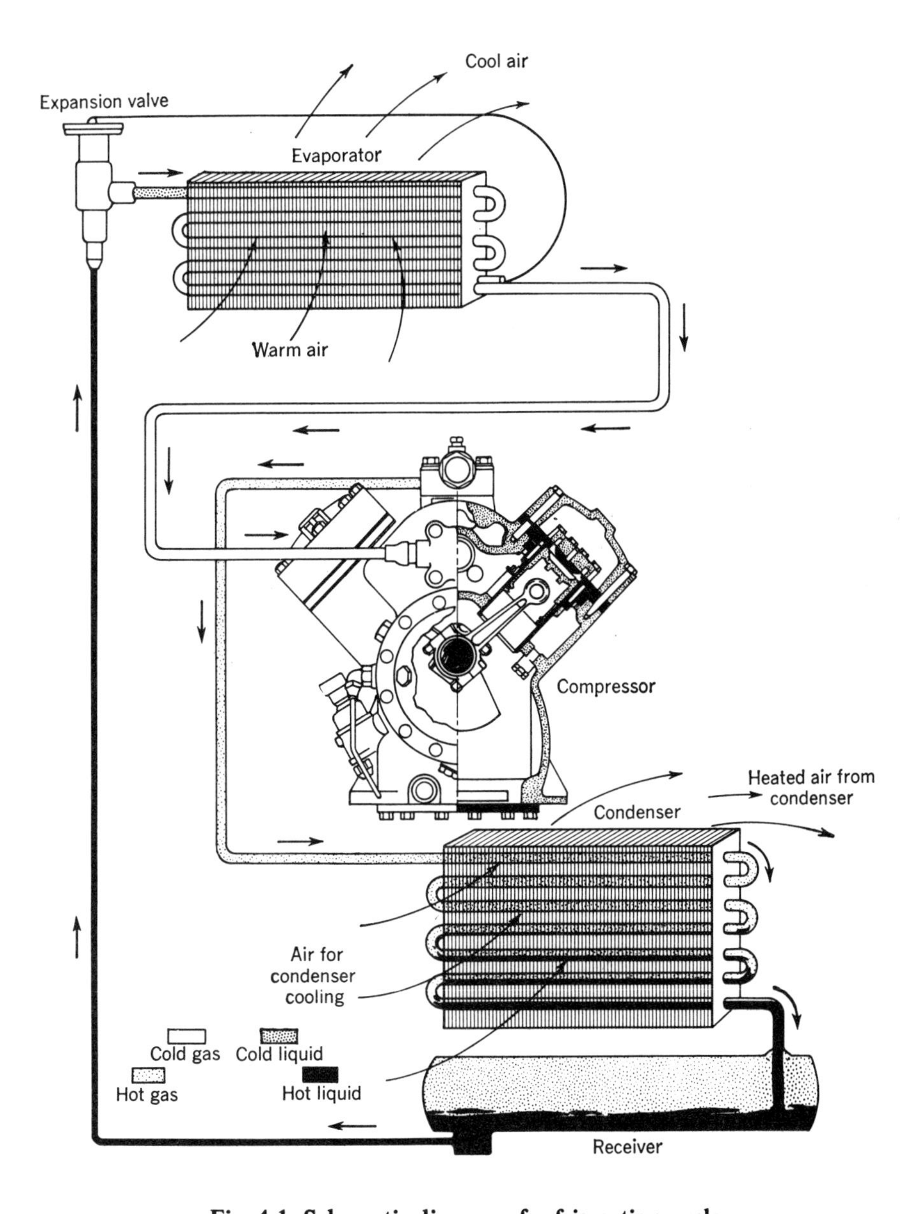

Fig. 4-1 Schematic diagram of refrigeration cycle

The cold, low-pressure liquid next flows into the cooling coil. The pressure in the cooling coil is low enough to allow the refrigerant to boil and vaporize as it again absorbs heat from the air passing over the coil surface. The cycle then repeats itself as the heated refrigerant vapor is drawn into the compressor.

COMPONENTS IN REFRIGERATION CYCLE

The following paragraphs describe briefly the purpose and function of each of the components of the refrigeration cycle as shown in figure 4-1. More detailed coverage of the refrigeration cycle is provided in the publication *Principles of Refrigeration* by Marsh and Olivo, Delmar Publishers.

The Cooling Coil (Evaporation)

The cooling coil is the only component that is common to both the air cycle and the refrigeration cycle. The main purpose of this coil is to provide a surface over which air from the room can flow. At the same time, the cooling coil provides a passage through which the refrigerant flows. The combination of warm air flowing over the cold refrigerant causes the air to lose heat and the refrigerant to gain heat. Actually, although the temperature of the refrigerant does not change, the heat picked up from the air is needed to vaporize the liquid refrigerant at the same temperature. In this sense, then, the refrigerant is heated.

By supplying a cool surface to the air, the cooling coil serves as a heat transfer device. Heat is transferred from the air to the coil surface, and then to the refrigerant in the coil. In effect, heat is transferred from the air to the refrigerant through the coil (heat transfer) surface.

The Compressor

The compressor serves two purposes. First, it draws the refrigerant from the cooling coil and forces it into the condenser. Secondly, the compressor increases the pressure of the refrigerant.

Suction. By drawing the refrigerant from the cooling coil, figure 4-2, the compressor reduces the pressure in the cooling coil. The compressor keeps the pressure at a low level to permit the refrigerant to boil or vaporize and absorb heat in the process. As stated previously, refrigerant boils at a relatively low temperature when the pressure on the refrigerant is reduced.

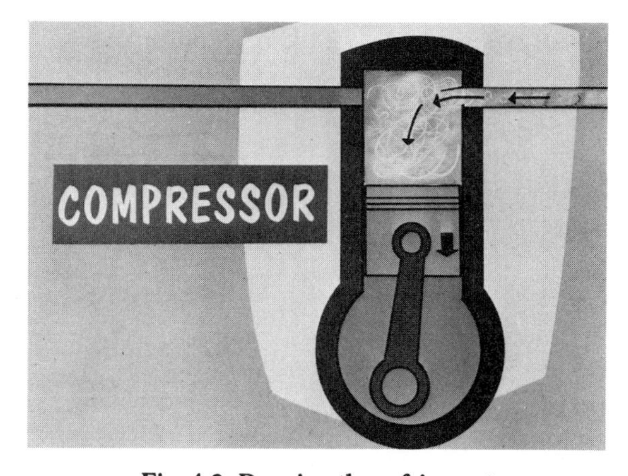

Fig. 4-2 Drawing the refrigerant

Discharge. The compressor then discharges or forces the refrigerant vapor into the condenser. During the discharge process, figure 4-3, the compressor increases the pressure of the refrigerant and also increases the refrigerant vapor temperature. As a result, it is easier for the condenser to do its job.

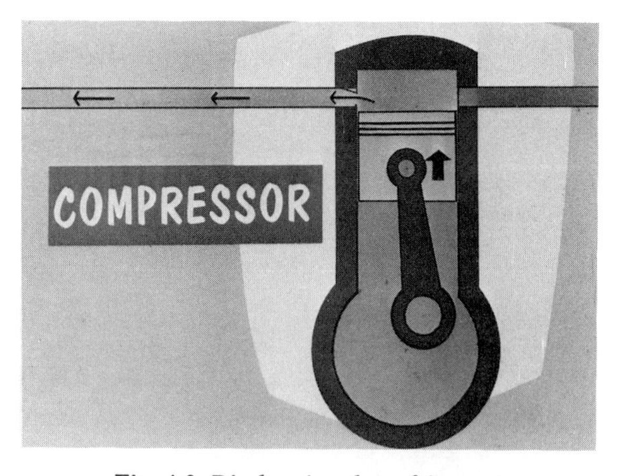

Fig. 4-3 Discharging the refrigerant

The Condenser

The condenser serves two important functions:

- It removes the heat picked up by the refrigerant in the cooling coil (evaporator).
- It condenses the refrigerant vapor to a liquid.

The heat removal and condensing processes can take place in either of two ways.

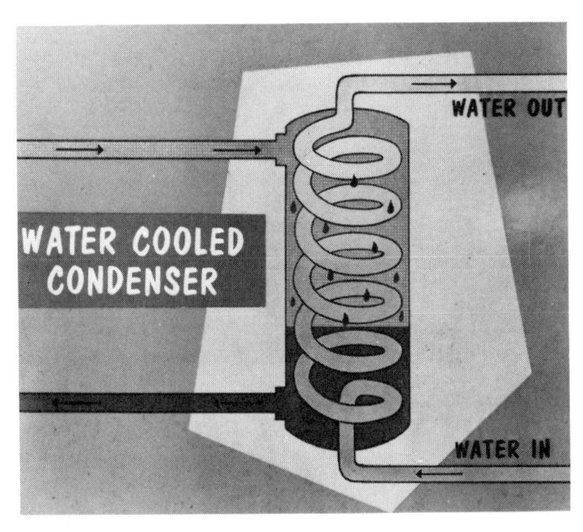

Fig. 4-4 The water-cooled process

Water-cooling Process. If water is used to remove the heat, figure 4-4, the refrigerant vapor is passed through a coil that is submerged in a container filled with water. The water supplied to the container is maintained at a lower temperature than that of the refrigerant vapor. Heat from the refrigerant is transferred to the water through the coil walls. The water then carries the heat from the container through a discharge or drain line.

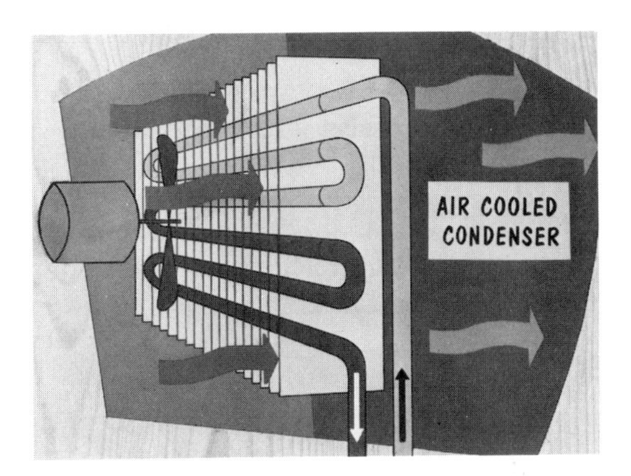

Fig. 4-5 The air-cooled process

Air-cooling Process. If air is used to remove the heat, the refrigerant vapor is passed through a fan-cooled coil. The air blown over the coil is cooler than the refrigerant vapor. Thus, the vapor gives off heat to the air through the coil walls.

The air is then blown to the outdoors and the heat is dissipated in the atmosphere. In the condensing process, the air (or water) picks up heat from the refrigerant. In the cooling process, the refrigerant picks up heat from the room air. In both instances, a coil is the heat transfer surface.

To summarize, the refrigerant enters the condenser as a hot vapor and leaves as a hot liquid. The condenser removes enough heat to change the refrigerant from a vapor to a liquid.

The Expansion Valve

The thermostatic expansion valve reduces the pressure of the refrigerant liquid and thus cools the liquid. Refrigerant enters the valve under pressure. As the refrigerant passes through the valve port, it enters the low-pressure area of the cooling coil. The valve port acts as a metering device between a high-pressure area (the condenser) and a low-pressure area (the cooling coil). Since the boiling point of a liquid is reduced when the pressure of the liquid is reduced, the liquid refrigerant begins to vaporize once it passes through the valve port into the low-pressure area.

The Heat Pump

A heat pump is an air-conditioning unit that consists of a combination of components arranged to supply heating in one cycle and cooling when the cycle is reversed.

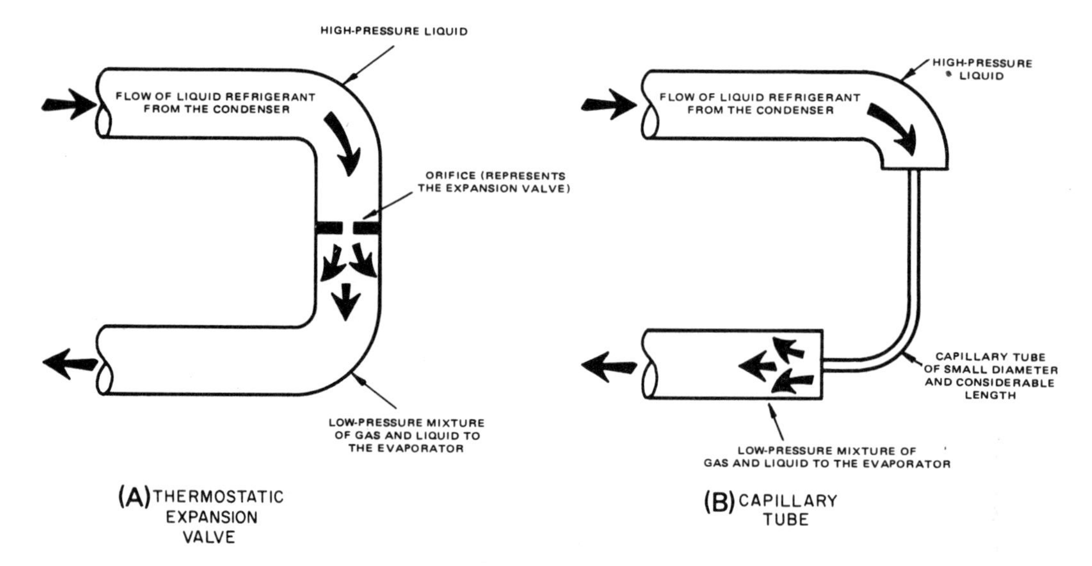

Fig. 4-6 Methods of metering refrigerant flow to the cooling coil

Heat pumps can be effective not only in moderate or warm climates where the heating demand is low, but also in cooler northern climates where the heating demand is considerably higher.

Heat Sources. Earth, air, and water are natural heat sources. At an eight-foot depth the earth temperature remains relatively constant, even in the winter. Thus, although water may be ice at the surface, it is still a liquid at a certain depth beneath the surface. Even cold winter air contains heat.

It is possible to use the heat from these natural sources to heat an area if it can be pumped, moved, or transferred in some manner from the source to the area. The most common method of heat transfer is air to air transfer.

Heating Cycle (Air to Air). In this method heat is taken from the outside air and is released to the inside air, figure 4-7.

Cooling Cycle. The heat pump cooling cycle takes heat from the inside air and releases it to outside air, figure 4-8. The components remain the same for the heat pump heating and cooling cycles, but the refrigerant flow is reversed. This reverse flow is readily achieved through the coils, but not through the expansion valve and the compressor. Thus, in figure 4-8, the expansion valve and the compressor are shown reversed from their positions in the heating cycle.

Heating and Cooling Cycle. In actual practice, the component arrangement shown in figure 4-9, page 24, is acceptable for both the heating and cooling functions of the heat pump.

In the system shown in figure 4-9, two expansion valves are installed so that the required flow direction can be achieved for both cycles. However, since both valves are in the same refrigerant line, one or the other will impede the refrigerant flow depending upon the cycle in operation. To solve this problem, a bypass is installed around each valve to cut it out of the system during the cycle in which it is not functioning.

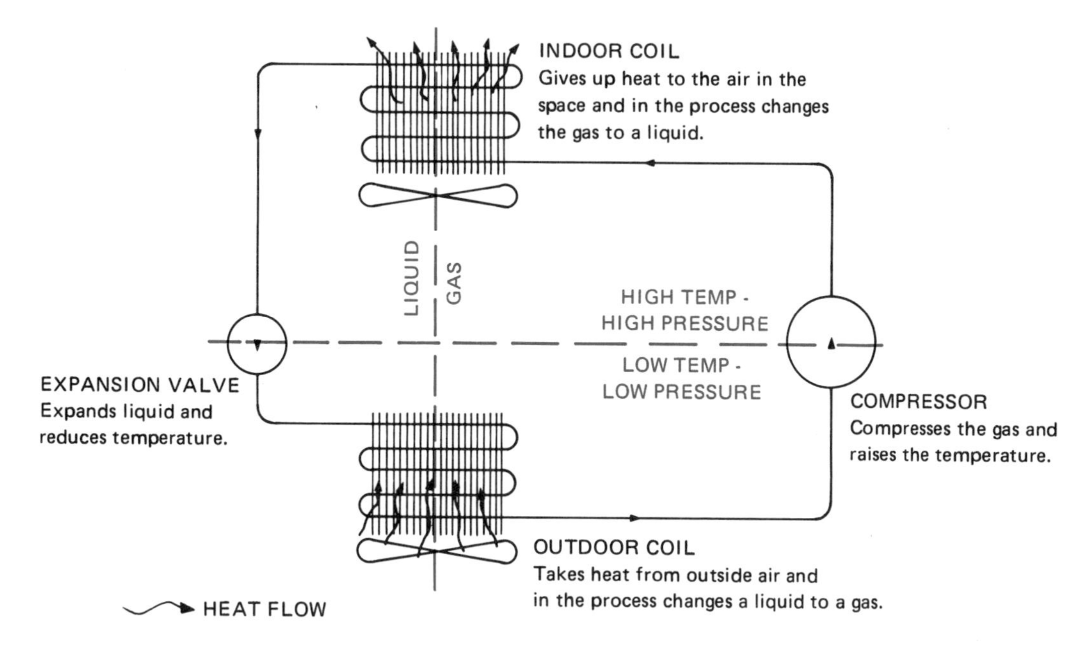

Fig. 4-7 Heat pump heating cycle

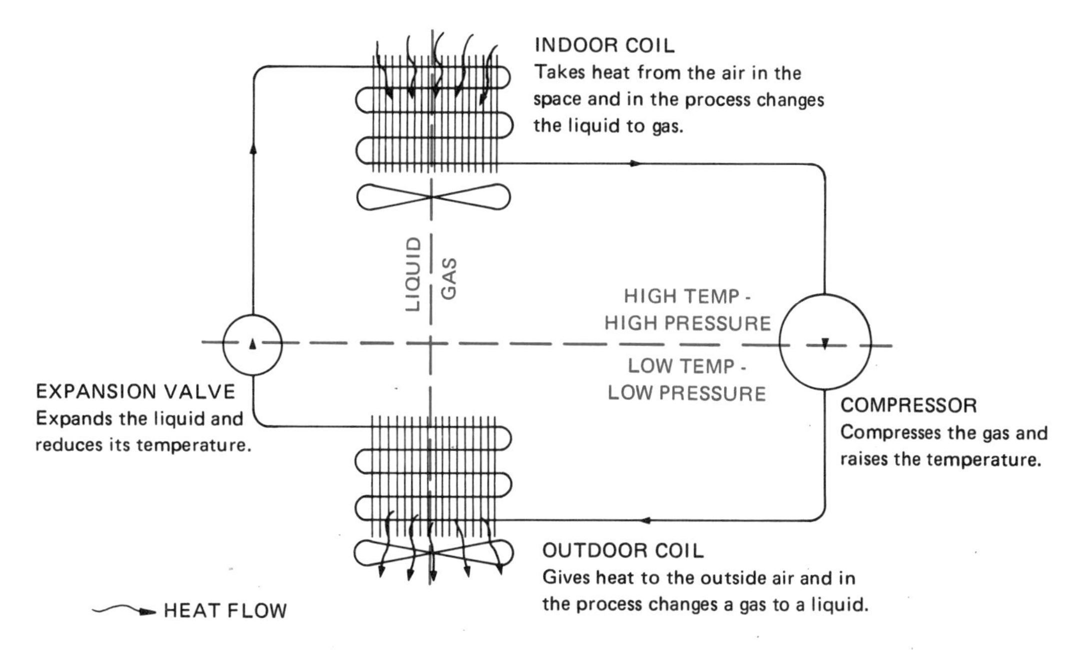

Fig. 4-8 Heat pump cooling cycle

To insure the correct expansion valve action for each cycle, check valves are also installed. These valves are installed so that the refrigerant flows through the heating cycle expansion valve only during the heating cycle and flows through the cooling cycle expansion valve only during the cooling cycle.

In addition, a four-way valve and the suitable piping insures a one-way flow through the compressor. The valve as shown is set for the heating cycle. A simple repositioning makes this valve and the compressor piping suitable for cooling cycle operation. (See the inset at the top right of figure 4-9.)

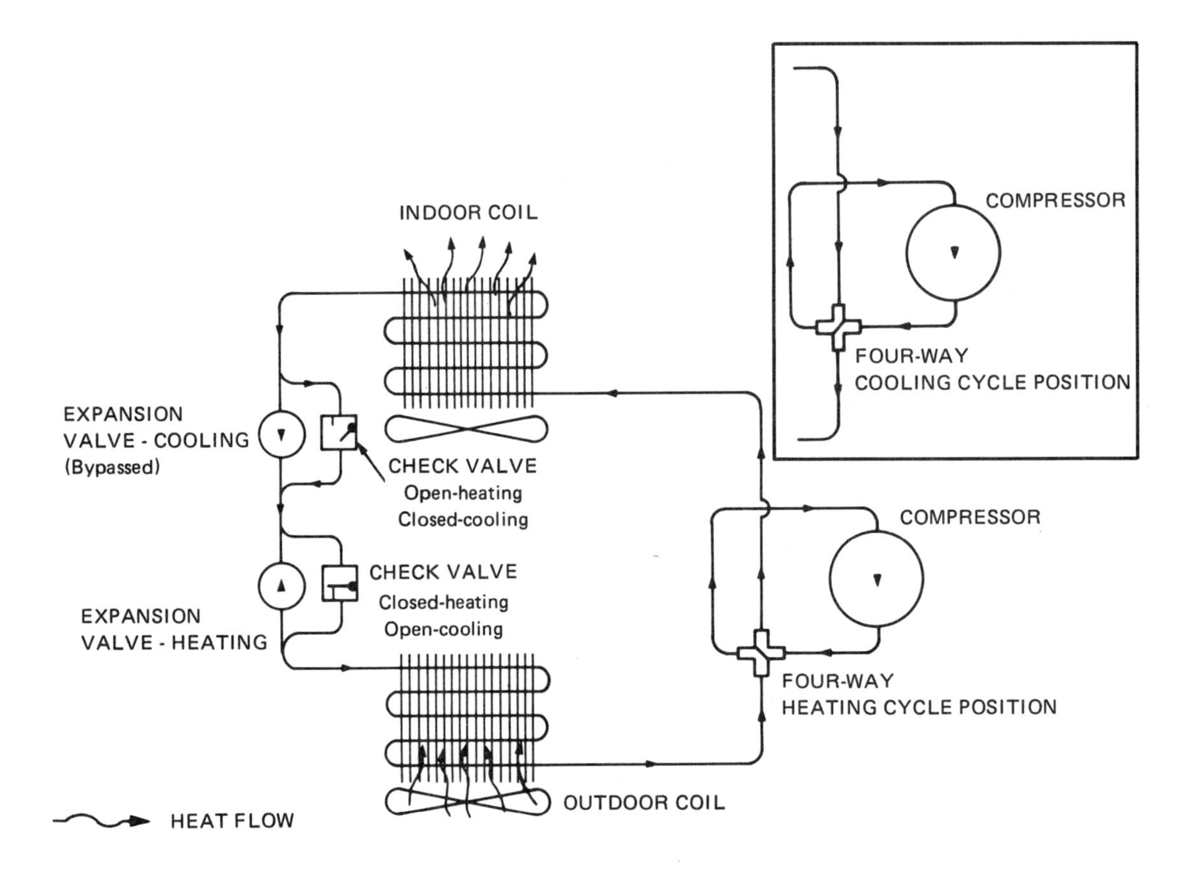

Fig. 4-9 Heat pump heating and cooling cycle

SUMMARY

- The cooling coil, or evaporator, is the key part of both the air cycle and the refrigeration cycle since it is at this point that heat is removed from the air and transferred to the refrigerant.
- When liquid changes to a vapor, it can absorb large quantities of heat.
- The boiling point of a liquid can be changed by changing the pressure exerted on the liquid.
- The refrigerant changes from a liquid to a vapor in the cooling coil and absorbs heat in the process.
- The compressor maintains low-pressure conditions in the cooling coil and creates high-pressure conditions in the condenser.
- The condenser removes heat from the refrigerant and changes the refrigerant to a liquid.
- The expansion valve meters the liquid refrigerant with a resulting reduction in the pressure and cooling of the refrigerant.

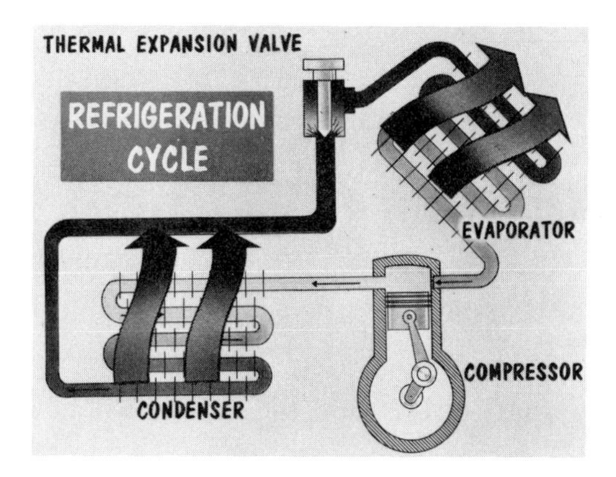

Fig. 4-10 The refrigeration cycle

REVIEW

Select the correct word or phrase to complete the statements 1-5.

1. Refrigeration is the process of ________________ (adding) (transferring) (removing) heat.
2. Evaporation takes place in refrigeration when there is a change of state from ________________ (gas to liquid) (gas to solid) (liquid to gas).
3. The unit in the refrigeration cycle in which heat is rejected is the ____________ (condenser) (coil) (compressor).
4. The dividing point between the high-pressure and low-pressure sides of the refrigeration cycle occurs at the ____________ (cooling coil) (expansion valve) (compressor).
5. Increasing the pressure of a refrigerant ____________________ (decreases its temperature) (increases its temperature) (does not affect its temperature).
6. On the schematic diagram shown in figure 4-11, identify the components of the refrigeration system. Briefly state the function of each component. Indicate the direction of the flow of the refrigerant. Fill in, using a solid color, the piping where the refrigerant is a liquid; leave the piping as shown where the refrigerant is a gas. Use labels to indicate the high and low pressures throughout the cycle.

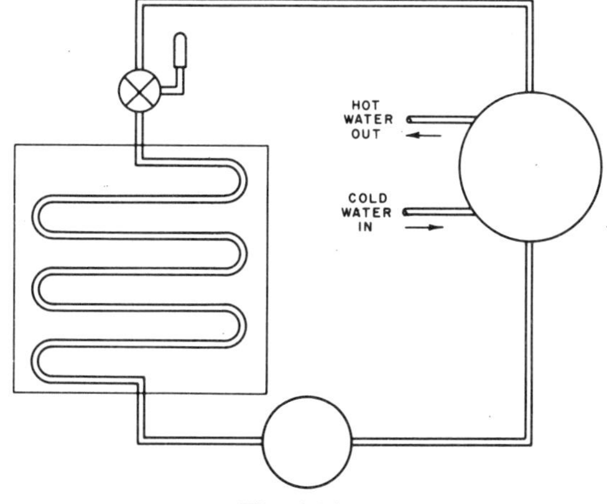

Fig. 4-11

UNIT 5
ABSORPTION-REFRIGERATION SYSTEM

OBJECTIVES

After completing the study of this unit, the student will be able to

- describe the basic absorption – refrigeration cycle.
- list the purpose and function of each of the basic components of the absorption system.
- describe the three absorption – refrigeration cycles: refrigerant, adsorbent, and chilled water.
- discuss the ammonia-water system and the water-lithium-bromide system.

An absorption-refrigeration system uses refrigerant, an adsorbent, and heat, figure 5-1, to create a cooling effect. Unlike the compression refrigeration system, the basic absorption cycle uses no moving parts. The cycle depends on the action and reaction between the refrigerant and the adsorbent under various pressure and temperature conditions in a vacuum. The system makes use of the cooling effect that results when liquids flash to a gaseous state and the condensing effect (changing the gas to a liquid) that results when heat is removed. The operation of the absorption-refrigeration cycle depends on heat applied at a key point and on the strong attraction of refrigerant moisture to an adsorbent medium.

	REFRIGERANT		ADSORBENT		HEAT
	WATER	+	LITHIUM BROMIDE	+	
OR	AMMONIA	+	WATER	+	

Fig. 5-1 Absorption-refrigeration system

ABSORPTION-REFRIGERANT CYCLE

Keep in mind the following points: (1) liquids boil (vaporize) more quickly at a low pressure and in a vacuum, and (2) the adsorbent has the ability to absorb moisture rapidly.

As shown in figure 5-2, when the refrigerant flows into the low-pressure evaporator (chiller), some of it flashes to a vapor. As a result, the remaining liquid refrigerant is chilled and is able to remove heat from the water passing through the evaporator chamber. The pickup of heat causes more refrigerant to vaporize, resulting in more cooling.

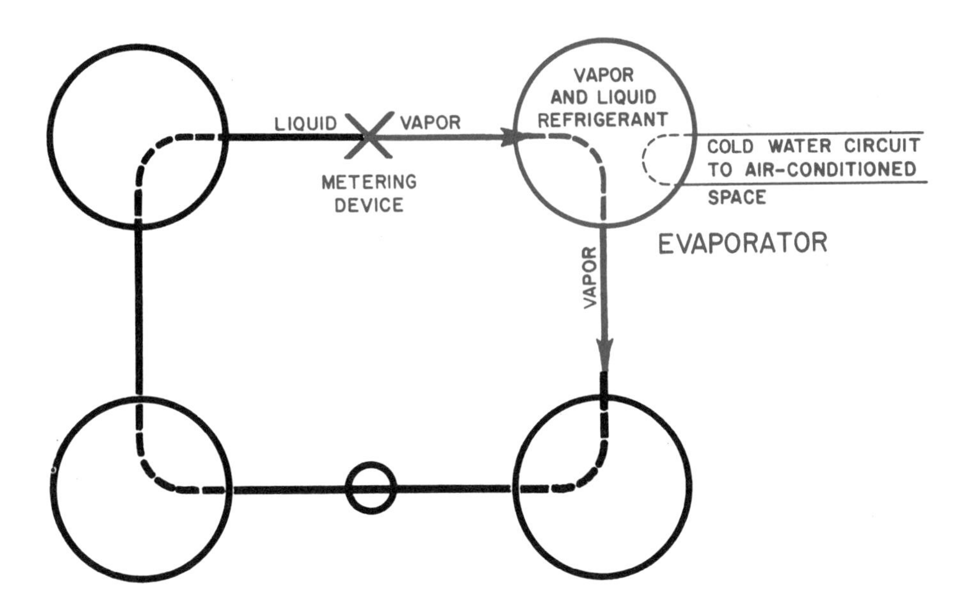

Fig. 5-2 Absorption-refrigeration cycle – evaporator segment

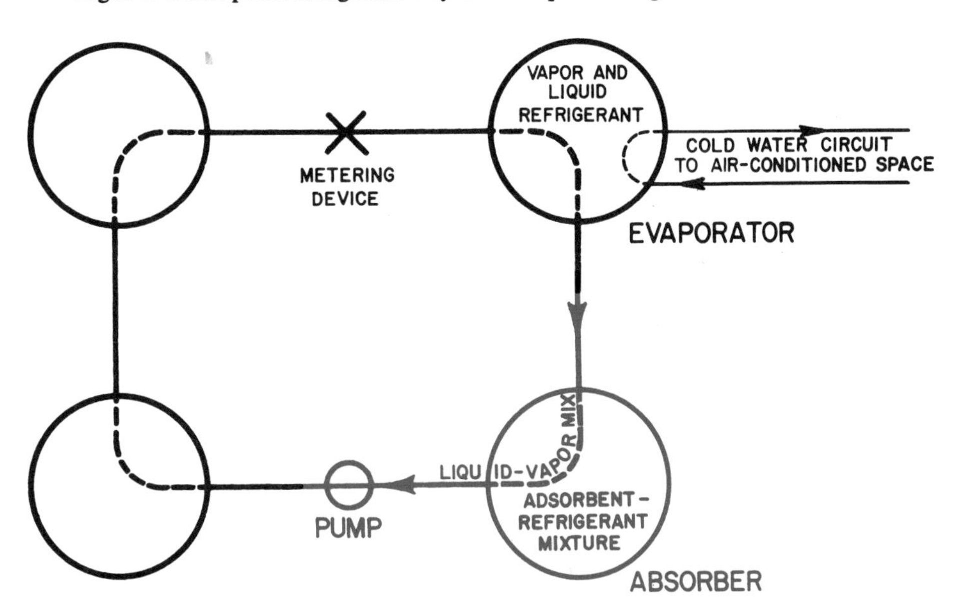

Fig. 5-3 Absorption-refrigeration cycle – absorber segment

After the refrigerant vapor picks up heat, it flows to the absorber, figure 5-3, as a result of the drawing action of the adsorbent. In the absorber, the vapor mixes with, or is dissolved in, the liquid adsorbent. The drawing and absorbing action of the adsorbent causes a pressure difference between the absorber and the evaporator. The pressure in the absorber is slightly lower than that of the evaporator.

The liquid adsorbent and refrigerant vapor mixture flows from the absorber to the generator, figure 5-4, page 28, where it is heated until the adsorbent and refrigerant percolate, or boil. As a result, the refrigerant vapor separates from the liquid adsorbent. The heat added in the generator increases the pressure of the refrigerant vapor.

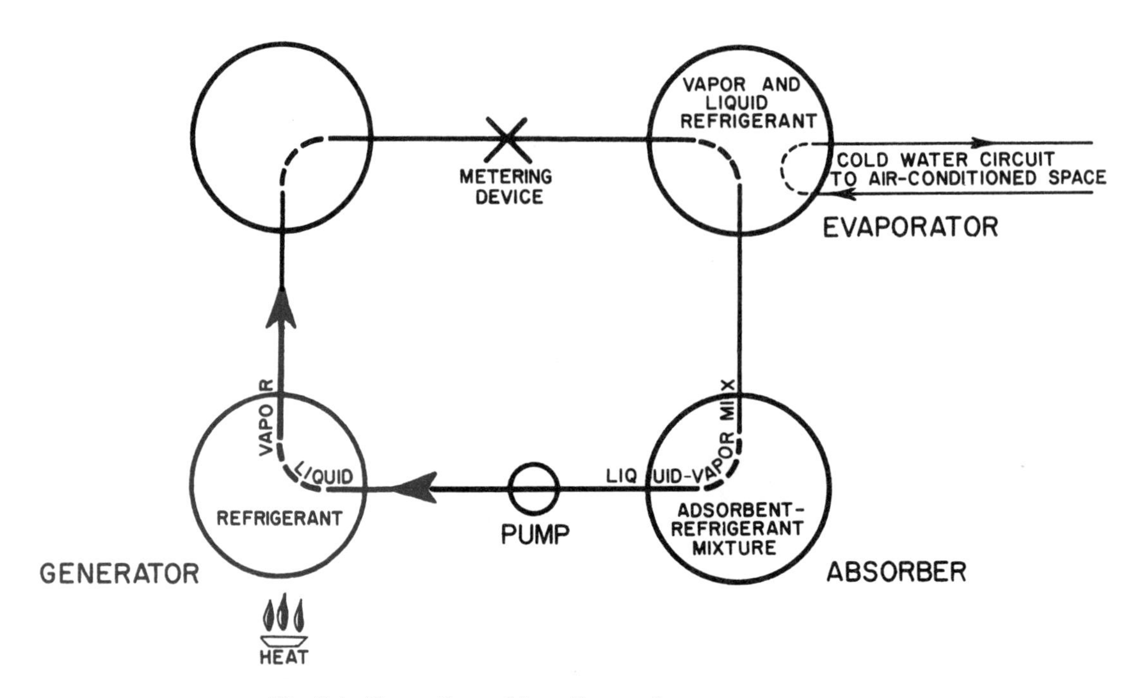

Fig. 5-4 Absorption-refrigeration cycle – generator segment

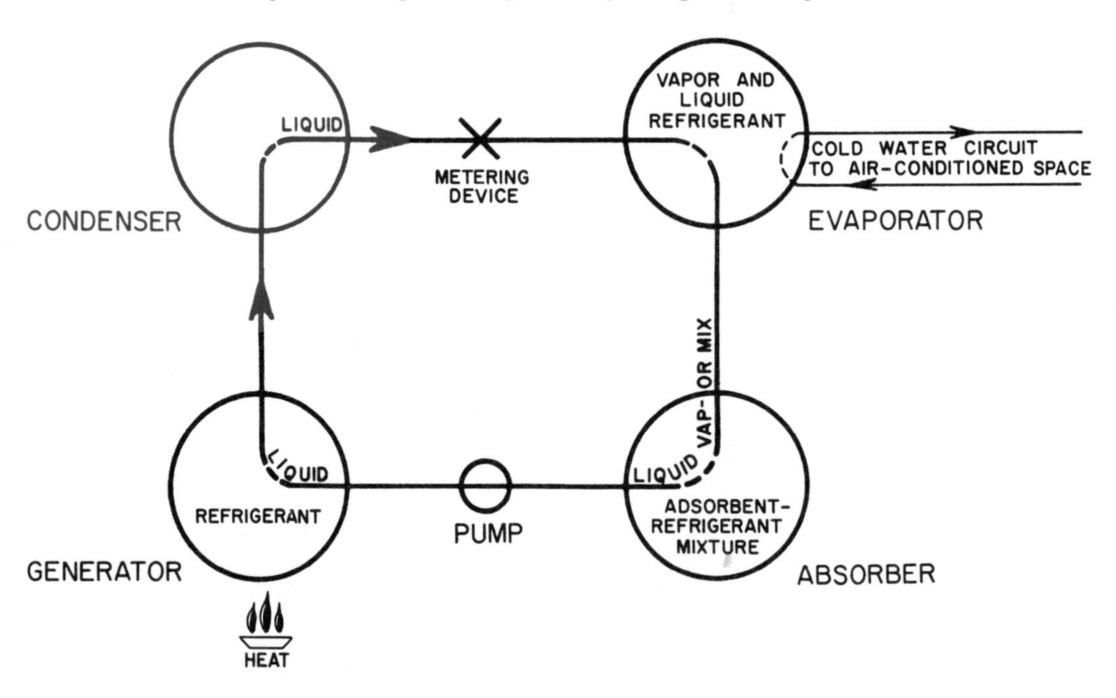

Fig. 5-5 Absorption-refrigeration cycle –condenser segment

The refrigerant is now heated and in a gaseous state. In this form, it passes to the condenser, figure 5-5, where heat is removed. As the condenser removes heat, the refrigerant vapor changes to a liquid. It can be seen that the functions of the condenser are similar for both the absorption system and the compressor system. Liquid refrigerant then flows through a metering device and flashes into the low-pressure evaporator where the cycle begins again.

In summary, the basic equipment in the absorption-refrigeration system includes the evaporator (chiller), absorber, generator and heat source, condenser, and metering device. The basic function and purpose of each component of the system are described in the following sections.

Evaporator

Basically, the evaporator is a heat transfer chamber in which heat flows from water to the refrigerant. The heat originates in the space to be conditioned. The heat is then carried to the evaporator via the chilled-water cycle (which is described on page 31). In the evaporator, the heat is transferred from the water through the coil surface to the refrigerant.

Some evaporators may contain a spray head to break up the liquid refrigerant into smaller particles to accelerate the heat transfer process. Refrigerant vapor is attracted by the adsorbent. In this manner, the heat is carried from the evaporator to the absorber.

In some absorption systems, the heat is transferred directly from the supply air to the refrigerant in the evaporator. In this type of system, the evaporator is a coil located in the supply airstream rather than an evaporator chamber.

Absorber

This component of the system contains an adsorbent that attracts refrigerant vapor from the evaporator and carries it to the generator in a mixed liquid-vapor state. Refrigerant is usually absorbed so rapidly that the action causes a low-pressure condition in the evaporator. Since the refrigerant vapor is changed to a liquid in the process of being absorbed, heat is generated in the absorber. When this is the case, the refrigerant used is generally ammonia.

Adsorbent Types. Some absorption systems use water as the adsorbent medium. Water has a strong attraction for ammonia and yet separates from it readily when enough heat is applied.

Other absorbent systems use lithium bromide as the adsorbent medium and water as the refrigerant. Since lithium bromide is a salt, it has a strong attraction for water. A liquid mixture of water and lithium bromide readily separates in a vacuum at low pressure when sufficient heat is applied.

An absorber using lithium bromide may contain a separate water coil to remove some of the heat of absorption generated when the lithium bromide absorbs water vapor from the evaporator.

Some absorbers contain spray heads. When lithium bromide is returned to the absorber from the generator, it is sprayed into the absorber. As a result, more surface area is exposed and the absorbing power of the lithium bromide is increased. A spray head in the absorber tends to increase the amount of refrigerant vapor absorbed from the evaporator. Thus, in addition to the adsorbent medium, an absorber may contain a heat-removal water coil and an adsorbent spray head.

Generator

The generator of an absorption system is the heat energy source that begins and maintains the operating cycle. The main function of the generator is to maintain the conditions in which the adsorbent readily separates from the refrigerant. The energy source may be steam, gas, or hot water. In general, gas is used for a system having a small capacity; steam and hot water may be used for systems having large capacities.

The functions performed by the absorber and the generator can be considered to be similar to the action of a compressor. That is, the absorber, in a sense, compresses the refrigerant and the generator heats the refrigerant with a resulting high-pressure condition.

Condenser

The condenser of an absorption system performs the same function as the condenser in a compression system. That is, it removes heat from the refrigerant vapor causing the vapor to condense to a liquid.

Some condensers use air blowing across the condenser coil surface to remove heat from the refrigerant through the coil wall. The refrigerant vapor is cooled and condenses to a liquid. The liquid is now under high pressure. It passes through a restriction and flashes into the low-pressure evaporator. Other condensers use water to remove heat from the refrigerant vapor.

ABSORPTION-ADSORBENT CYCLE

Adsorbent in a liquid state attracts refrigerant vapor from the evaporator, figure 5-6. Because of this attraction, the vapor mixes with the adsorbent. The mixture flows to the generator with the aid of a pump to insure the proper flow.

Heat is applied to the adsorbent-refrigerant mixture in the generator. The mixture is brought to a boil and the refrigerant vaporizes, but the adsorbent tends to remain liquid. Thus, the adsorbent separates from the refrigerant and returns to the absorber. The return flow is regulated by a flow-control device.

Once the adsorbent releases the refrigerant vapor in the generator, it can absorb more moisture from the evaporator. Therefore, the adsorbent again attracts refrigerant vapor from the evaporator and the adsorbent cycle repeats.

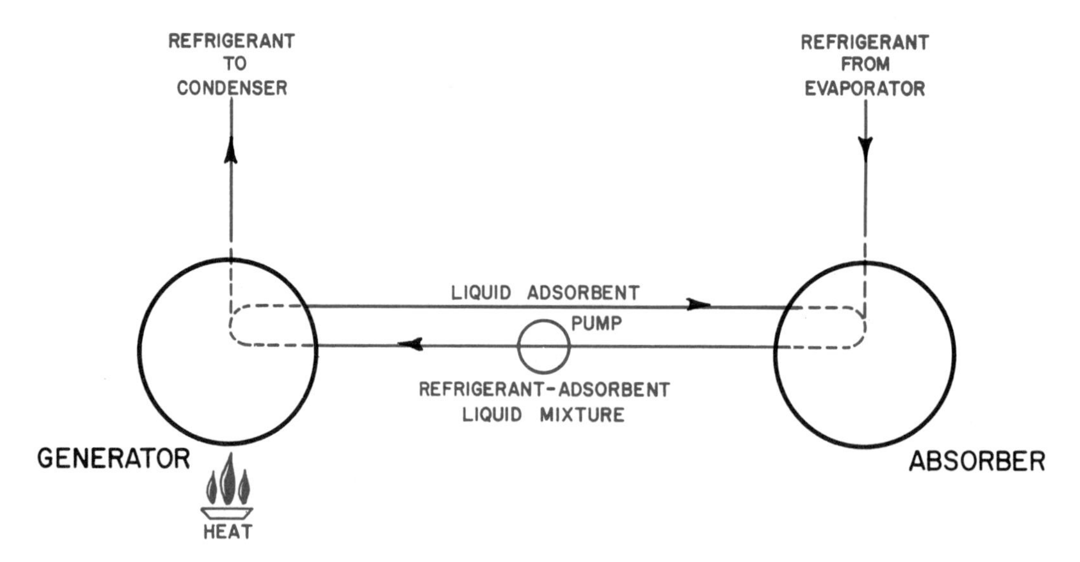

Fig. 5-6 Adsorbent cycle

CHILLED-WATER CYCLE

A third cycle in an absorption cooling system removes heat from the air-conditioned space and releases it to the evaporator, figure 5-7. Chilled water flows to a coil located in the supply airstream. The air is cooled before it enters the conditioned space. Heat absorbed from the air by the water flows to the coil in the evaporator. The refrigerant in the evaporator removes the heat from the water. The chilled water returns to the coil to cool the supply air again. The air picks up more heat from the conditioned space and the cycle repeats itself.

AMMONIA-WATER SYSTEM

In an ammonia-water absorption system, ammonia is the refrigerant and water is the adsorbent.

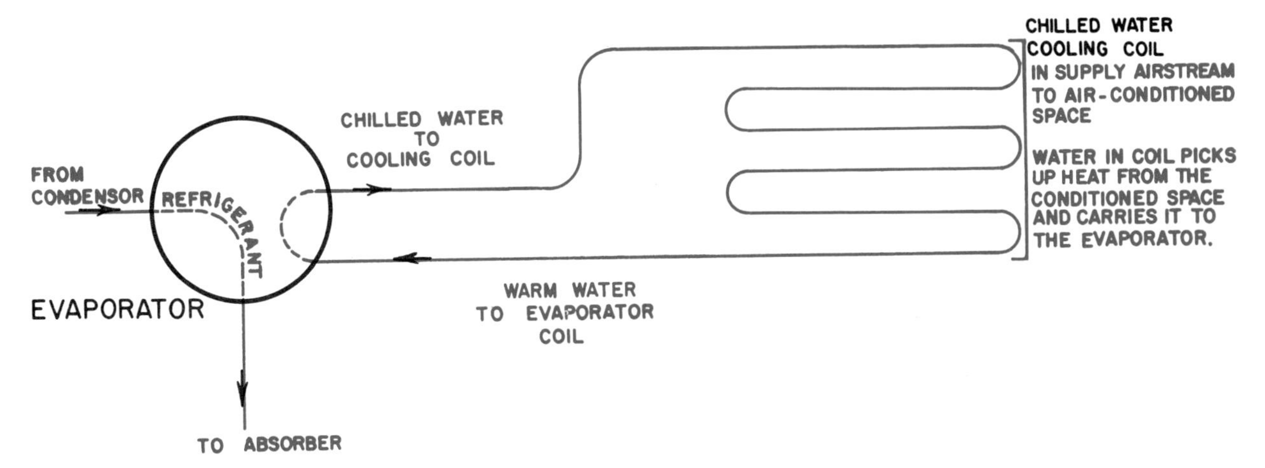

Fig. 5-7 Chilled-water cycle

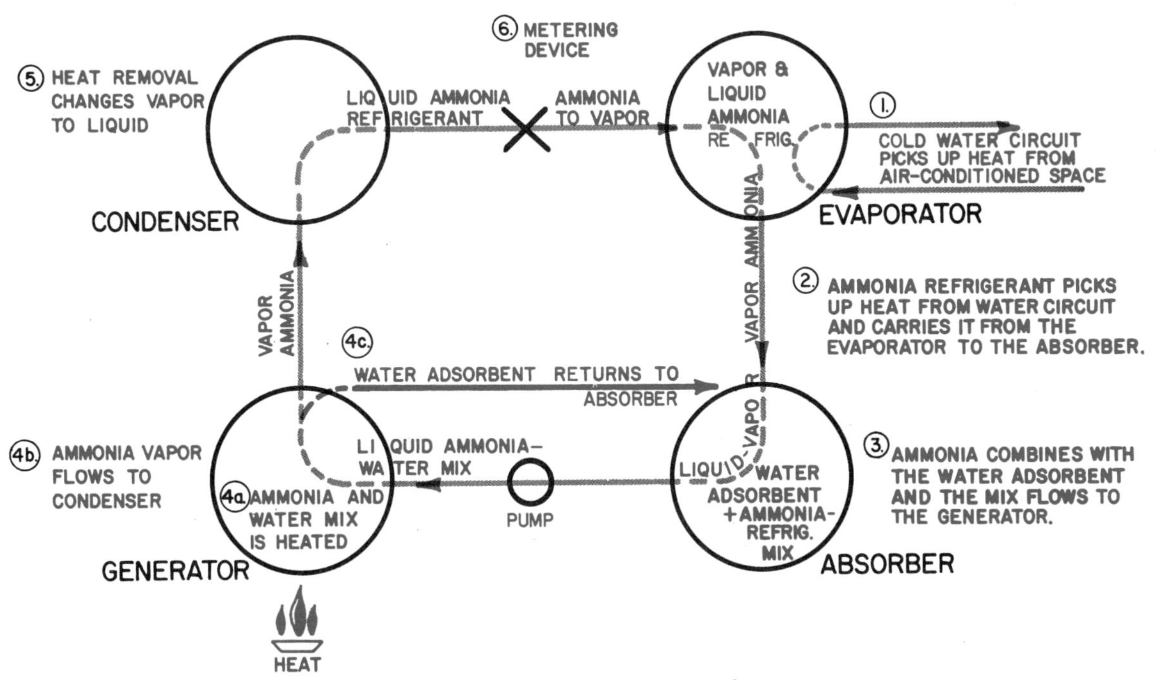

Fig. 5-8 Ammonia – water system cycle

The Cycle

As the liquid ammonia flows into the low-pressure evaporator, some of the ammonia flashes to a vapor. As a result, the remaining ammonia liquid is chilled and begins to remove heat from the water passing through the evaporator. The resulting cool water is then used to cool the space to be air conditioned. The pickup of heat from the water causes more ammonia to vaporize, resulting in more cooling.

After the ammonia picks up heat in the evaporator, ammonia vapor is attracted to the absorber by the pull of the water. The ammonia vapor combines with the adsorbent water to form a liquid mixture. This liquid mixture of ammonia and water flows to the generator.

In the generator, the mixture is heated until it percolates. The ammonia refrigerant is now in the vapor state and separates from the water adsorbent. In a sense, ammonia is driven out of the water by the heat added in the generator. That is, the ammonia is under a higher pressure than it was before it was heated.

The water adsorbent returns to the absorber. The pressurized ammonia vapor flows to the condenser. As heat is removed from the ammonia vapor due to the air blowing across the condenser coil, the vapor is cooled and returns to a liquid form while still under pressure. The liquid ammonia then flows through a metering device and flashes into the low-pressure chamber. The ammonia refrigerant is again a vapor and the cycle begins again.

WATER-LITHIUM-BROMIDE SYSTEM

A water-lithium-bromide absorption system uses water as the refrigerant and lithium bromide as the adsorbent.

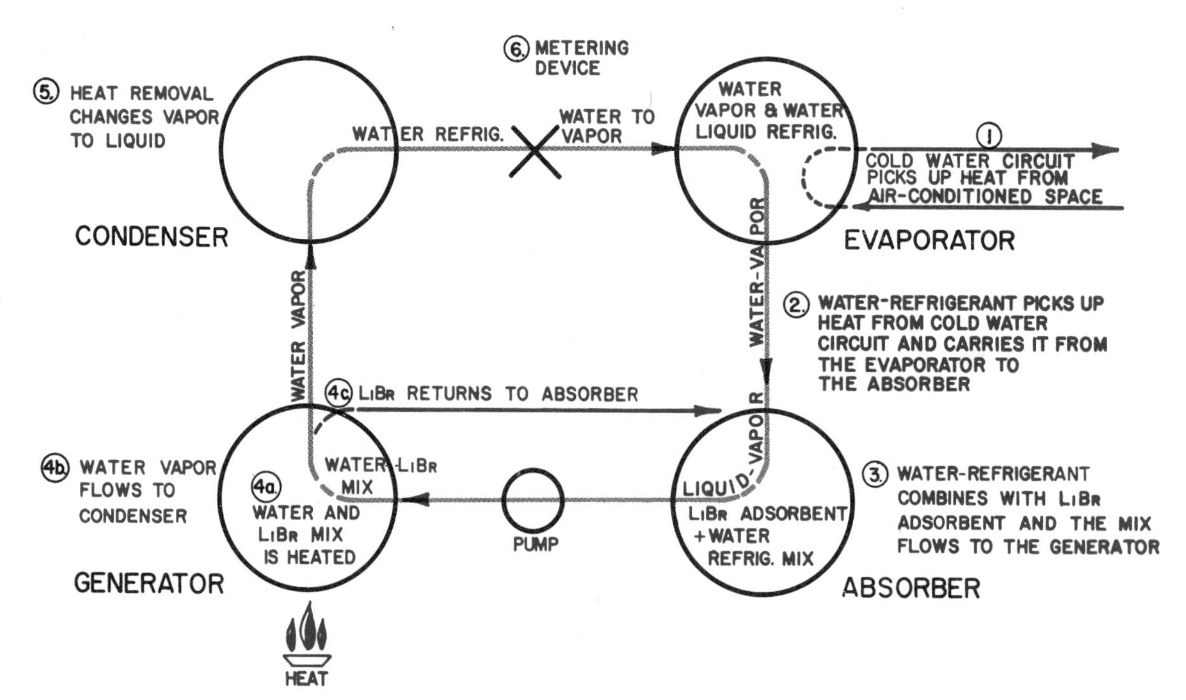

Fig. 5-9 Water – lithium bromide (Li Br) system cycle

The Cycle

As the refrigerant water flows into the low-pressure evaporator, some of the water flashes to the vapor form. As a result, the remaining refrigerant water is chilled and heat transfer takes place. The chilled refrigerant water removes heat from the water coil passing through the evaporator. The cool water is used to cool the space to be air conditioned. The heat picked up by the refrigerant water causes more water to evaporate, resulting in more cooling.

Once the refrigerant water vapor picks up heat in the evaporator, it flows to the absorber. The vapor is drawn there by the lithium bromide adsorbent. The refrigerant vapor is adsorbed by the lithium bromide. In the process, the vapor changes back to water. The water-lithium liquid mixture then flows to the generator.

The mixture is heated in the generator until the refrigerant separates from the lithium bromide adsorbent. The adsorbent returns to the absorber and the refrigerant water vapor flows under pressure to the condenser.

As heat is removed from the water vapor by the condenser, the water is cooled and returns to its liquid state. The refrigerant water is still under pressure. It flows through a metering device and flashes into the low-pressure evaporator. Some of the refrigerant water is vaporized and the cooling cycle begins again.

SUMMARY

- The basic absorption-refrigeration cycle uses no moving parts, but relies on the action and reaction between the adsorbent and the refrigerant in a vacuum.
- Adsorbents absorb moisture rapidly and then give it up again when enough heat is applied.
- The evaporator, absorber, generator, condenser, and metering device are the basic components in an absorption system.
- In absorption systems, ammonia or water is used as the refrigerant; lithium bromide or water is used as the adsorbent.
- The action of the generator and the absorber is sometimes considered to be similar to the work done by the compressor.
- An absorption system contains a refrigeration cycle, an adsorbent cycle, and a chilled-water cycle.

REVIEW

Select the one best answer for each of the following questions.

1. A basic absorption cycle functions properly

   a. because it has no moving parts.
   b. because of the action and reaction between adsorbent and refrigerant.
   c. because it is cylindrical.
   d. because it uses no heat.

2. Heat in the cycle should be applied

   a. in the cooler.
   b. in the absorber.
   c. in the generator.
   d. with a torch.

3. Adsorbent absorbs

   a. rapidly, if properly heated.
   b. slowly.
   c. rarely.
   d. only in total darkness.

4. Adsorbents separate from the refrigerant

   a. only when sufficiently heated.
   b. only when cooled.
   c. by means of a filter.
   d. by using a separator.

Select the word or phrase which best completes each statement.

5. The vacuum in an absorption system __________the action and______________ the reaction between the refrigerants and adsorbents.

   a. assists
   b. hinders

6. The basic components of an absorption-refrigeration cycle include __________

   a. evaporator, absorber, generator, condenser, metering device
   b. evaporator, absorber, generator, condenser, compressor

7. Commonly used refrigerants in an absorption system are __________

   a. ammonia, lithium bromide, water
   b. ammonia, water

8. Commonly used adsorbents in an absorption system are __________

   a. lithium bromide, water
   b. ammonia, lithium bromide, water

9. The basic cycles in an absorption system are __________

   a. absorption-refrigeration cycle, absorption-adsorbent cycle, chilled-water cycle
   b. refrigeration cycle, chilled-water cycle

Supply the word or phrase which best completes the following statements.

10. In an ammonia-water absorption system, ____________________ is used as the adsorbent.

11. In a lithium-bromide-water absorption system, ________________ is used as the refrigerant.

SECTION 2
PSYCHROMETRICS

UNIT 6

PSYCHROMETRICS AND THE PSYCHROMETRIC CHART

OBJECTIVES

After completing the study of this unit, the student will be able to

- define psychrometrics.
- define the terms dry-bulb temperature, wet-bulb temperature, relative humidity, grains of moisture, and dewpoint temperature.
- identify the lines and scales which represent these terms on the psychrometric chart.
- use a psychrometric chart to determine the conditions of air.

Psychrometrics is the determination and measurement of the properties of air. As applied to air conditioning, psychrometrics is concerned with the properties of the outside air and air that is present in the conditioned room or building. The principles of psychrometrics are also used to determine the air conditions that are most comfortable in a given air-conditioning application.

PSYCHROMETRIC CHART

The psychrometric chart is a tool which simplifies the measurement of air properties. The chart also eliminates many time-consuming and tedious calculations. Various air-conditioning equipment manufacturers provide slightly different forms of the chart. The differences between the charts usually concern the location of information. All psychrometric charts are basically a graphic representation of the conditions or properties of air. These conditions include temperature, humidity, and condensation point (or dewpoint). A full-size psychrometric chart is included in the envelope attached to the back cover of the text.

PSYCHROMETRIC TERMS

The properties that can be determined from the psychrometric chart are: dry-bulb temperature, wet-bulb temperature, relative humidity, dewpoint, and grains of moisture.

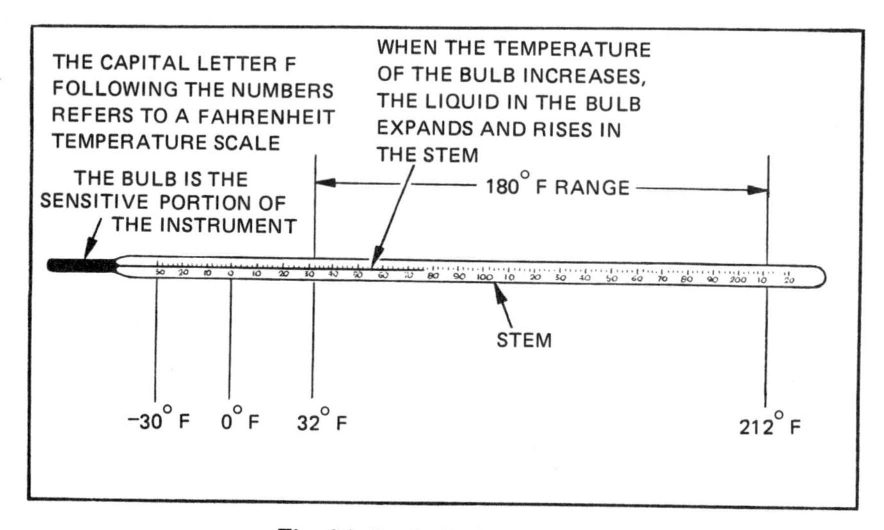

Fig. 6-1 Dry-bulb thermometer

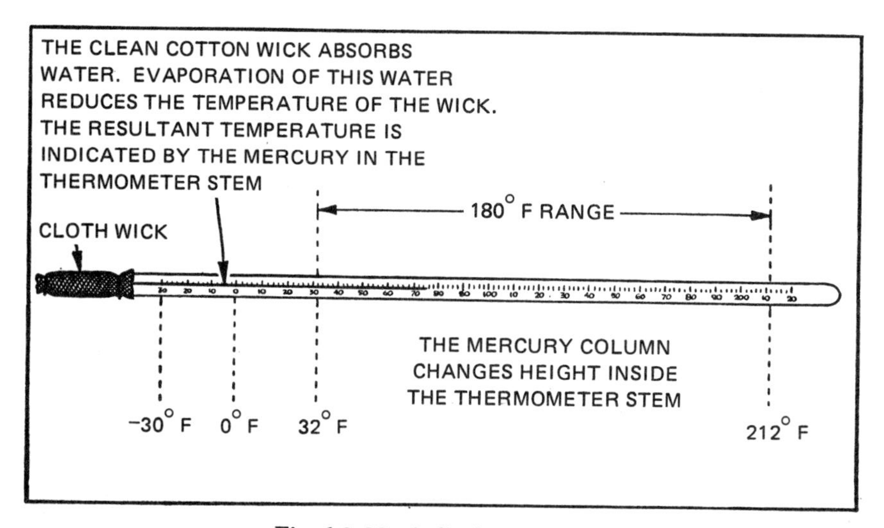

Fig. 6-2 Wet-bulb thermometer

The *dry-bulb temperature* is the temperature of the air as measured by an ordinary household thermometer, figure 6-1.

The *wet-bulb temperature* is the temperature of the air as measured by an ordinary thermometer whose glass bulb is covered by a cloth sock, figure 6-2. The sock-covered bulb is then dipped into water so that the bulb is wet when the temperature reading is taken. The temperature is recorded after the thermometer is moved rapidly in the air. Except for the sock, a wet-bulb thermometer is the same as a dry-bulb or ordinary thermometer. To measure wet- or dry-bulb temperatures, a sling psychrometer is used, figure 6-3.

The two thermometers of the psychrometer are mounted on a base plate as shown in figure 6-3. The thermometer with the sock over the bulb is the wet-bulb thermometer. The wet-bulb thermometer extends below the dry-bulb thermometer so that the sock can be dipped in water without wetting the dry-bulb thermometer.

After the sock is dipped in water, the two thermometers are moved quickly through the air until the water evaporates from the sock. Although the air passing

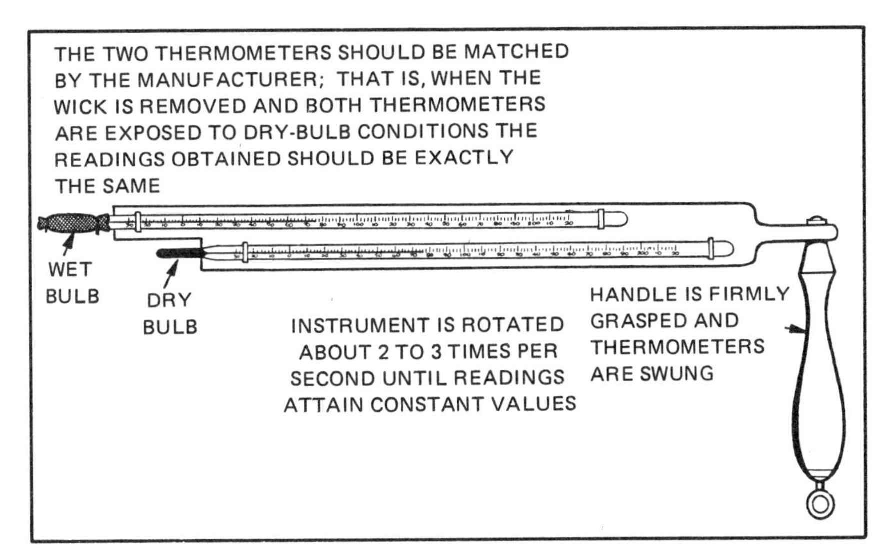

Fig. 6-3 Sling psychrometer

over the dry-bulb has the same temperature as the air passing over the wet-bulb, the temperature registered by the two thermometers is not the same. The dry-bulb thermometer always registers the actual air temperature. The wet-bulb thermometer registers a temperature that is lower than the dry-bulb reading.

The key to the difference in the temperature readings of the thermometers is the word evaporation. Unit 2 described the evaporation process and its effects. Recall from the discussion in unit 2 that as moisture evaporates from a surface, heat is removed from that surface.

In the case of the sling psychrometer, moisture evaporates from the wet sock of the wet-bulb thermometer. Thus, the surface of the thermometer bulb is cooled, resulting in a lower wet-bulb thermometer reading.

The temperature spread between the dry-bulb and wet-bulb readings depends on the amount of moisture in the air. If the moisture content is high, water evaporation from the sock on the wet-bulb thermometer occurs at a slower rate. Consequently, less heat is removed and the wet-bulb reading is high (closer to the dry-bulb reading). If the moisture content of the air is low, the air is dry and can readily absorb moisture. Therefore, evaporation from the wet sock takes place at a rapid pace and more heat is removed. As a result, the surface of the wet-bulb cools rapidly and the reading on the wet-bulb thermometer is lower than the reading for air containing more moisture.

Dry air or air with a low moisture content has a low wet-bulb temperature. Humid air or air with a high moisture content, has a high wet-bulb temperature. Thus, it can be seen that when the moisture content reaches 100% (100% relative humidity), the wet-bulb temperature is the same as the dry-bulb temperature. This condition is readily apparent on the psychrometric chart. At 100% relative humidity, evaporation stops because the air can absorb no more moisture. Therefore, it is not possible to remove heat by evaporation from the sock on the wet-bulb thermometer, and the two thermometers register the same temperature.

Relative humidity is the actual amount of moisture in the air compared to the total or maximum moisture the air can hold.

Grains of moisture is the unit of measurement of the amount of moisture in the air.

Dewpoint temperature is the temperature at which moisture condenses on a surface.

When related to the psychrometric chart, the terms defined in the preceding paragraphs can give a great deal of information about the condition of the air. Consider the following situations:

- If the dry-bulb and wet-bulb temperatures are known, the relative humidity can be read from the chart.
- If the dry-bulb temperature and the relative humidity are known, the wet-bulb temperature can be determined.
- If the wet-bulb temperature and the relative humidity are known, the dry-bulb temperature can be found.
- If the wet-bulb and dry-bulb temperatures are known, the dewpoint can be found.
- If the wet-bulb temperature and the relative humidity are known, the dewpoint can be read from the chart.
- If the dry-bulb temperature and the relative humidity are known, the dewpoint can be found.

The grains of moisture in the air can be determined if any of the following combinations of readings are known:

Dry-bulb temperature and relative humidity (RH)
Dry-bulb temperature and dewpoint
Wet-bulb temperature and relative humidity
Wet-bulb temperature and dewpoint
Dry-bulb and wet-bulb temperatures
Dewpoint alone

IDENTIFICATION OF THE SCALES ON THE PSYCHROMETRIC CHART

The psychrometric chart can be pictured as a shoe with the toe on the left and the heel on the right, figure 6-4.

The dry-bulb temperature scale, figure 6-5A, extends along the sole from the toe to the heel. The dry-bulb lines are perpendicular to the sole. Each line represents one degree of temperature change.

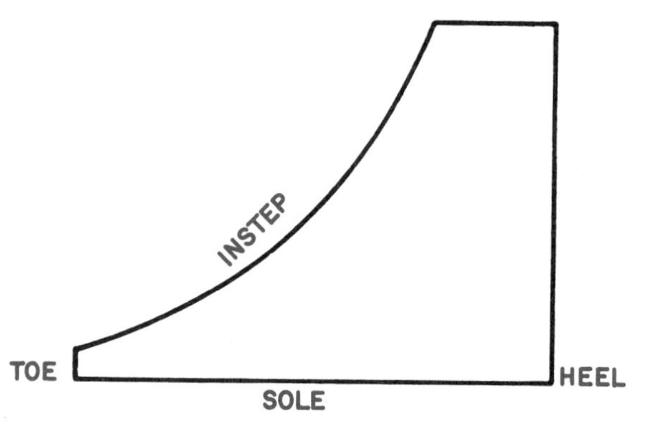

Fig. 6-4

The wet-bulb scale, figure 6-5B, extends along the instep from the toe to the top of the shoe. The wet-bulb lines extend diagonally downward to the sole and the back of the shoe. Each line represents one degree of temperature change.

The condensation or dewpoint scale, figure 6-5C, extends from the sole to the top of the shoe. Each line is parallel to the sole and represents one dewpoint temperature.

The relative humidity lines, figure 6-5D, follow approximately the same curve as the instep. The instep line itself is actually the line representing 100% relative humidity.

The grains of moisture lines, figure 6-5E, are the same as the dewpoint lines. However, the scale for the grains of moisture line, as shown on the right of the chart, are different.

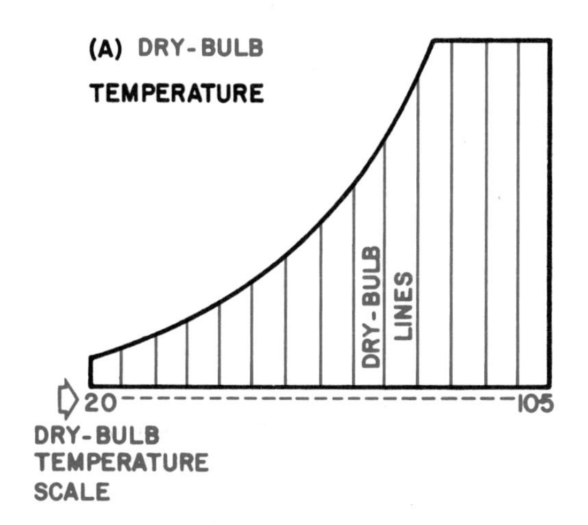
(A) DRY-BULB
TEMPERATURE
DRY-BULB
LINES
20
105
DRY-BULB
TEMPERATURE
SCALE

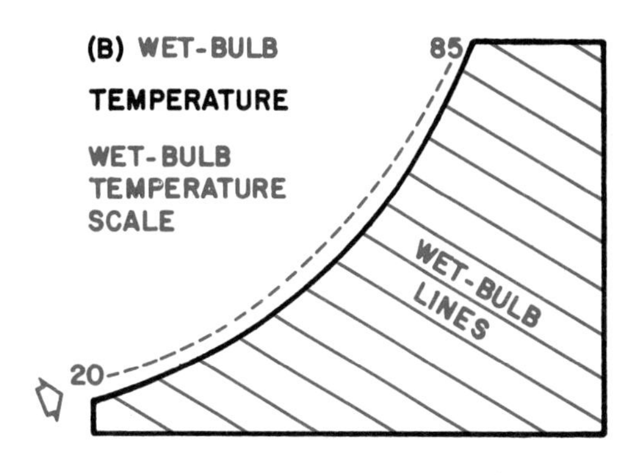
(B) WET-BULB
TEMPERATURE
WET-BULB
TEMPERATURE
SCALE
85
20
WET-BULB
LINES

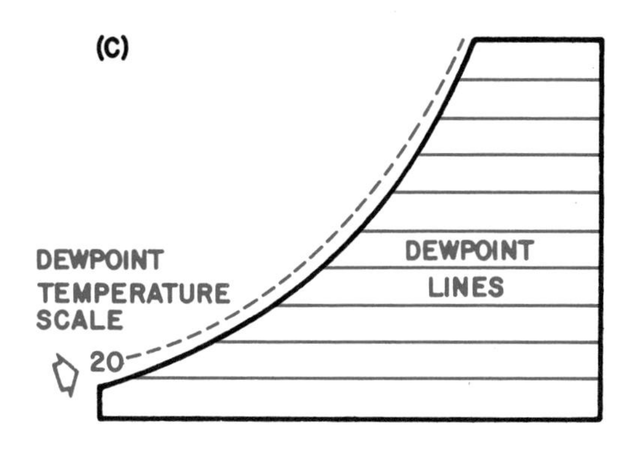
(C)
DEWPOINT
TEMPERATURE
SCALE
20
DEWPOINT
LINES

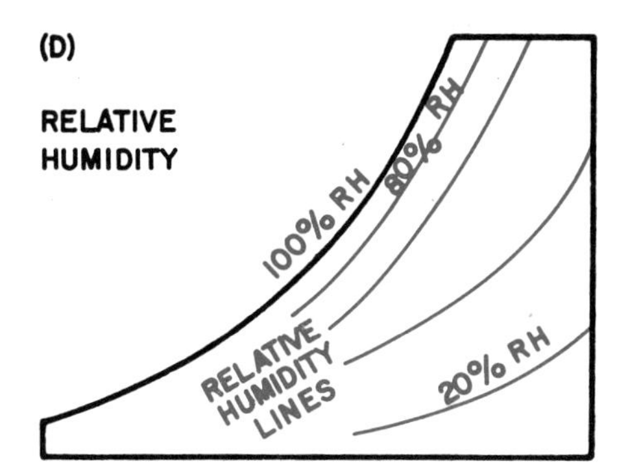
(D)
RELATIVE
HUMIDITY
100% RH
80% RH
RELATIVE
HUMIDITY
LINES
20% RH

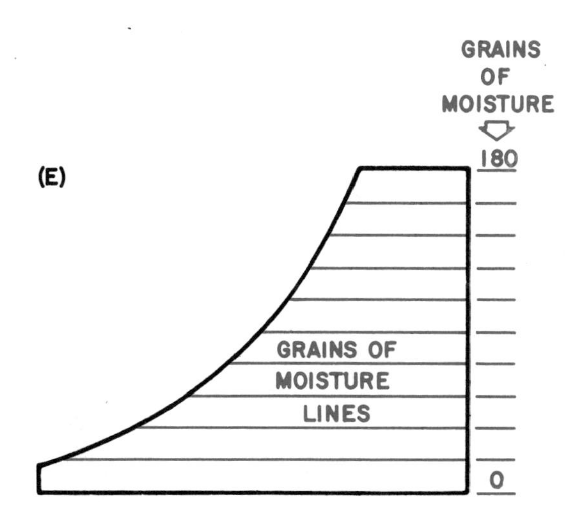
GRAINS
OF
MOISTURE
(E)
180
GRAINS OF
MOISTURE
LINES
0

Fig. 6-5

SOLVING PROBLEMS USING THE PSYCHROMETRIC CHART

Problems 1-7 illustrate the various relationships between the wet-bulb temperature, the dry-bulb temperature, the dewpoint temperature, relative humidity, and the grains of moisture. The psychrometric chart in the back cover envelope of this text is to be used to follow the solution to each problem.

PROBLEM 1

Dry-bulb and Wet-bulb Temperatures – Relative Humidity

Given

Dry-bulb temperature 78°F   Wet-bulb temperature 65°F

Find

Relative humidity

Solution

1. Locate 78°F on the dry-bulb scale at the bottom of the chart, figure 6-6A.
2. At 78°F, draw a line perpendicular to the sole until it meets the curved line (instep), figure 6-6B.
3. Move down the instep (wet-bulb scale) to 65°F, figure 6-6C.
4. Draw a line diagonally along the 65°F wet-bulb line until it crosses the 78°F dry-bulb line, figure 6-6D.
5. Read 50% relative humidity at the intersection of the dry- and wet-bulb lines, figure 6-6E.

Thus, at the conditions of 78°F dry-bulb and 65°F wet-bulb temperatures, the relative humidity is 50%, figure 6-6F.

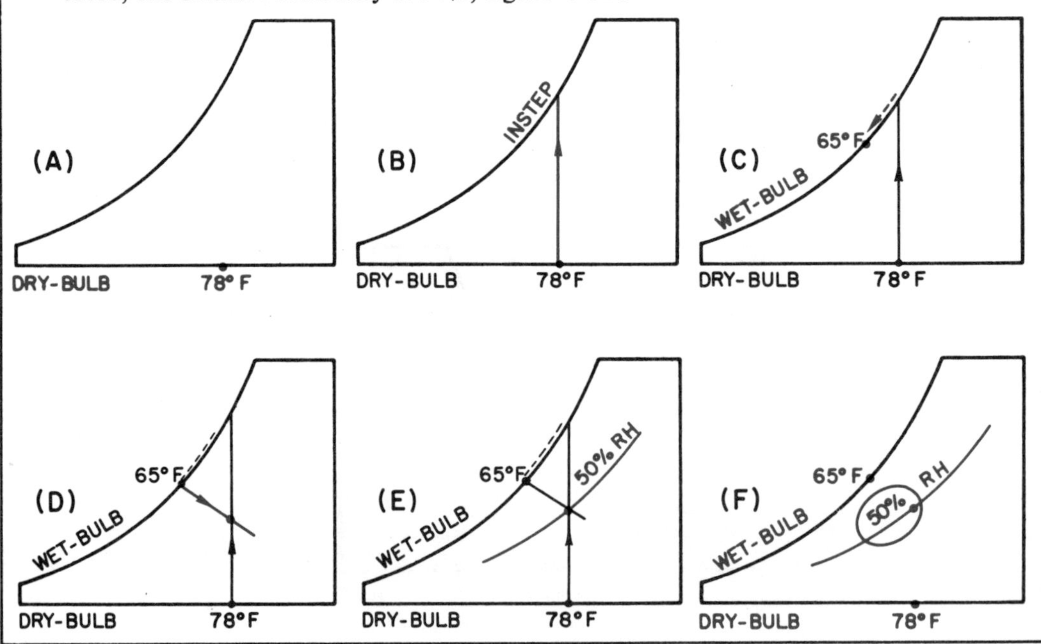

Fig. 6-6 Problem 1

PROBLEM 2

Dry-bulb Temperature, Relative Humidity – Wet-bulb Temperature

Given

Dry-bulb temperature 78°F   Relative humidity 50%

Find

Wet-bulb temperature

Solution

1. Locate 78°F on the dry-bulb scale at the bottom of the chart, figure 6-7A.
2. Draw a line from the 78°F line perpendicular to the sole until it intersects with the 50% RH line, figure 6-7B.
3. Locate the wet-bulb line that is closest to the intersection of the 78°F dry-bulb line and the 50% RH line, figure 6-7C.
4. Follow the wet-bulb line diagonally upward to the wet-bulb scale on the instep, figure 6-7D.
5. The wet-bulb temperature at that point is 65°F, figure 6-7E.

Thus, at a dry-bulb temperature of 78°F and 50% RH, the wet-bulb temperature is 65°F, figure 6-7F.

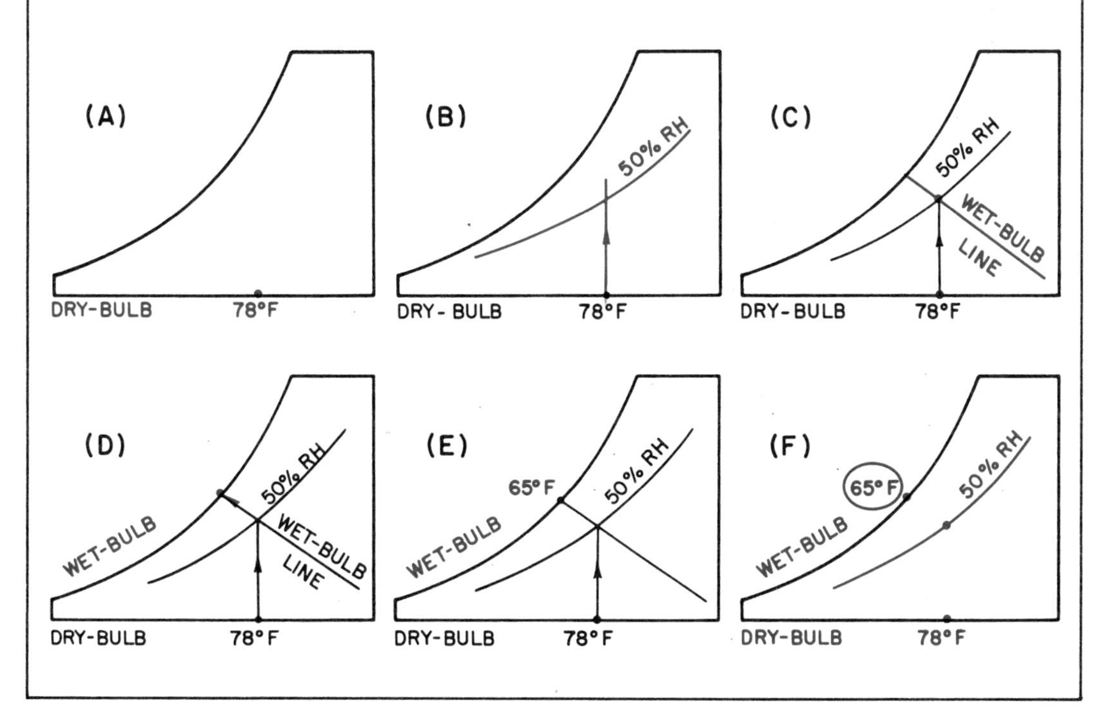

Fig. 6-7 Problem 2

PROBLEM 3

Wet-bulb Temperature, Relative Humidity – Dry-bulb Temperature

Given

Wet-bulb temperature 65°F

Relative humidity 50%

Find

Dry-bulb temperature

Solution

1. Locate 65°F on the wet-bulb scale, figure 6-8A.
2. Draw a line diagonally downward to the 50% RH line, figure 6-8B.
3. Locate the dry-bulb line that is closest to the intersection of the 65°F wet-bulb line and the 50% RH line, figure 6-8C.
4. Follow the dry-bulb line down to the dry-bulb scale, figure 6-8D.
5. The dry-bulb temperature at that point is 78°F, figure 6-8E.

Thus, at the conditions of a 65°F wet-bulb temperature and 50% RH, the dry-bulb temperature is 78°F, figure 6-8F.

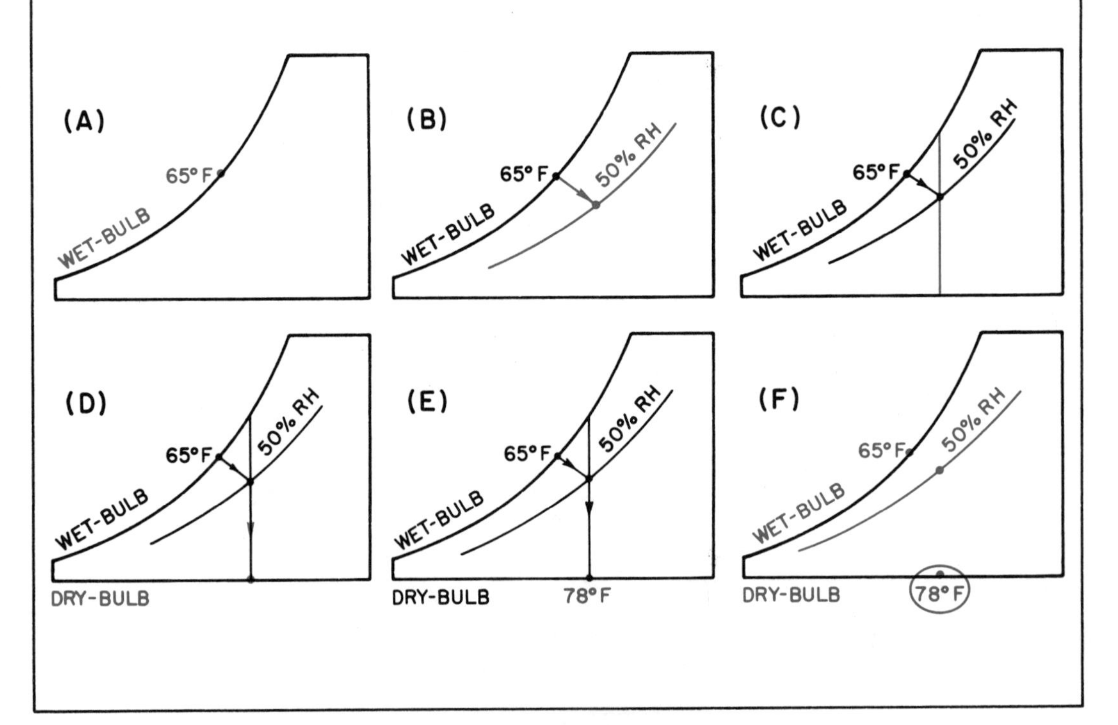

Fig. 6-8 Problem 3

PROBLEM 4

Dry-bulb Temperature, Wet-bulb Temperature – Dewpoint

Given

Dry-bulb temperature 78°F

Wet-bulb temperature 65°F

Find

Dewpoint

Solution

1. Find the intersection of the 78°F dry-bulb line and the 65°F wet-bulb line, figure 6-9A.
2. Follow the horizontal dewpoint line from the intersection to the in-step line, figure 6-9B.
3. The dewpoint temperature is read as 58°F, figure 6-9C.

Thus, at the conditions of a 78°F dry-bulb temperature and a 65°F wet-bulb temperature, the dewpoint temperature is 58°F.

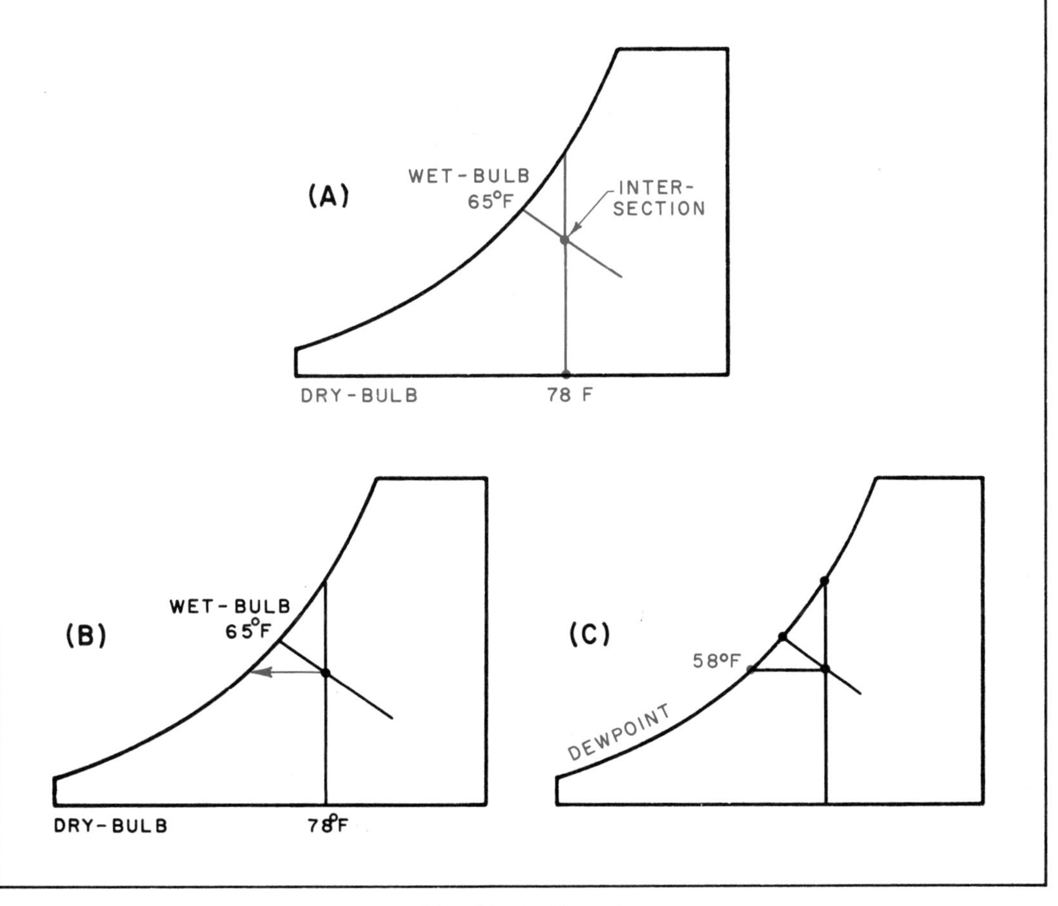

Fig. 6-9 Problem 4

PROBLEM 5

Wet-bulb Temperature, Relative Humidity – Dewpoint

Given

Wet-bulb temperature 65°F   Relative humidity 50%

Find

Dewpoint

Solution

1. Locate 65°F on the wet-bulb scale, figure 6-10A.
2. Follow the wet-bulb temperature line diagonally downward until it intersects with the 50% RH line, figure 6-10B.
3. Locate the dewpoint line at the intersection of the wet-bulb and relative humidity lines, figure 6-10C.
4. Follow the horizontal dewpoint line to the instep line and read the dewpoint temperature as 58°F, figure 6-10D.

Thus, at the conditions of a 65°F wet-bulb temperature and 50% RH, the dewpoint temperature is 58°F.

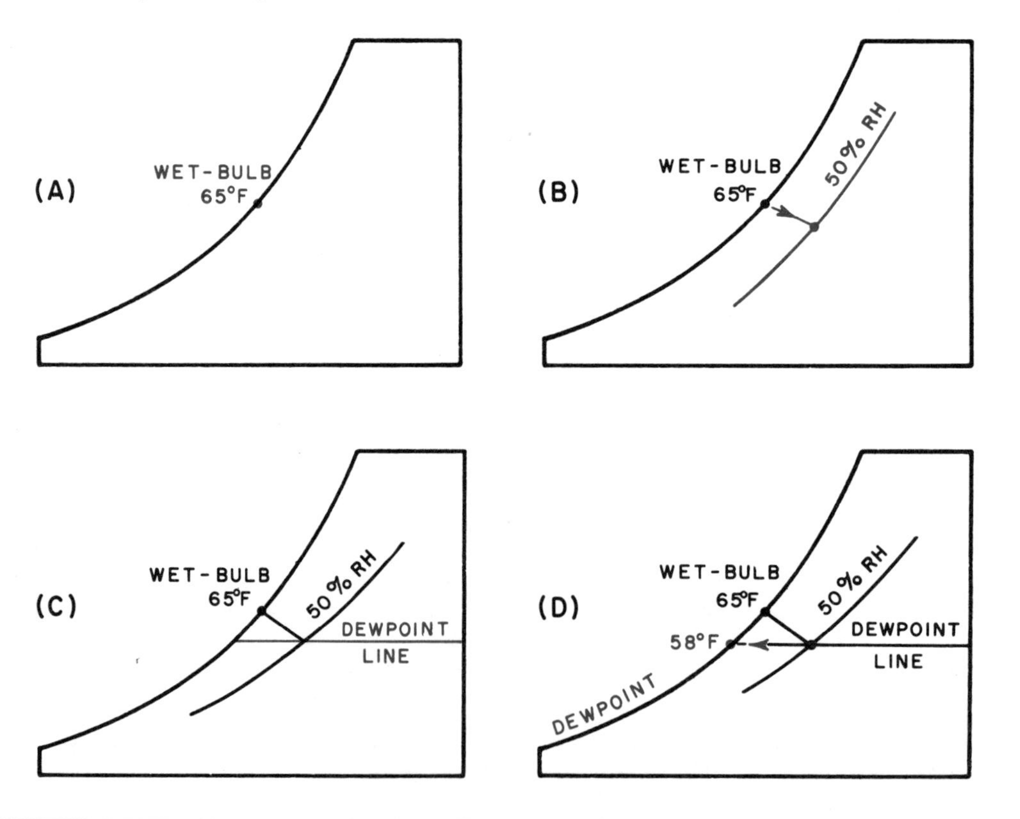

Fig. 6-10 Problem 5

As shown in Problem 3, the same conditions of 65°F wet-bulb temperature and 50% RH result in a dry-bulb temperature of 78°F. Thus, the same set of conditions can be used to determine more than one other condition. So far, the wet-bulb temperature and the relative humidity have been used to find the dry-bulb and dewpoint temperatures.

PROBLEM 6

Dry-bulb Temperature, Relative Humidity – Dewpoint

Given

Dry-bulb temperature 78°F

Relative humidity 50%

Find

Dewpoint

Solution

1. Find the intersection of the 78°F dry-bulb line and the 50% RH line, figure 6-11A.
2. Follow the horizontal dewpoint line to the instep line, figure 6-11B.
3. Read the dewpoint temperature, 58°F, figure 6-11C.

Thus, at the conditions of a 78°F dry-bulb temperature and 50% RH, the dewpoint temperature is 58°F.

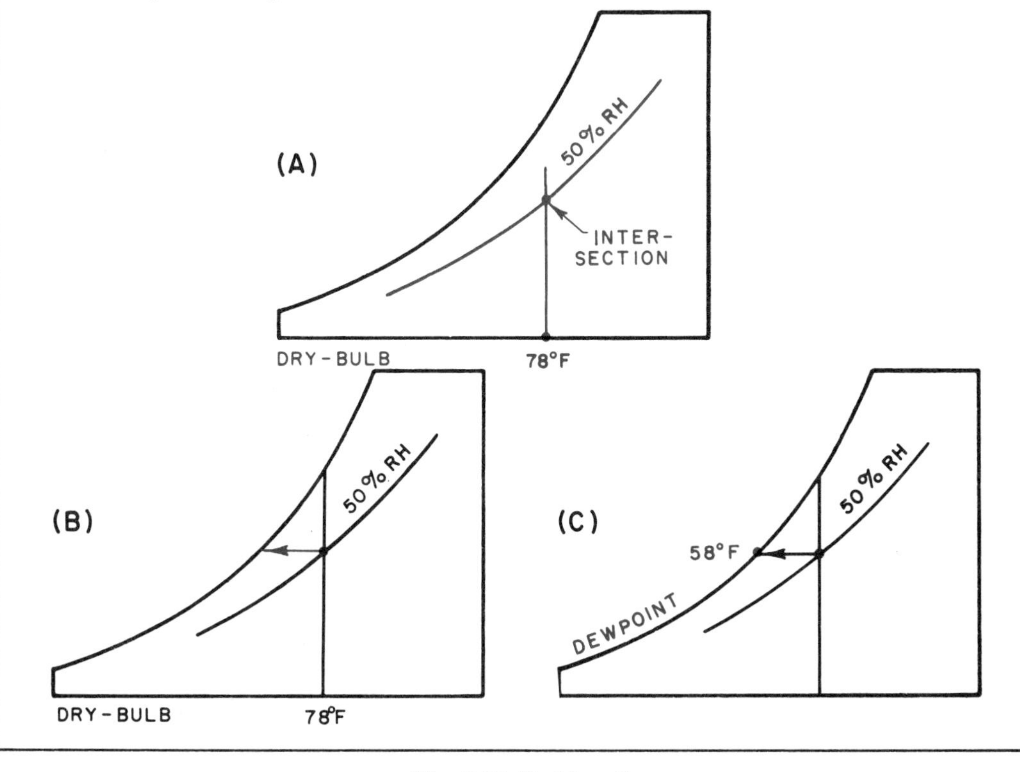

Fig. 6-11 Problem 6

As shown in Problem 2, the same conditions of 78°F dry-bulb temperature and 50% RH result in a wet-bulb temperature of 65°F. Thus, the same temperature conditions can be used to find more than one other condition. To this point, the dry-bulb temperature and relative humidity have been used to find the wet-bulb and dewpoint temperatures.

PROBLEM 7

Dry-bulb and Wet-bulb Temperatures – Grains of Moisture

Given

Dry-bulb temperature 78°F — Wet-bulb temperature 65°F

Find

Grains of Moisture

Solution

1. Find the intersection of the 78°F dry-bulb and 65°F wet-bulb lines, figure 6-12A.
2. Follow the horizontal grains of moisture line to the right (back of the shoe), figure 6-12B.
3. Stop at the first column of numbers which is the grains of moisture scale. Read the value of 72 grains of moisture, figure 6-12C.

Thus, at the conditions of 78°F dry-bulb and 65°F wet-bulb temperatures, the moisture in the air is 72 grains.

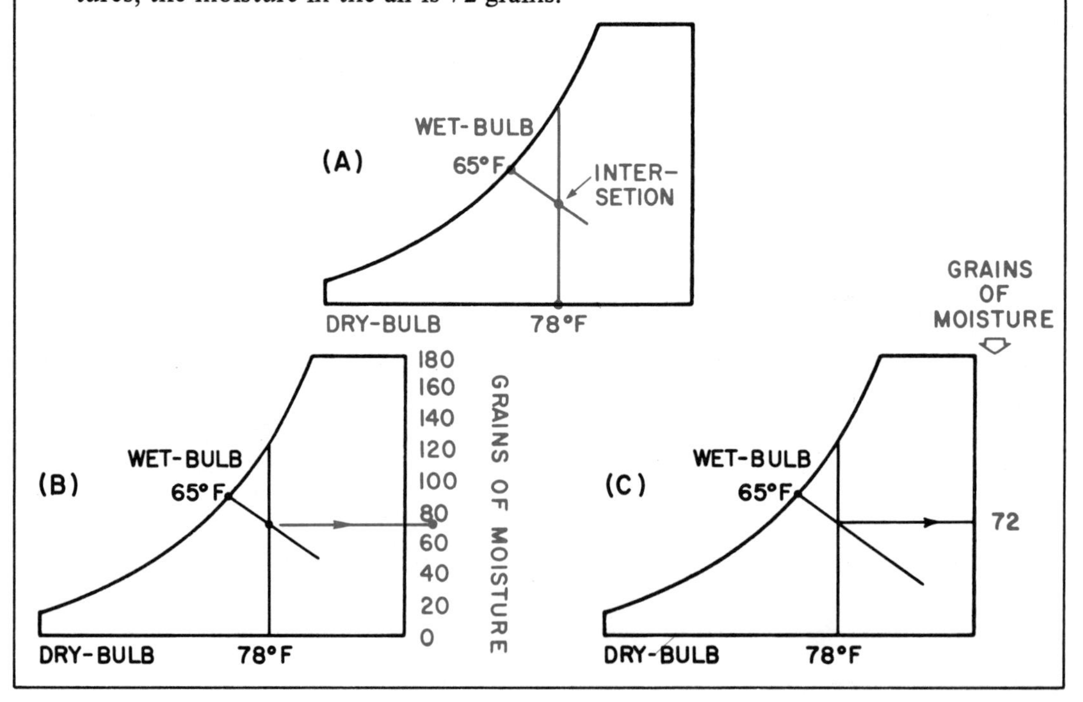

Fig. 6-12 Problem 7

Problem 7 shows how to find the grains of moisture in the air using the dry-bulb and wet-bulb temperatures. The value for the grains of moisture can also be found using other combinations of psychrometric properties. The combinations are listed as follows.

- dry-bulb temperature and relative humidity
- dry-bulb and dewpoint temperatures
- wet-bulb temperature and relative humidity
- wet-bulb and dewpoint temperatures

For each combination, simply find the intersection of the lines given and follow the point across the chart to the grains of moisture scale.

Grains of Moisture Per Pound of Dry Air or Per Cubic Foot of Air

Refer to figure 6-13A and note that the words "grains of moisture per pound of dry air" appear at the top of the scale. This means that at the conditions of a 78°F dry-bulb temperature and 65°F wet-bulb temperature, the air holds 72 grains of moisture per pound.

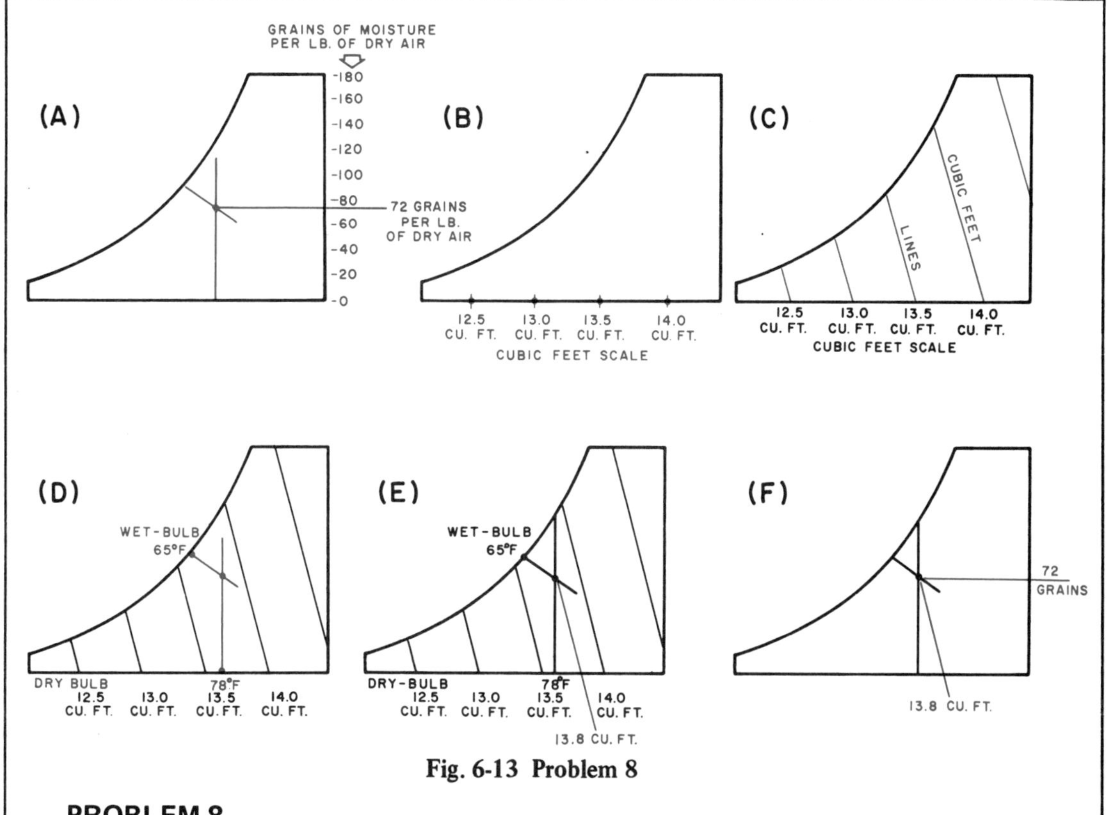

Fig. 6-13 Problem 8

PROBLEM 8

Moisture in the air can be measured per pound of air or per cubic foot of air. To find the moisture in a cubic foot of air, use the conditions given (78°F dry-bulb and 65°F wet-bulb temperatures) and proceed as follows:

1. Find the intersection of the 78°F dry-bulb and the 65°F wet-bulb temperatures, figure 6-13A.

2. Follow the horizontal grains of moisture line to the right to the grains of moisture scale.
3. Read 72 grains at this point.
4. Find the cubic foot scale along the sole of the shoe, figure 6-13B. The scale starts at 12.5 cubic feet and ends at 14.0 cubic feet. The cu. ft. lines extend diagonally from the sole of the shoe to the instep, figure 6-13C.
5. Again locate the intersection of the 78°F dry-bulb and 65°F wet-bulb lines, figure 6-13D.
6. Draw a line parallel to the cu. ft. line from the point of intersection located in step 5 to the sole of the shoe, figure 6-13E. The line crosses the sole at a point that is more than halfway between 13.5 and 14 on the cubic foot scale. Assume that the reading is 13.8 cubic feet.
7. Divide 72 grains by 13.8 cubic feet, figure 6-13F.

Thus, the moisture in the air = 72 ÷ 13.8 = 5 grains per cubic foot (approximately).

At 78°F dry-bulb temperature and 65°F wet-bulb temperature, the moisture in the air can be read as 72 grains per pound or 5 grains per cubic foot.

SUMMARY

- Psychrometrics is the study of the properties of air.
- The psychrometric chart simplifies the measurement of air properties.
- The psychrometric chart is a graphic representation of air properties and conditions, figure 6-14.

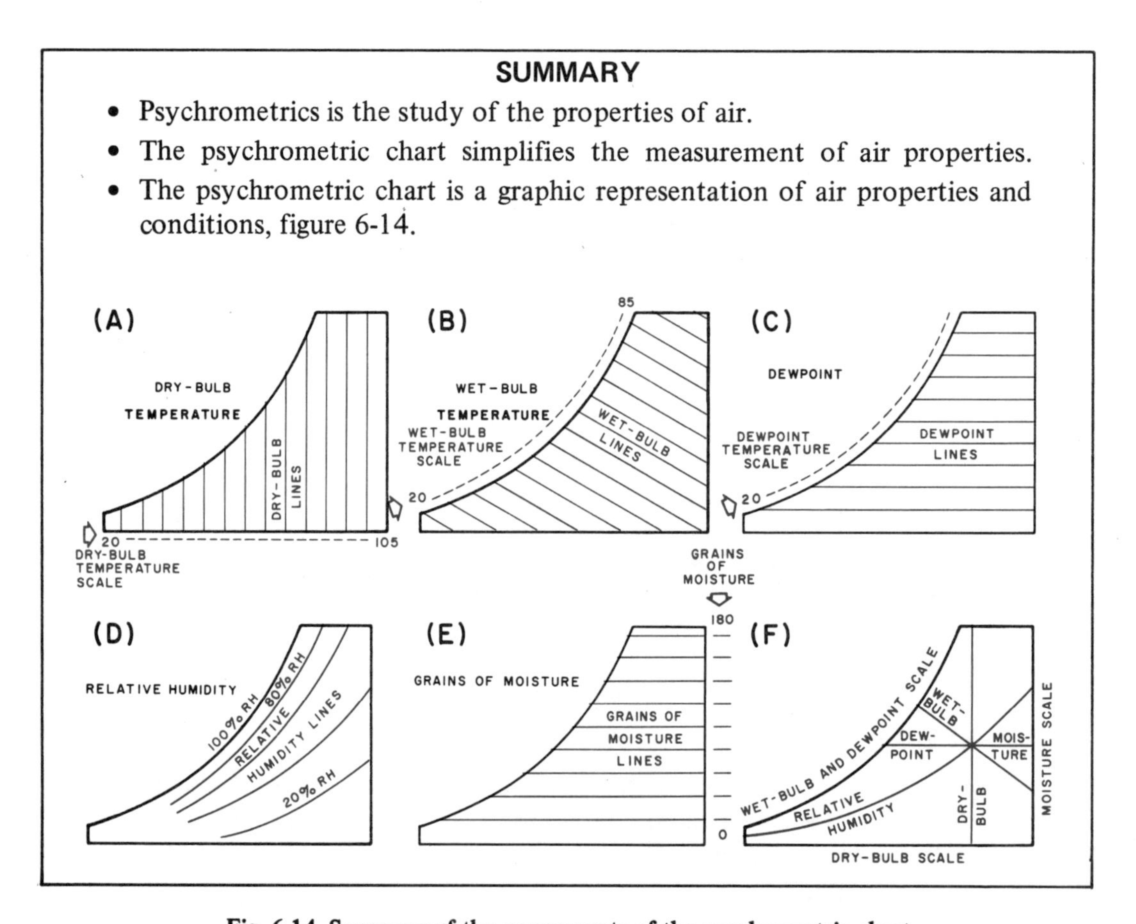

Fig. 6-14 Summary of the components of the pyschrometric chart

- Dry-bulb and wet-bulb temperatures, relative humidity, dewpoint, and grains of moisture are common psychrometric terms.
- If the values of any two of the psychrometric terms are known, the value of any other term can be found on the psychrometric chart.
- The psychrometric chart is shaped like a boot, figure 6-14. The sole is the dry-bulb temperature scale; the instep curve is the wet-bulb temperature and dewpoint temperature scale; the humidity lines are approximately parallel to the instep curve and follow along the side of the shoe; the grains of moisture scale is along the back of the shoe.
- The dry-bulb temperature lines are vertical on the chart; the wet-bulb temperature lines are diagonal; the dewpoint and grains of moisture lines are horizontal.
- For the wet-bulb thermometer, the bulb end of the thermometer is kept moist by a wet sock when the temperature reading is taken.
- A wet-bulb thermometer registers a lower temperature than a dry-bulb thermometer except at 100% RH.
- The bulb surface of a wet-bulb thermometer is cooled by the effect of moisture evaporating from the wet sock.
- The wet-bulb temperature is higher in wet air than it is in dry air at the same temperature.
- The amount of cooling that takes place at the bulb depends upon the amount of moisture in the air. At 100% RH, cooling ceases because the air is saturated and can no longer absorb moisture from the wet sock.

REVIEW

Use the psychrometric chart provided in the back cover pocket of this text to solve the following problems.

1. Find the relative humidity

	DB	WB	RH
a.	60°F	50°F	
b.	70°F	60°F	
c.	79°F	69°F	
d.	79°F	70.5°F	
e.	80°F	56.5°F	

2. Find the wet-bulb temperature

	DB	RH	WB
a.	72°F	34%	
b.	78°F	96%	
c.	79°F	62%	
d.	70°F	9%	
e.	66°F	40%	

3. Find the dry-bulb temperature.

	WB	RH	DB
a.	70°F	80%	
b.	40°F	10%	
c.	75°F	50%	
d.	30°F	20%	
e.	62°F	65%	

4. Find the dewpoint.

	DB	WB	DP
a.	70°F	61°F	
b.	90°F	81°F	
c.	55°F	50°F	
d.	101°F	62°F	
e.	80°F	71°F	

5. Find the grains of moisture

	DB	WB	Gr./lb.
a.	80°F	65°F	
b.	70°F	61°F	
c.	101°F	62°F	
d.	60°F	60°F	
e.	45°F	40°F	

6. Supply the missing values.

	DB	WB	RH	DP	Gr./lb.
a.	72°F				59 gr.
b.		61°F	45%		
c.	85°F		60%		
d.	68°F			24°F	
e.		71°F			63 gr.
f.			12%		10 gr.
g.			90%	68°F	
h.	98°F				180 gr.
i.	83°F	53°F			
j.			30%		10 gr.

Solve the following problems using the psychrometric chart as needed.

7. a. How many grains of water vapor equal 1 lb.?
   b. 0.01 lb.?
   c. 0.001 lb.?
8. How many pounds do the following quantities of moisture equal?
   a. 14 grains
   b. 70 grains
   c. 3,500 grains
   d. 70,000 grains
9. Under saturated conditions, how much more water vapor can 60°F air hold than 30°F air?
10. A room air conditioner is started when the temperature is 75°F and the RH is 70%.
    a. How much moisture does the air contain?
    b. After three hours of operation, the temperature is 70°F and the RH is 50%. How much moisture was removed from the air?
11. A homeowner complains of dryness resulting from the hot air heating system. When checked, the dry-bulb temperature is 72°F and the wet-bulb temperature is 48°F.
    a. What is the relative humidity?
    b. How much moisture must be added to the system to bring the RH to 50%?

UNIT 7
APPLICATION OF PSYCHROMETRIC TERMS

OBJECTIVES

After completing the study of this unit, the student will be able to

- apply and use the psychrometric terms humidity, dewpoint, and wet bulb.
- given a set of conditions, determine the remaining psychometric conditions.

Unit 6 described the basic psychrometric terms and pointed out their locations and relationships on the psychrometric chart. Unit 7 shows how these terms and the psychrometric chart can be put to practical use.

If the chart is to be used efficiently, the technician must understand the working uses of several of the psychrometric terms that can be determined from it. These terms are humidity, dewpoint, and wet-bulb temperature.

PRACTICAL APPLICATION OF THE TERM HUMIDITY

As stated previously, relative humidity indicates the amount of moisture in the air. This condition is an important factor in comfort air conditioning. The phrase comfort air conditioning is another way of describing air conditioning that provides comfort for the human body rather than air conditioning that is used for an industrial process.

Experiments have shown that certain combinations of moisture and air temperature within narrow ranges are more comfortable than other combinations. In winter, for example, 30% to 35% relative humidity at 72° to 75°F represents the indoor combination of moisture and temperature that is comfortable for most people. In summer, comfortable conditions occur at 45% to 50% relative humidity and approximately 75° to 78°F.

By applying this knowledge of the psychrometric chart, it is possible to determine what must be done to the outside air before it can be supplied to the room or conditioned space to maintain the most comfortable combination of moisture and temperature. The following problems show how air is conditioned to achieve the desired comfort conditions.

Problem 1 describes a simple winter heating operation in which a furnace, a boiler, or a heating coil adds heat. At the same time, a humidifier, water pan, or water spray adds moisture to the air.

PROBLEM 1 WINTER AIR CONDITIONING

Given

Outdoor dry-bulb temperature	30°F
Outdoor relative humidity	20%

Find

(a) The combination of moisture and dry-bulb temperature that falls within the indoor comfort conditions for winter (30% to 35% RH and 72°F to 75°F).

(b) The treatment required to change the outdoor air to meet the inside comfort conditions.

Solution

1. Place a dot on the psychrometric chart at the intersection of the dry-bulb temperature (30°F) and the 20% RH line, figure 7-1A.
2. Place a dot at the intersection of a dry-bulb line and a relative humidity line that fall within the indoor comfort condition range for winter (30% RH and 72°F), figure 7-1B.
3. Draw a line between the two dots, figure 7-1C.
4. By following the line from the first to the second intersection, it is evident that several changes must be made to bring the air to the desired temperature and relative humidity, figure 7-1D.
   (a) The relative humidity must be increased by adding 20% to 30% moisture.
   (b) Heat must be added to bring the dry-bulb temperature from 30°F to 72°F.

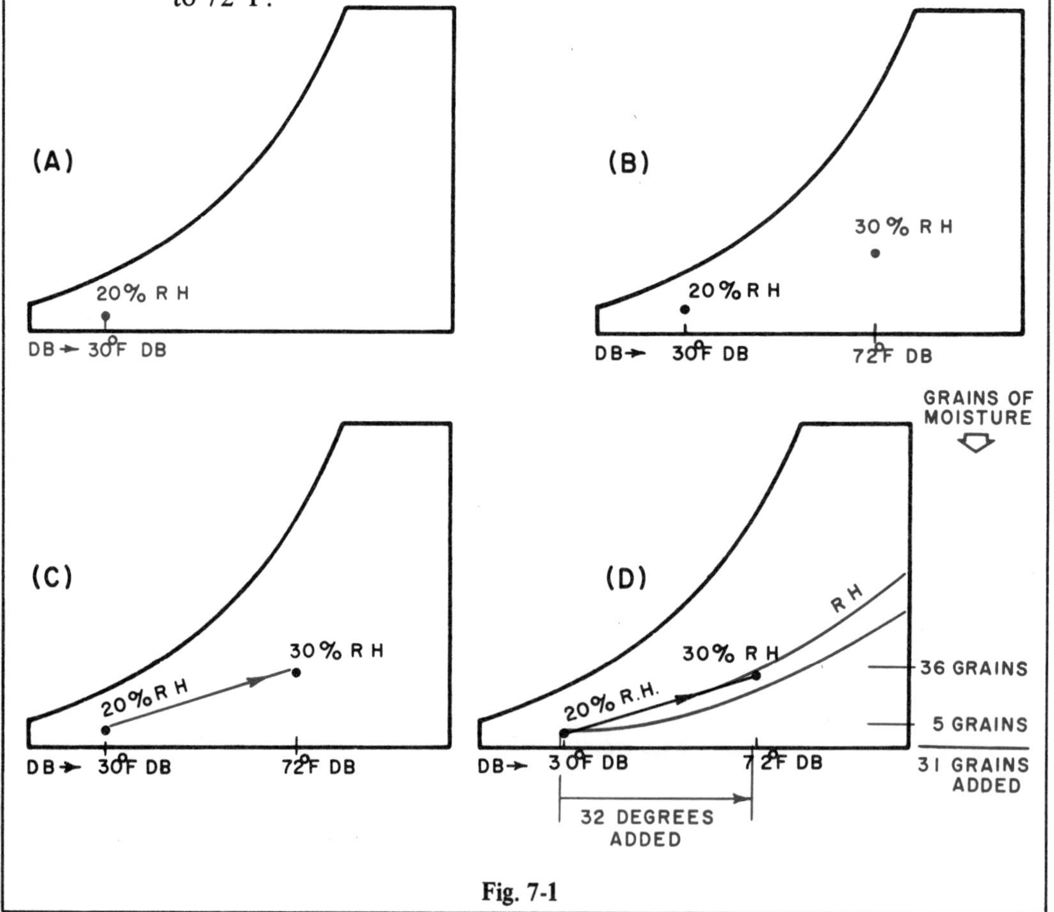

Fig. 7-1

Problem 2 shows a simple summer cooling operation in which a cooling coil removes both moisture and heat from the air.

PROBLEM 2 SUMMER AIR CONDITIONING

Given

Outdoor dry-bulb temperature 85°F
Outdoor relative humidity 70%

Find

(a) The combination of moisture (RH) and dry-bulb temperature that falls within the indoor comfort conditions for summer (45% to 50% RH and 75°F to 78°F).

(b) The treatment that is required to change the outside air to meet the inside comfort conditions.

Solution

1. Place a dot on the psychrometric chart at the intersection of the 70% RH line and the 85°F dry-bulb temperature, figure 7-2A.
2. Place a dot at the intersection of values for the relative humidity and the dry-bulb temperature that fall within the indoor comfort range for summer (50% RH and 75°F), figure 7-2B.

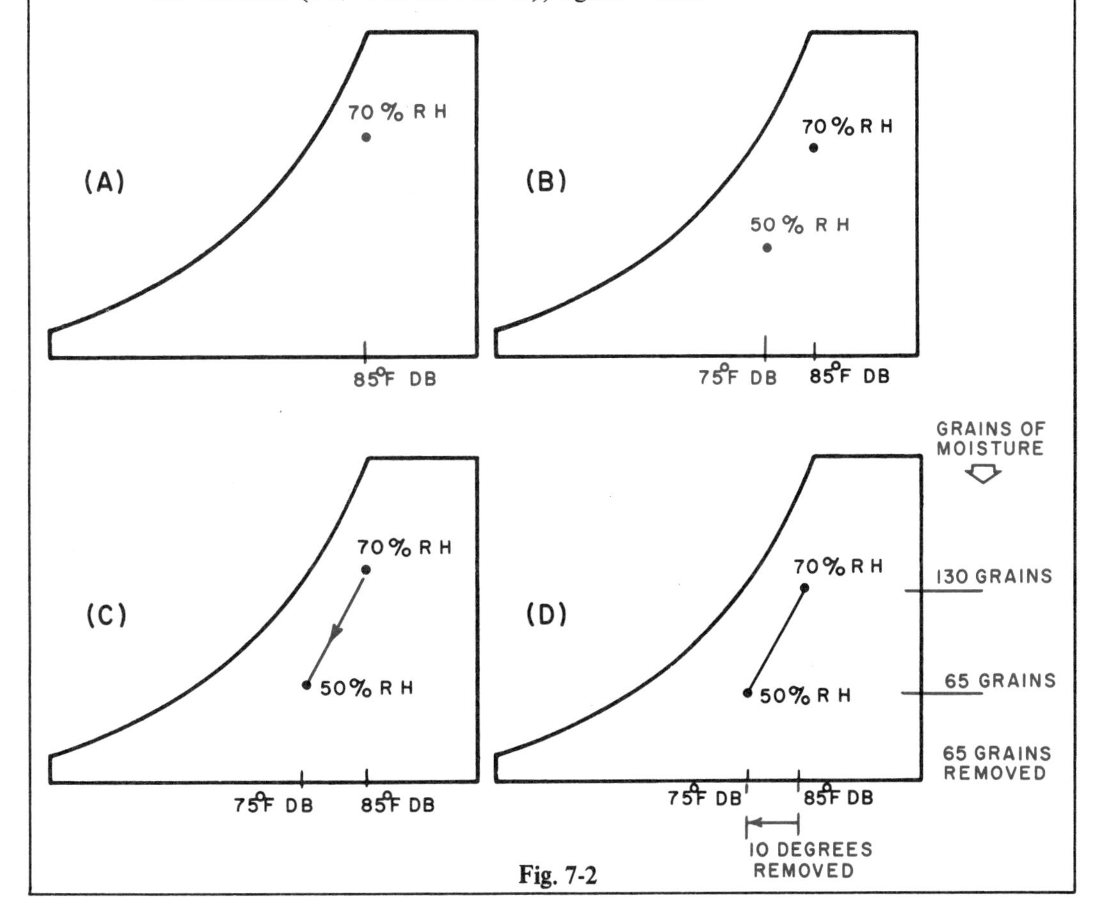

Fig. 7-2

3. Draw a line between the two dots, figure 7-2C.
4. Follow the line from the first to the second intersection points and note that several changes must be made to bring the air to the desired inside condition, figure 7-2D.
   (a) Since the relative humidity decreases from 70% to 50%, moisture must be removed from the air.
   (b) Since the temperature decreases from 85°F to 75°F, heat must be removed.

Problems 1 and 2 show the working relationship between relative humidity and dry-bulb temperature. If the relative humidity is maintained within the established comfort range (30% to 35% for winter and 45% to 50% for summer), the occupant in the air-conditioned space will be comfortable.

Using given temperature conditions, the psychrometric chart can be used, as in Problem 3, to find the value of relative humidity at which condensation forms on cold surfaces.

PROBLEM 3 CONDENSATION IN WINTER

Given

Window surface temperature	30°F
Indoor temperature	72°F

Find

The relative humidity at which condensation will *not* form on windows.

Solution

1. Use the window temperature as the dewpoint temperature, and locate 30°F on the dewpoint scale, figure 7-3A.
2. Find the intersection of the 30°F dewpoint line and the 72°F dry-bulb temperature line, figure 7-3B.
3. Read the relative humidity at the point of intersection: the relative humidity is approximately 20%, figure 7-3C. In other words, at 72°F, a relative humidity up to 20% keeps the windows dry. If the humidity is above 20%, moisture forms. In fact, under these conditions, moisture forms on any surface at a temperature of 30°F.

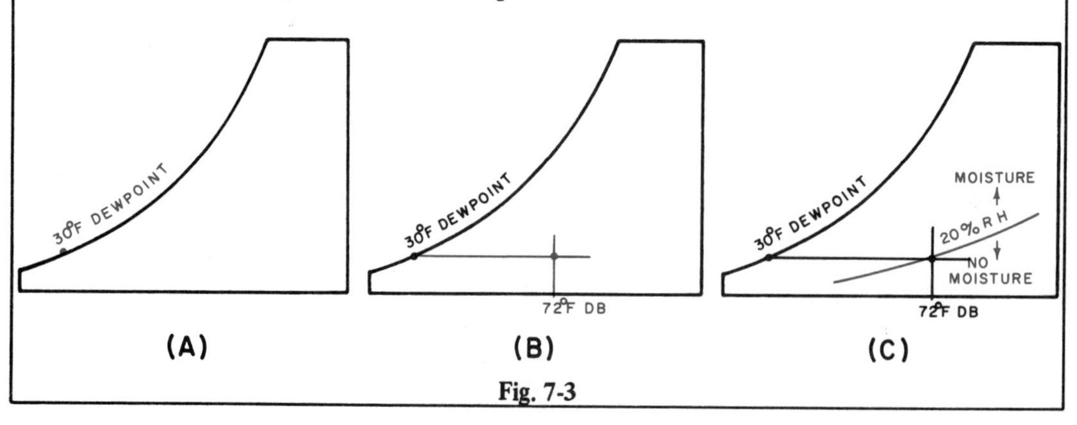

Fig. 7-3

Problem 1 states that a combination of 30% relative humidity and 72°F results in an acceptable comfort condition. However, Problem 3 shows that at 72°F, the maximum relative humidity that is permitted to prevent condensation is only 20%. This value is 10% less than that required for the acceptable comfort condition.

There are two possible alternatives to correct this lack of adequate moisture in the air.

Solution 1. It was shown previously that moisture can be removed from the air or can be prevented from building up in the air by air motion. Thus, blowing warm air over the surface of the window maintains a higher relative humidity without causing condensation. In addition, the warm air tends to increase the temperature of the window surface.

Solution 2. A second window surface (storm window) can be added or a window with a double layer of glass (thermopane), can be used to prevent heat loss and thus increase the surface temperature of the inside pane (above 30°F). As a result, the relative humidity can be increased to a more comfortable level.

In most instances, modern structures use both approaches to prevent condensation.

PRACTICAL APPLICATION OF THE DEWPOINT

In Problem 3, the relationship between dewpoint, relative humidity, and dry-bulb temperature is applied to an indoor comfort situation. In addition to illustrating the practical use of relative humidity, the problem also shows that the dewpoint temperature plays an important part in establishing and maintaining indoor conditions to prevent condensation from forming on cold surfaces such as windows.

A knowledge of dewpoint also has a practical application in those areas where the air is not conditioned. Air-conditioning supply ducts (carrying cold air) that run through unconditioned furnace rooms, equipment rooms, storage areas, or other unconditioned space, are likely to be covered with condensation.

PROBLEM 4 CONDENSATION IN UNCONDITIONED SPACES

Given

Unconditioned space dry-bulb temperature 90°F
Unconditioned space wet-bulb temperature 75°F
Cold air supply duct surface temperature 60°F

Find

The dewpoint temperature; determine if condensation will form on the duct.

Solution

1. To find the dewpoint temperature for the conditions given, locate the intersection of the 90°F dry-bulb and 75°F wet-bulb temperatures. Move along the horizontal line to the dewpoint scale. The dewpoint is approximately 69°F, figure 7-4A.
2. Use the duct surface temperature as the duct surface dewpoint or condensation temperature. The duct surface condensation temperature is 60°F, figure 7-4B.
3. The temperature at which condensation begins to form on the duct is 69°F. Any duct temperature below 69°F causes condensation. Since the duct temperature is 60°F, moisture condenses on the duct surface, figure 7-4C.

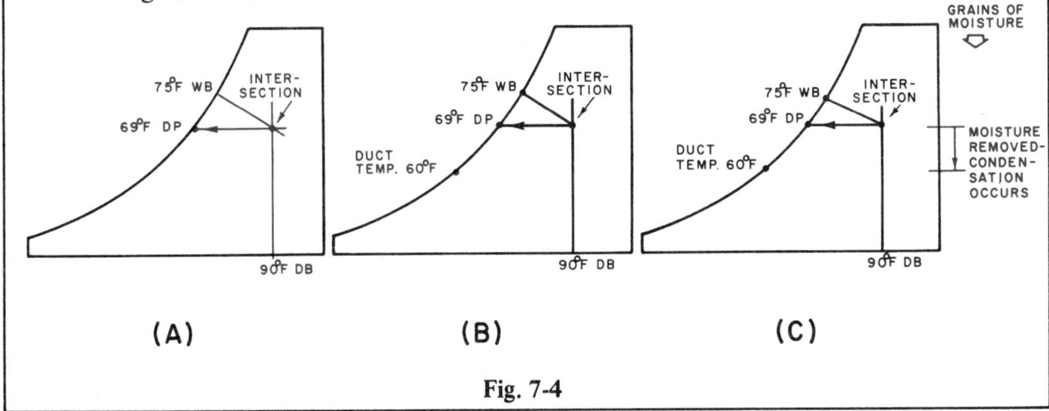

Fig. 7-4

In many instances, the water (condensate) that drops from the duct is harmful to the area below the duct. This is especially true if merchandise is stored in the area or if machinery is located below the duct. To prevent the harmful effects of condensation, it is necessary to prevent the formation of the moisture. The most common method of preventing condensate is to wrap the duct with insulation and then apply a vapor barrier, figure 7-5. The insulation must be thick enough to maintain the duct surface at a temperature that prevents condensation.

In summary, the dewpoint is used primarily to determine the temperature combinations of the air and the duct, wall, and window surfaces that produce condensation.

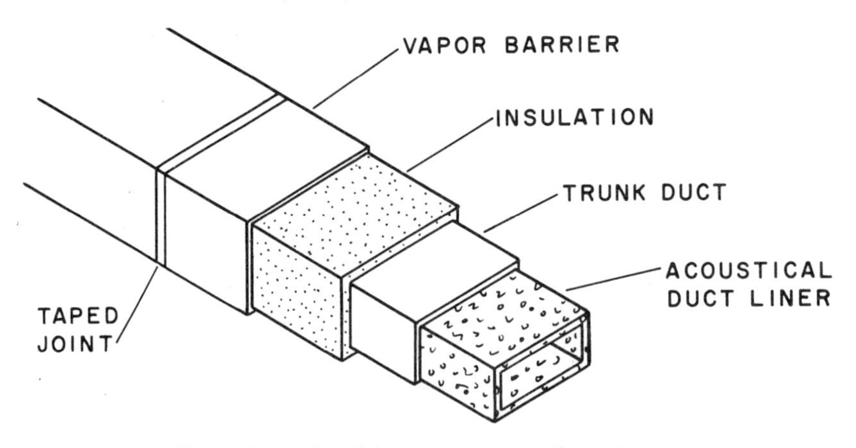

Fig. 7-5 Insulated duct prevents condensation

PRACTICAL APPLICATION OF THE WET-BULB TEMPERATURE

The wet-bulb temperature is used with the psychrometric chart to determine the moisture content in the air. It was shown previously that a direct relationship exists between the wet-bulb temperature and the moisture content of the air. Because of this relationship and because it is a simple process to determine the wet-bulb temperature, the wet-bulb values are important in determining the values of other psychrometric terms.

Problem 3 lists the wet-bulb temperature as one of the given conditions. In the same problem, the wet-bulb and dry-bulb temperatures are used to determine the dewpoint of the air in an unconditioned space. If these temperatures are not known, they can be found with a sling psychrometer. The dewpoint can then be found on the psychrometric chart using the temperature values just determined. Once the dewpoint of the air is known, it is possible to determine the surfaces on which condensation is likely to occur. In this instance, then, knowing the wet-bulb temperature simplifies the process of determining the temperature at which condensation may occur.

SUMMARY

- Certain combinations of moisture and temperature are more comfortable than other combinations.
- For winter indoor comfort, the preferred combination of relative humidity and dry-bulb temperature is 30% to 35% RH and 72°F to 75°F.
- For summer indoor comfort, the preferred combination of relative humidity and dry-bulb temperature is 45% to 50% RH and approximately 75°F to 78°F.
- The psychrometric chart can be used to determine the treatment that must be given to outside air before it can be delivered to a room or space.
- The psychrometric chart can illustrate a winter heating operation that requires the addition of heat and moisture, and a summer cooling operation that requires the removal of heat and moisture.
- Relative humidity is used to determine comfort conditions and to establish the conditions at which condensation will form on cold surfaces.
- A comfortable relative humidity can be maintained in winter by blowing warm air over the window surface or by installing double-pane glass (storm sash or thermopane), or both. As a result, the formation of condensate on the window is prevented.
- Dewpoint is used in conditioned or unconditioned areas to determine the temperature combinations at which condensation is likely to occur on cold surfaces.
- The most common method of preventing condensation on ducts in unconditioned spaces is the application of insulation and a vapor barrier to the duct surface.

- The wet-bulb temperature is important in simplifying the process of determining other pyschrometric values when the dry-bulb or ordinary temperature is known. The wet-bulb temperature is also important in simplifying the process of determining the temperature at which condensation may occur.

REVIEW

1. When the outdoor conditions are a dry-bulb temperature of 40°F and a relative humidity of 20%, it is desired to treat the air to arrive at the following indoor conditions: 75°F dry-bulb temperature and 35% RH. Determine:
   a. the amount of moisture that must be added or removed. ________
   b. the amount of heat that must be added or removed. ________
2. When the outdoor conditions are a dry-bulb temperature of 50°F and a wet-bulb temperature of 40°F, the desired indoor conditions are a 72°F dry-bulb temperature and 40% RH. Determine the amount of moisture that must be added or removed. ________
3. When the outdoor conditions are a dry-bulb reading of 30°F and a wet-bulb reading of 25°F, an indoor condition of 74°F and 40% RH is desired. Determine:
   a. the amount of heat to be supplied by a heating coil. ________
   b. the amount of moisture to be supplied by a humidifier. ________
4. The outdoor dry-bulb temperature is 90°F and the outdoor RH is 70%. The desired indoor condition is 75°F at 50% RH. Determine:
   a. the amount of moisture to be added or removed. ________
   b. the amount of heat to be added or removed. ________
5. The outdoor dry-bulb temperature is 92°F, the outdoor wet-bulb temperature is 70°F. The desired indoor conditions are 78°F and 50% RH. Determine:
   a. the amount of moisture to be added or removed. ________
   b. the amount of heat to be added or removed. ________
6. The dry-bulb temperature on the inside surface of a window is 30°F. The indoor air dry-bulb reading is 75°F at 30% RH.
   a. Will condensation form on the window? ________
   b. At what RH can condensation be prevented? ________
   c. If the indoor dry-bulb temperature rises to 80°F and the window surface temperature remains at 30°F, what affect is there on the relative humidity at which condensation can be prevented? ________
   ____
   ____
   ____

7. In the basement of a house, the dry-bulb temperature is 70°F and the wet-bulb temperature is 60°F. The surface of the basement walls has a dry-bulb temperature of 50°F.

   a. Will condensation form on the walls? ________

   b. At what RH can condensation be prevented? ________

   c. How much moisture must a dehumidifier remove to keep the walls dry? ________

   d. Will the moisture removal affect the dry-bulb air temperature? ________

8. A large cold water pipe runs through a furnace room where the dry-bulb reading is 90°F and the relative humidity is 50%.

   a. If the dry-bulb temperature of the surface of the pipe is 50°F, will the pipe sweat? ________

   b. What is the maximum pipe surface temperature required to prevent sweating? ________

   c. What is a common method of preventing pipe sweating (formation of condensate)? ________

UNIT 8
PSYCHROMETRIC PROCESSES

OBJECTIVES

After completing the study of this unit, the student will be able to

- relate the psychrometric terms to the conditions of the air as it passes through an air-conditioning system.
- define latent heat, sensible heating and cooling, sensible heat factor, evaporative cooling, air mixture, bypass air, and apparatus dewpoint.

The psychrometric processes described in this unit show in a simple, graphic manner the relationship of the psychrometric terms to the changes that take place in the condition of the air as it passes through a typical air-conditioning process. These changes involve heating or cooling and the addition or removal of moisture.

A TYPICAL AIR-CONDITIONING PATTERN

Many air-conditioning systems take air from the room and return it to the air-conditioning apparatus where it is reconditioned and supplied again to the room. For added ventilation, most systems mix the return air from the room with outdoor air.

An example of a typical psychrometric pattern describing the use of room air and outdoor air is illustrated in figure 8-1.

1. Air is taken from the room at 75°F and is returned to the air-conditioning apparatus, figure 8-1A.
2. The 75°F indoor air is mixed with 85°F outdoor air, figure 8-1B.

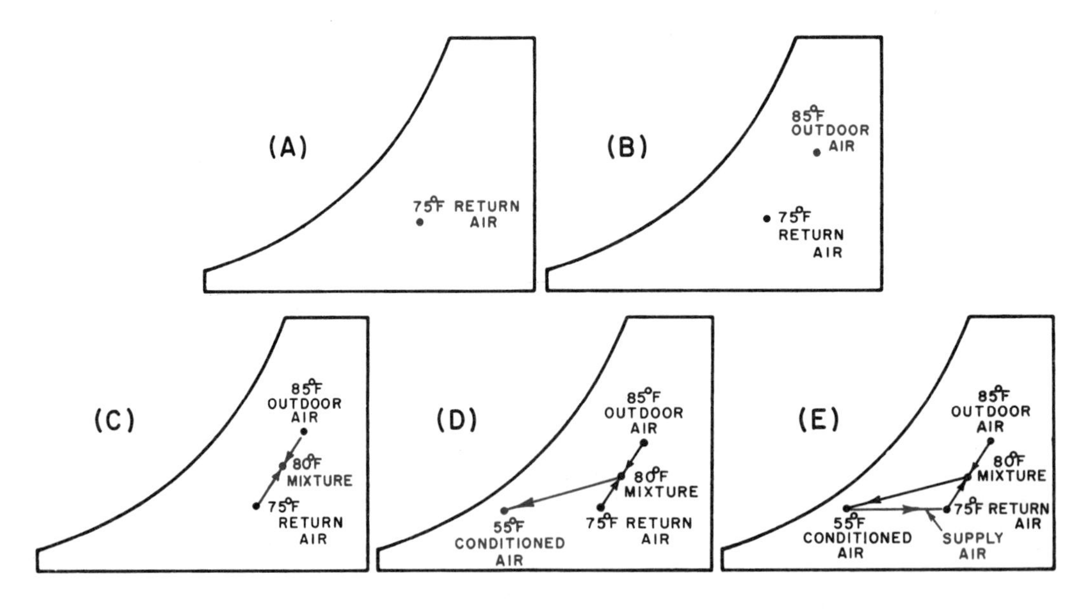

Fig. 8-1

3. The temperature of the resulting air mixture is 80°F, figure 8-1C.
4. The air mixture enters the conditioning apparatus at 80°F and is conditioned to 55°F, figure 8-1D.
5. The 55°F air is then supplied to the room where it picks up heat from the room and reaches a temperature of 75°F before it is again returned to the apparatus and mixed with outdoor air, figure 8-1E.

The psychrometric chart shows the following processes for an air-conditioning cycle.

- The room air picks up heat and moisture when it is mixed with outdoor air.
- Heat and moisture are removed as the mixture passes through the conditioning apparatus.
- The air mixture is supplied to the room at a condition that is dry enough and cool enough to maintain the required room temperature and humidity conditions.

PSYCHROMETRIC PATTERNS FOR HEATING AND COOLING PROCESSES

The psychrometric pattern for a typical air-conditioning process shows that air is heated or cooled and moisture is added or removed. To establish an accurate understanding of each of the psychrometric processes, it is necessary to know the basic types of heating and cooling changes that can be interpreted on the psychrometric chart. These changes are known as latent heating and cooling and sensible heating and cooling.

LATENT HEAT

The general term *latent heat* may refer to either the latent heat of vaporization or the latent heat of fusion. The *latent heat of vaporization* is the heat required to change a liquid to a vapor without increasing the temperature of the fluid. For example, water can be heated to its boiling point or 212°F. If more heat is added, the temperature of the water does not increase but the water begins to boil and vaporize. Thus, the latent heat of vaporization in this case is the heat required to change water at 212°F to a vapor at 212°F.

The *latent heat of fusion* is the amount of heat that must be removed to change a liquid to a solid at the same temperature. For example, assume that water is cooled to the freezing point, 32°F. If more heat is removed, the water changes to ice. Thus, the latent heat of fusion actually reflects a cooling process, because heat is removed from the water.

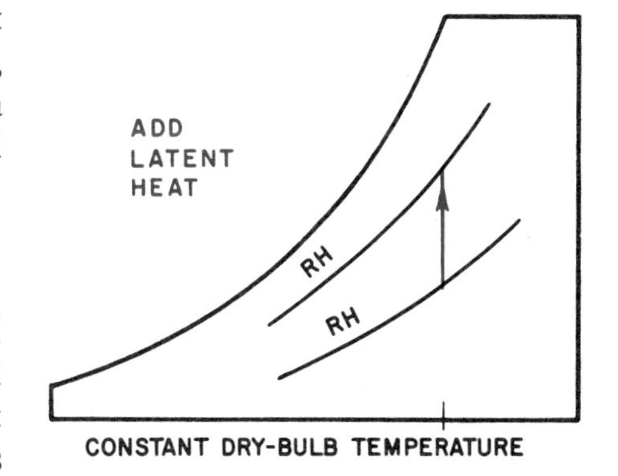

Fig. 8-2

Latent Heat Applied to Air

When the latent heat principle is applied to air, the changes that occur are related to the moisture content of the air. As latent heat is added, the moisture content increases but the air temperature does not change.

Any condition in which evaporation takes place without changing the air temperature, adds latent heat to the air.

This effect is shown on the psychrometric chart as a straight dry-bulb line indicating an increase in humidity.

As latent heat is removed, the moisture content decreases but the air temperature remains unchanged.

Any condition which causes condensation without changing the air temperature, removes latent heat.

This effect is shown on the psychrometric chart as a straight dry-bulb line indicating a decrease in humidity, figure 8-3.

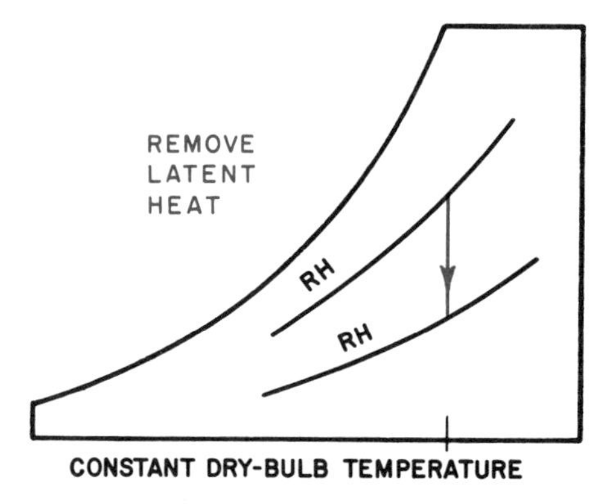

Fig. 8-3

For air-conditioning processes, then, latent heat involves evaporation and condensation. Latent heat is also specifically related to the conditions and devices which cause either process to occur. Because the removal of latent heat is actually a cooling process, this change is sometimes called latent cooling.

SENSIBLE HEATING AND COOLING

Sensible heat is any heat that raises the temperature, but not the moisture content, of a substance. Because a temperature change is involved, sensible heat can be detected by human senses.

Examples of sensible heat include the heat required to raise the temperature of water and the heat removed to lower the temperature of water. When heat is removed to lower the temperature of the water, the process is also known as sensible cooling.

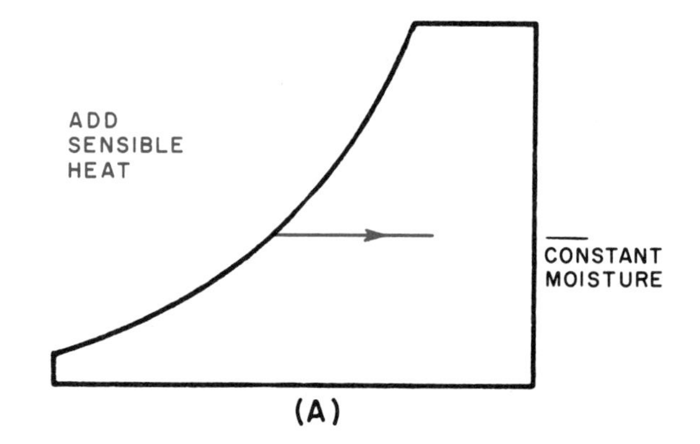

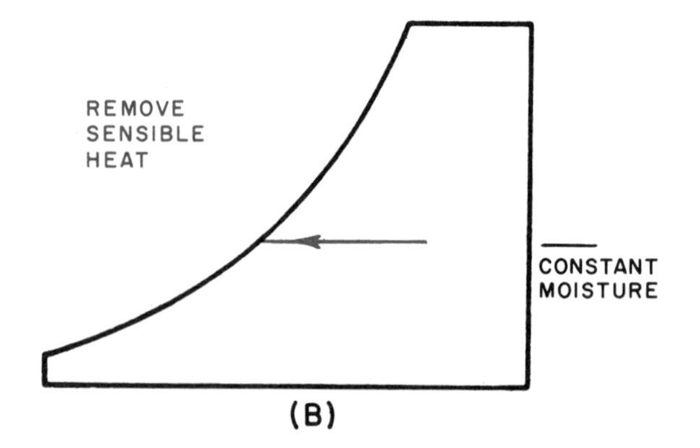

Fig. 8-4

Sensible Heat Applied to Air

When the sensible heat principle is applied to air, the changes that occur are related to the air temperature. As sensible heat is added, the air temperature increases;

however, there is no change in the moisture content of the air. This effect is shown on the psychrometric chart as a straight horizontal line starting at the left and extending to the right, figure 8-4A. For example, if the air temperature is 50°F, and 20 degrees of sensible heat are added, the air temperature rises to 70°F. Note in the figure that the line indicating this temperature rise remains at the same moisture content level.

As sensible heat is removed from the air, the air temperature decreases. Again, there is no change in the moisture content. This effect is shown on the psychrometric chart as a straight horizontal line starting at the right and extending to the left, figure 8-4B. For example, if the air temperature is 70°F and 20 degrees of sensible heat is removed, the air temperature drops to 50°F. Again note that the line illustrating the temperature decrease remains at the same moisture content level.

To this point, the psychrometric chart has shown that the sensible heating or cooling process changes the dry-bulb temperature but does not change the dewpoint temperature of the air. The chart can also show what happens to the wet-bulb temperature and the relative humidity during a sensible heating or cooling process, figure 8-5.

In this case, assume that the conditions given for figure 8-4A also are true for this example. Thus, the dry-bulb temperature increases 20 degrees to 70°F and the dewpoint temperature remains unchanged at 45°F when sensible heat is added to the air. In addition, the wet-bulb temperature increases from 47°F to 56°F and the relative humidity decreases from 80% to 40%. Figure 8-5 and Table 8-1 give a composite picture of the changes that take place when sensible heat is added to the air.

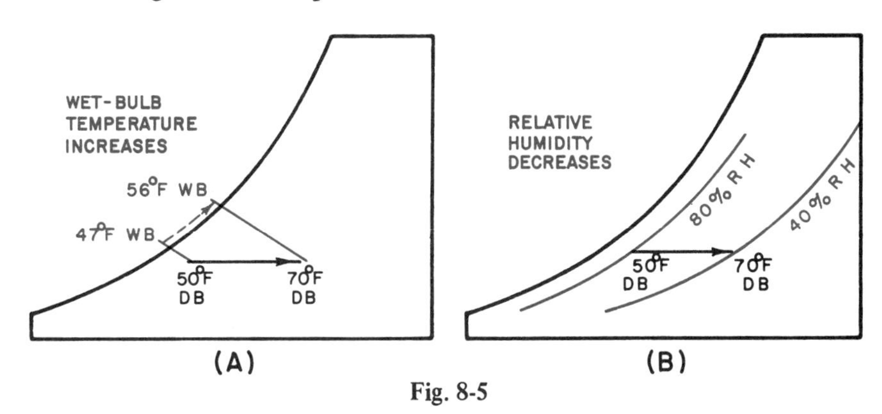

Fig. 8-5

Sensible Heat	Starting Conditions	Conditions after Adding Sensible Heat	Change
DB (dry-bulb temperature)	50°F	70°F	Increase
DP (dewpoint temperature)	45°F	45°F	No change
WB (wet-bulb temperature)	47°F	56°F	Increase
RH (relative humidity)	80%	40%	Decrease

Table 8-1 Adding sensible heat

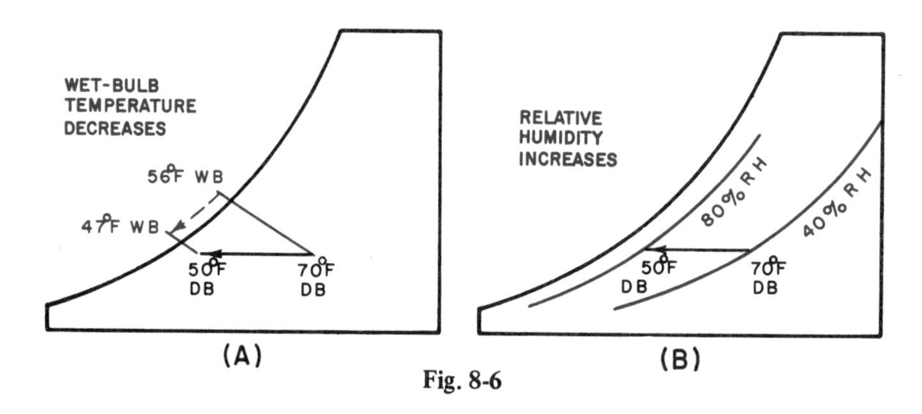

Fig. 8-6

Sensible Cooling	Starting Conditions	Conditions after Removing Sensible Heat	Change
DB (dry-bulb-temperature)	70°F	50°F	Decrease
DP (dewpoint temperature)	45°F	45°F	No change
WB (wet-bulb temperature)	56°F	47°F	Decrease
RH (relative humidity)	40%	81%	Increase

Table 8-2 Removing Sensible Heat

When sensible cooling takes place using the conditions given for figure 8-4B, the dry-bulb temperature decreases from 70°F to 50°F and the dewpoint temperature remains the same at 45°F. The wet-bulb temperature, however, drops from 56°F to 47°F. The relative humidity increases from 40% to 80%. Figure 8-6 and Table 8-2 give a composite picture of the changes that take place when sensible heat is removed from the air.

SENSIBLE AND LATENT HEATING AND COOLING

When sensible heat and latent heat are added at the same time in an air-conditioning process, the process provides heating and humidification. Examples of such a process occur in comfort air conditioning during the winter season when the air is usually colder and dryer. Typical examples are as follows:

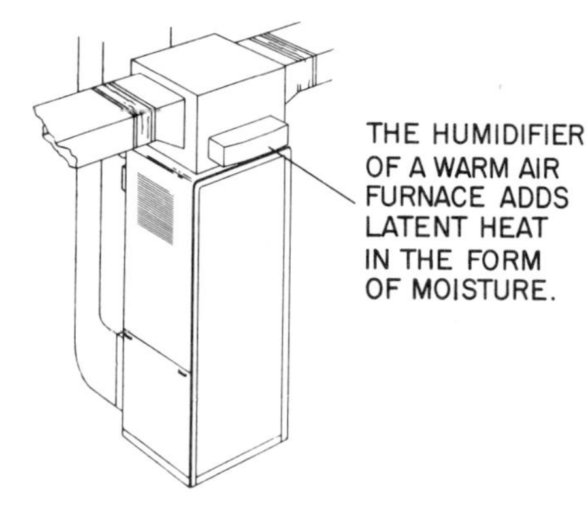

THE HUMIDIFIER
OF A WARM AIR
FURNACE ADDS
LATENT HEAT
IN THE FORM
OF MOISTURE.

Fig. 8-7

Residential Installation. The comfort air-conditioning process in winter requires sensible heat from a furnace source and latent heat from an evaporation pan system in the furnace. These sources will keep the air at a comfortable temperature and humidity level.

Commercial Installation. For large commercial structures such as office buildings, schools, hospitals, hotels, and churches, the comfort conditioning system requires heating coils to supply sensible heat and sprays to supply latent heat to keep the temperature at a comfortable level.

Sensible heat alone reduces the relative humidity and makes the air dry. The effect of dry air is noticed especially during the winter heating season. For example, people find that their nasal and throat linings are dry. In addition, the wooden parts of a structure, such as joists and beams, appear to be dry. Thus, latent heat must be used to add moisture and increase the relative humidity. An example of simultaneous heating and humidification is shown in figure 8-8.

Heating and Humidifying

Air that is to be heated and humidified must pass through a heating coil and a water spray coil. Assume that air enters the heating coil at 60°F DB (dry-bulb temperature), 45°F WB (wet-bulb temperature), 25°F DP (dewpoint), and 28% RH (relative humidity).

The same air leaves the spray coil at 80°F DB, 62°F WB, 50°F DP, and 35% RH. As a result, a reverse L psychrometric pattern occurs, as shown in figure 8-8A. In actual practice, however, heating ahd humidifying take place almost simultaneously. Therefore, the psychrometric pattern follows a straight diagonal line rather than a reverse L pattern, figure 8-8B.

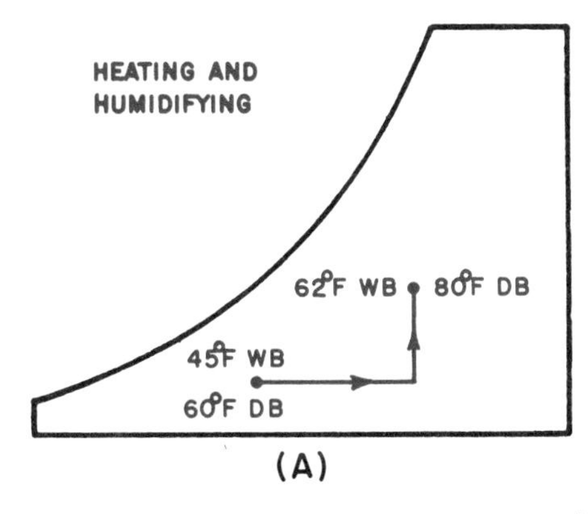

(A)

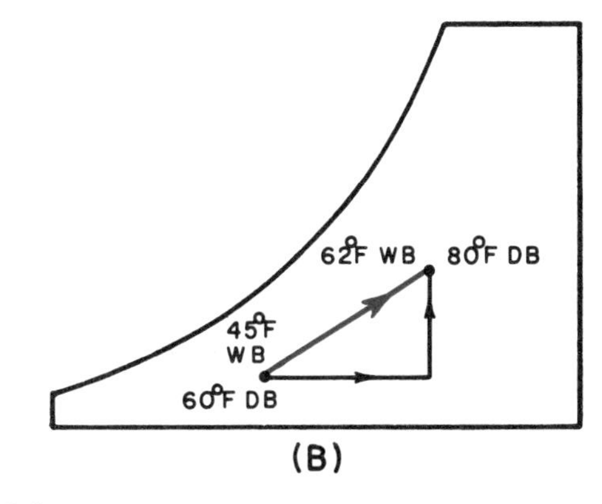

(B)

Fig. 8-8

Cooling and Dehumidifying

Air to be cooled and dehumidified must pass through a cooling coil. In this process, sensible heat is removed. As the air approaches its final temperature, latent heat is also removed. The cooling and dehumidifying process produces changes that are opposite to those produced by the heating and humidifying process. That is, sensible heat and latent heat are removed rather than added.

Assume that air enters the cooling coil at 80°F DB, 62°F WB, 50°F DP, and 35% RH. The same air leaves the cooling coil at 60°F DB, 45°F WB, 25°F DP, and 28% RH, figure 8-9A. In actual practice, the cooling and dehumidifying process takes place almost simultaneously. Therefore, the psychrometric pattern follows a straight diagonal line representing the simultaneous processes of sensible cooling and dehumidifying, figure 8-9B.

Sensible Heat Factor

The simultaneous cooling and dehumidifying process occurs so frequently in air conditioning that the psychrometric line representing this process is called the *sensible heat factor line.*

This line represents the change that takes place in the sensible heat and the latent heat. If the cooling process involves the removal of only sensible heat (no latent heat is removed), the sensible heat factor line is horizontal, figure 8-10A. The sensible heat factor, in numerical terms, is 1.0. If 50% sensible heat and 50% latent heat are removed, the sensible heat factor line slopes at an angle of approximately 45°, figure 8-10B. In this case, the numerical value of the sensible heat factor is 0.5. The 0.5 factor indicates that half of the sensible heat and half of the latent heat are removed in the process.

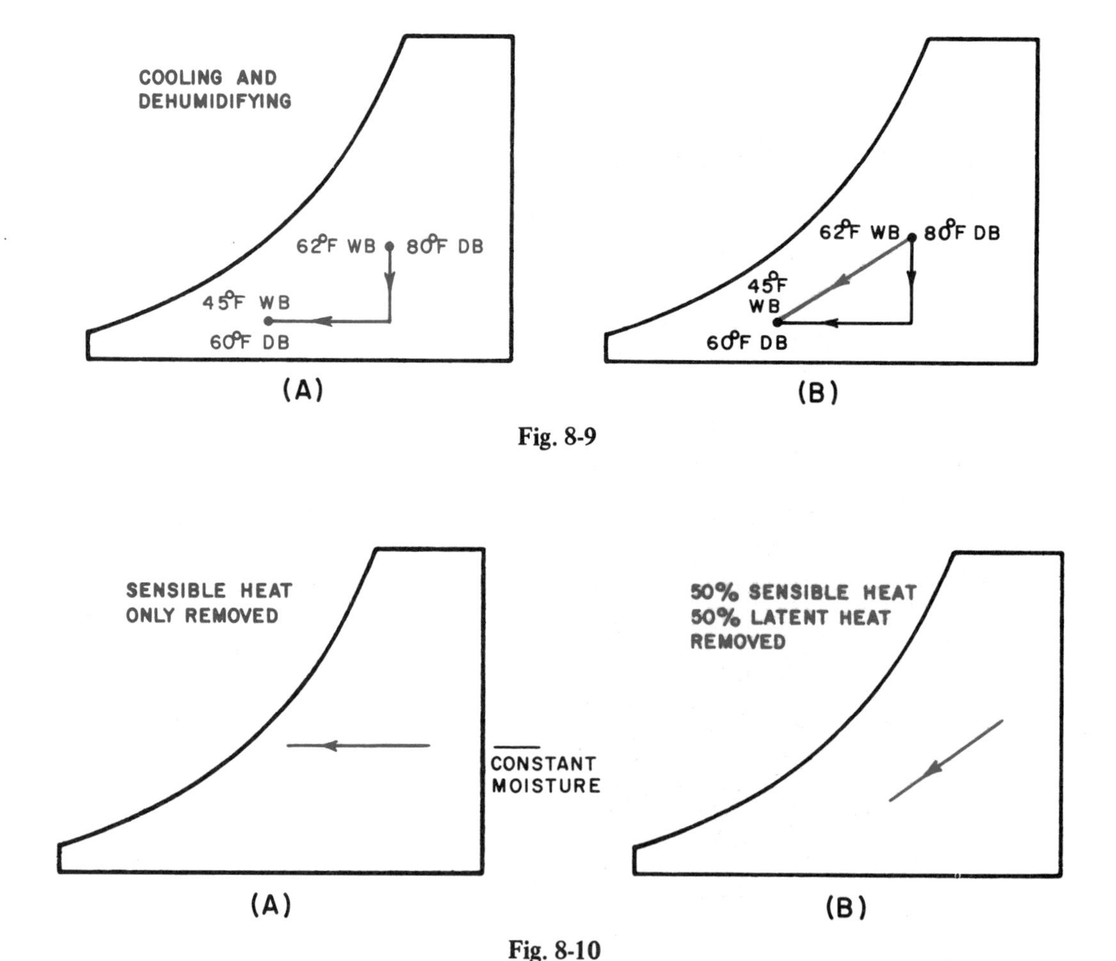

Fig. 8-9

Fig. 8-10

In most comfort conditions, the sensible heat factor has a value greater than 0.5. This condition is reasonable since most comfort cooling applications remove more sensible heat than latent heat. The percentages of the types of heat removed do vary, however.

A typical residence may have a sensible heat factor of 0.8. The sensible heat factor for a typical restaurant may be 0.6. Of course, the restaurant produces more latent heat in the form of steam and vapor because cooking and dishwashing are being done continually and also because of the concentration of people.

The 0.8 factor for the residence indicates that 80% (0.80) of the total heat change is due to a change in sensible heat; the remaining 20% is the change in latent heat. Thus, if the total cooling requirement is 10 tons, 8 tons of cooling are required to remove the sensible heat and 2 tons are required to remove the latent heat.

A practical use of the sensible heat factor is shown in Problem 1. The following conditions must be known for this problem: the total amount of cooling that is required, the conditions that are to be maintained in a room, and the condition of the air that is supplied to the room.

PROBLEM 1 SENSIBLE HEAT FACTOR

Given

Cooling load	10 tons
Desired room conditions	80°F DB temperature
	67°F WB temperature
Supply air conditions	60°F DB temperature
	58°F WB temperature

Find

Sensible heat factor

Solution

1. Plot the given conditions on a psychrometric chart and draw a straight line between the two points, figure 8-11A.
2. Extend the line to the sensible heat factor scale on the right-hand side of the chart, figure 8-11B.
3. Read the value 0.80 on the sensible heat factor scale. This value means that 80% of the heat loss is due to sensible heat.

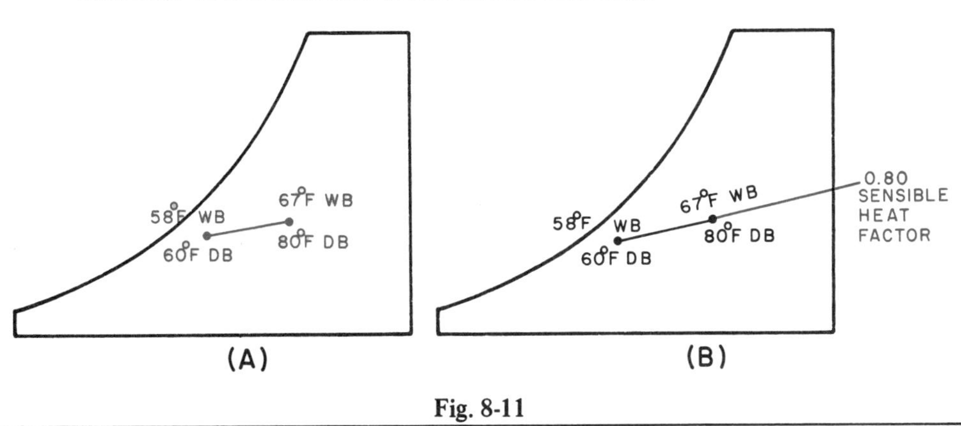

Fig. 8-11

For the conditions of Problem 1, the conditioning equipment must be capable of removing 8 tons of sensible heat and 2 tons of latent heat.

Evaporative Cooling

Evaporative cooling is the process by which sensible heat is removed from the air and latent heat (moisture) is added. The evaporative cooling process can be plotted on the psychrometric chart as a diagonal line extending upward and to the left, figure 8-12.

Evaporative cooling requires a spray coil. As air passes through the water spray, it gives up heat to the cooler water. Some of the water evaporates because of the heat picked up from the air. The resulting vapor is carried along in the airstream. As a result, the air is cooled and humidified. In the case shown in figure 8-12, the air enters the spray water at 95°F DB and is cooled to 75°F DB. The moisture content for this example increases from approximately 70 grains per pound to 102 grains per pound. The relative humidity increases from approximately 29% to 80%. The wet-bulb temperature remains constant at 70°F. As shown by the psychrometric chart in figure 8-12, most evaporative cooling processes travel diagonally upward to the left along the wet-bulb line.

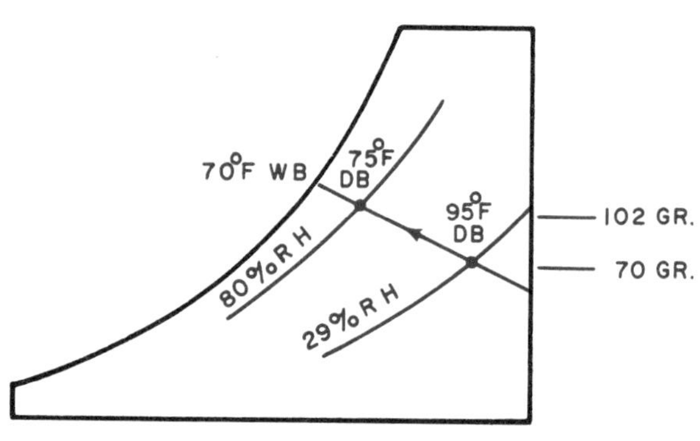

Fig. 8-12

Evaporative cooling is used primarily for industrial conditioning rather than for comfort air conditioning. Certain industrial processes that require high relative humidity (such as textile manufacturing) may use evaporative cooling.

Air Mixture

Air mixture is another air-conditioning process that can be plotted on the psychrometric chart. The air cycle described in Unit 3 indicates that return air from the room is reconditioned and supplied again to the room. In some air-conditioning applications, the supply air may consist of 100% return air and no outdoor air. In most cases, however, the supply air consists of return air from the room plus outdoor air. The outdoor air is required for ventilation purposes in that it serves to replenish the oxygen supply.

A mixture of outdoor air and return air can be plotted on the psychrometric chart. The resulting air mixture temperature can be determined immediately. Once this temperature is known, it is possible to determine the treatment required and the changes that must be made to maintain the desired room conditions. Problem 2 illustrates the procedure used to determine the the air mixture temperature.

PROBLEM 2 AIR MIXTURE

Given

Total air quantity	2,000 cfm (cubic feet per minute)
Return air quantity	1,000 cfm

Return air temperature	80°F DB and 62°F WB
Outdoor air quantity	1,000 cfm
Outdoor air temperature	90°F DB and 75°F WB

Find

Dry-bulb temperature of the air mixture
Wet-bulb temperature of the air mixture

Solution

1. Place a dot at the return air conditions (80°F DB and 62°F WB temperatures) on the psychrometric chart, figure 8-13.
2. Place a dot at the outdoor air conditions (90°F DB and 75°F WB temperatures) on the chart.
3. Connect the two dots.
4. Determine the percentage of return air used.
   Total air: 2,000 cfm
   Return air: 1,000 cfm
   Percent return air used: 1,000 ÷ 2,000 = 0.50 = 50%

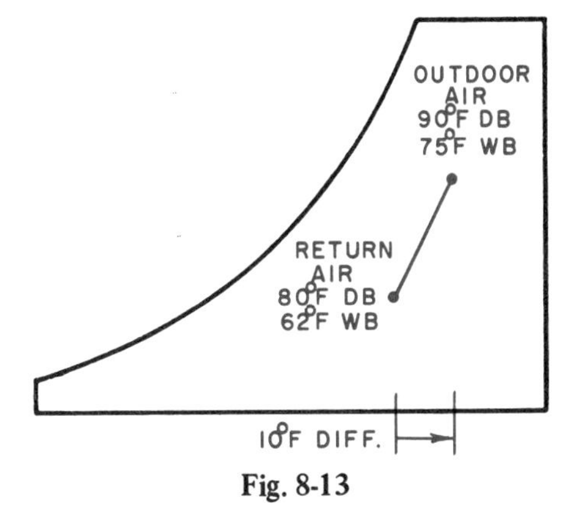

Fig. 8-13

5. Determine the percentage of outdoor air that is used.
   Total air: 2,000 cfm
   Outdoor air: 1,000 cfm
   Percent outdoor air used: 1,000 ÷ 2,000 = 0.50 = 50%
6. Determine the difference in dry-bulb temperature between the return air (80°F) and the outdoor air (90°F), figure 8-14.
   Difference in dry-bulb temperature = 90° - 80° = 10°F

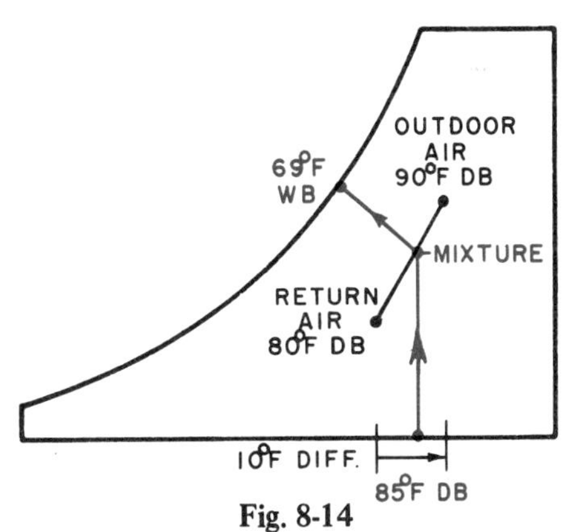

Fig. 8-14

7. Determine the dry-bulb temperature of the mixture:
   a. Multiply the temperature difference (10°F) by the percentage of the return air used (50%): 10 x 0.50 = 5.0.
   b. Subtract 5.0 from the supply air temperature (90°F): 90 – 5 = 85°F. Thus, the air mixture dry-bulb temperature = 85°F.
8. Determine the mixture wet-bulb temperature:
   a. Follow the 85°F DB line on the psychrometric chart until it intersects the line connecting the return air and outdoor air temperatures.
   b. Read the **WB** temperature at the point of intersection. The wet-bulb temperature of the air mixture is 69°F.

A typical comfort air-conditioning installation may require a mixture of 10% outdoor air and the return air from the room. The actual composition of the air mixture varies, however, depending upon the installation. An example of an air mixture that may be encountered in actual practice is illustrated in Problem 3.

PROBLEM 3 PRACTICAL AIR MIXTURE

Given

Total air quantity required	10,000 cfm
Return air quantity	9,000 cfm
Return air temperature	80°F DB and 62°F WB
Outdoor air quantity	1,000 cfm
Outdoor air temperature	90°F DB and 75°F WB

Find

Dry-bulb temperature of the air mixture
Wet-bulb temperature of the air mixture

Solution

1. Plot the return air and the outdoor air conditions on the psychrometric chart. Connect these points with a line as shown in figure 8-15.
2. Determine the percentage of the return air to be used.
   Total air: 10,000 cfm
   Return air: 9,000 cfm
   Percent return air used:
   $9{,}000 \div 10{,}000 = 0.90 = 90\%$
3. Determine the dry-bulb temperature difference between the return air and the outdoor air: 90°F - 80°F = 10°F
4. Determine the dry-bulb and wet-bulb temperatures of the air mixture.

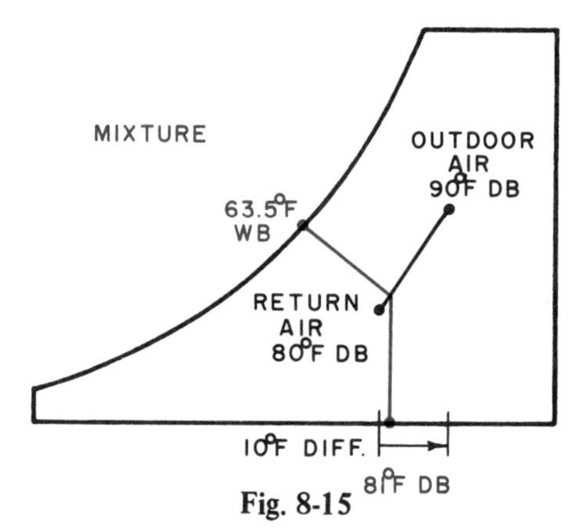

Fig. 8-15

   a. Multiply the dry-bulb temperature difference by the percentage of the return air: 10 x 0.90 = 9.0.
   b. Subtract 9.0 from the outdoor air dry-bulb temperature: 90 - 9.0 = 81. The dry-bulb temperature of the air mixture = 81°F.
   c. Follow the 81°F dry-bulb line until it intersects the line connecting the return air and outdoor air temperatures. The wet-bulb temperature of the air mixture at the point of intersection is 63.5°F.

Since 90% of the total air supply (return air) is at 80°F and only 10% (outdoor air) is at 90°F, it is evident that the temperature of the mixture must be closer to the return air temperature. That is, there is nine times as much return air at 80°F than there is outdoor air at 90°F.

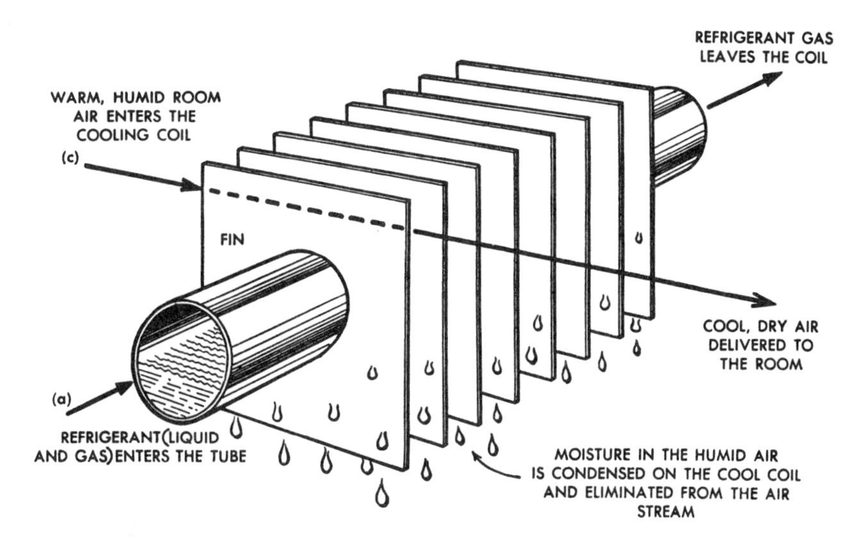

Fig. 8-16 Bypass air does not contact the coil surface

For comfort air conditioning, the dry-bulb temperature of the air mixture is usually closer to the dry-bulb temperature of the largest part of the air mixture.

With regard to air mixtures, then, the psychrometric chart is useful in determining the dry-bulb and wet-bulb temperatures. Once the dry-bulb and wet-bulb temperatures are known, the dewpoint, relative humidity, and moisture content can be determined.

Bypass Air

Bypass air is the air that flows through a coil without making contact with the coil surface. The amount of bypass air depends basically on the construction of the coil and the velocity of the air flow. If the coil tubing has one fin per inch, more air bypasses the coil surface than if the tubing has seven fins per inch. If the velocity of the air passing over the coil surface is low, more of the air contacts the coil surface than it does when its velocity is high.

The bypass process is described in terms of a *bypass factor*. The use of this factor is shown in Problem 4.

PROBLEM 4 BYPASS FACTOR

Given

Coil temperature	50°F
Temperature of the air entering the coil	80°F
Temperature of the air leaving the coil	60°F

Find

Bypass factor

Solution

1. Subtract the coil temperature from the temperature of the air leaving:

$$60 - 50 = 10$$

2. Subtract the coil temperature from temperature of the air entering:

$$80 - 50 = 30$$

3. Divide the first answer by the second:

$$10 \div 30 = 0.33$$

$$\text{Bypass factor} = 0.33$$

In Problem 4, a bypass factor of 0.33 (1/3) indicates that the temperature of the air leaving the coil (60°F) drops 2/3 of the total temperature difference between the air entering the coil (80°F) and the coil surface temperature (50°F).

In actual practice, bypass factors for cooling coils are established through tests conducted by the manufacturers. Thus, the factors are known before a coil is installed. As a result, it is possible to predict the coil performance and select the optimum coil for the job to be done. The following conditions are true for various values of the bypass factor.

1. If the bypass factor is high, the following requirements are necessary to produce the required temperature of the air leaving the coil.
   - More air at slower velocities.
   - Larger ducts to handle the larger air quantities.
   - A larger fan and fan motor to supply the larger air quantities.
2. A low bypass factor produces advantages as follows:
   - Less air is required because the temperature of the air leaving the coil is lower.
   - Smaller ducts are required to handle the smaller air quantities.
   - A smaller fan and fan motor are required to handle the smaller air quantities.

   The following disadvantages are the result of a low coil bypass factor:
   - A low supply air temperature may cause discomfort in small rooms.
   - Larger coils may be required.
   - Since a greater coil surface area is contacted by the air, the coil picks up heat faster; therefore, larger refrigeration equipment may be needed to maintain the low coil temperature.
   - Because of the low temperature in the supply duct to the room, insulation and a vapor seal may be needed.

It is necessary to evaluate a number of factors to determine the best combination of equipment and comfort conditions. The average bypass factor for comfort air conditioning varies according to the optimum combination of conditions and equipment; the bypass factor usually ranges between 0.10 and 0.30.

Apparatus Dewpoint

Problem 4 demonstrates that the coil surface temperature is used to help determine the bypass factor. In this problem, the air temperature drops from 80°F to 60°F as the air passes through the coil; the coil surface temperature is 50°F. If a lower temperature for the air leaving the coil is required, more coil surface must be provided. The coil surface can be increased by adding a second row of coil tubing.

The 60°F air leaving the first row of coils then passes through the second row of tubing and the air temperature drops to some point between 60°F and the coil surface temperature (50°F). As more rows of coil are added, the temperature of the air is lowered to a value closer to the surface temperature of the coil.

If enough coil surface is provided, the temperature of the air leaving the coil becomes the same as the coil surface temperature. When this condition is achieved, the air is at the dewpoint temperature and it is saturated.

Since the cooling coil is located in the air-conditioning apparatus, and since the air reaches its dewpoint when it is cooled to the surface temperature of the coil, the coil surface temperature is also known as the *apparatus dewpoint.* Many cooling coils are rated according to the value of the apparatus dewpoint.

SUMMARY

- Psychrometric processes involve the heating or cooling of the air and the addition or removal of moisture.
- A typical comfort air-conditioning process uses a mixture of outdoor air and return air from the room.
- For a typical comfort air-conditioning process, the psychrometric chart can show:
    1. That the room air picks up heat and moisture when it is mixed with outdoor air.
    2. That heat and moisture are removed as the mixture passes through the conditioning apparatus.
    3. At what temperature and humidity the air mixture must be supplied to the room to maintain comfort conditions.
- Latent and sensible heating and cooling are the basic heating and cooling processes that can be interpreted on the psychrometric chart.

Latent Heat

- Latent heat of vaporization is the heat required to change a liquid to a vapor (with no change in temperature).
- Latent heat of fusion is the heat that must be removed to change a liquid to a solid (with no change in temperature).
- As it is applied to air conditioning, the latent heat principle refers to changes in the moisture content.
- An increase in latent heat is an increase in the moisture content; a decrease in latent heat is a decrease in the moisture content. In both instances, there is no change in temperature.
- An increase or decrease in latent heat is shown as a straight vertical line on the psychrometric chart.
- Latent heat is concerned with the processes of evaporation and condensation.

Sensible Heating and Cooling

- Sensible heat increases the temperature but does not change the moisture content of the air. Sensible cooling lowers the temperature but does not change the moisture content of the air.
- Sensible heat can be detected by the senses.
- Sensible heating or cooling is shown as a straight, horizontal line on the psychrometric chart.
- When sensible heat is added to the air:
  1. the dry-bulb temperature increases.
  2. the dewpoint does not change.
  3. the wet-bulb temperature increases.
  4. the relative humidity decreases.
- The simultaneous addition of sensible heat and latent heat represents a heating and humidifying process.
- A heating coil and a water spray coil are required for a heating and humidifying process.
- The heating and humidifying process is shown as a diagonal line beginning at the lower left and moving to the upper right of the psychrometric chart.
- A cooling and dehumidifying process removes sensible heat and latent heat. This process is shown as a diagonal line running from the upper right to the lower left of the psychrometric chart.

Sensible Heat Factor

- The cooling and dehumidifying line is also called the sensible heat factor line.
- The slope of the sensible heat factor line changes according to the amount of sensible or latent heat removed.
- A typical residence may have a sensible heat factor of 0.8. Thus, 80% of the total heat change is due to sensible heat and 20% is due to latent heat.

Evaporative Cooling

- The evaporative cooling process removes sensible heat and adds latent heat.
- This process is shown as a diagonal line extending from the lower right to the upper left of the psychrometric chart.
- Evaporative cooling requires a spray water coil.

- Most evaporative cooling processes proceed along the wet-bulb temperature line.
- Evaporative cooling is used primarily in industrial processes rather than for comfort air conditioning.

Air Mixture

- Most air-conditioning applications require a mixture of return air from the room and outdoor air.
- The psychrometric chart can be used to determine the dry-bulb temperature, wet-bulb temperature, dewpoint temperature, relative humidity, and moisture content of the air mixture.

Bypass Air

- Bypass air flows through a cooling coil but does not contact the coil surface.
- The amount of air that bypasses the coil surface depends upon the coil construction and the air velocity.
- The bypass process is measured in terms of a bypass factor.
- The bypass factor makes it possible to predict coil performance and to select the optimum coil to meet specific requirements.
- A low bypass factor can mean that less air is required and that smaller ducts, fans, and motors are necessary.
- The average bypass factor for comfort air conditioning varies, but it is usually in the range of 0.10 and 0.30.

Apparatus Dewpoint

The coil surface temperature is the apparatus dewpoint.

REVIEW

Answer the following problems using the simplified psychrometric charts provided.

1. In a sensible heating process, what change takes place in the:
   a. dry-bulb temperature ____________
   b. wet-bulb temperature ____________
   c. dewpoint ____________
   d. relative humidity ____________
2. In a sensible cooling process, what change takes place in the:
   a. dry-bulb temperature ____________
   b. wet-bulb temperature ____________
   c. dewpoint ____________
   d. relative humidity ____________

3. Given the following initial conditions, sketch a heating and humidifying process on the psychrometric chart.

   DB 60°F
   WB 45°F
   DP 25°F
   RH 28%

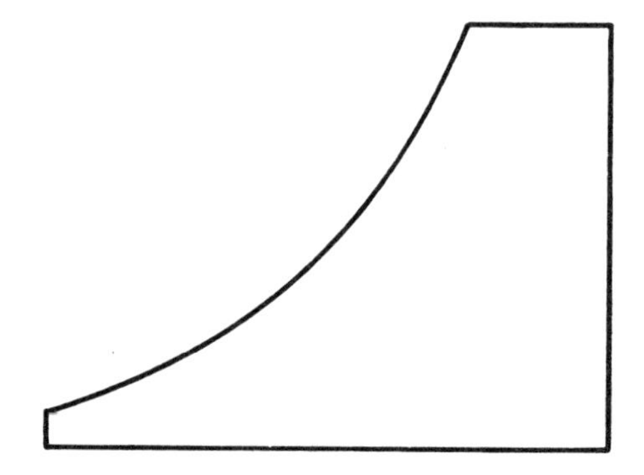

4. Given the following initial air conditions, sketch a cooling and dehumidifying process on the psychrometric chart.

   DB 80°F
   WB 67°F
   DP 61°F
   RH 50%

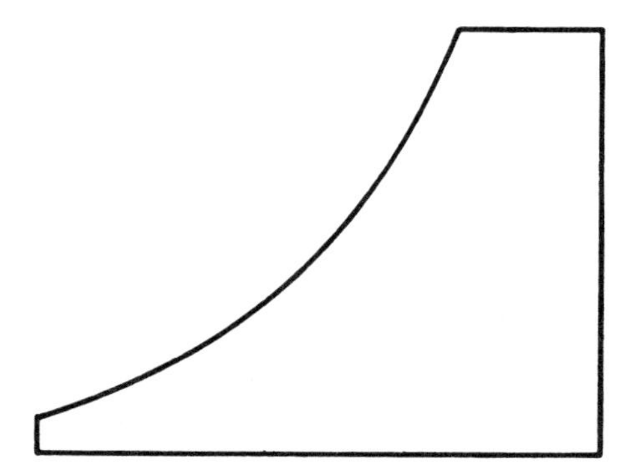

5. Sketch a sensible heat factor line on a psychrometric chart.

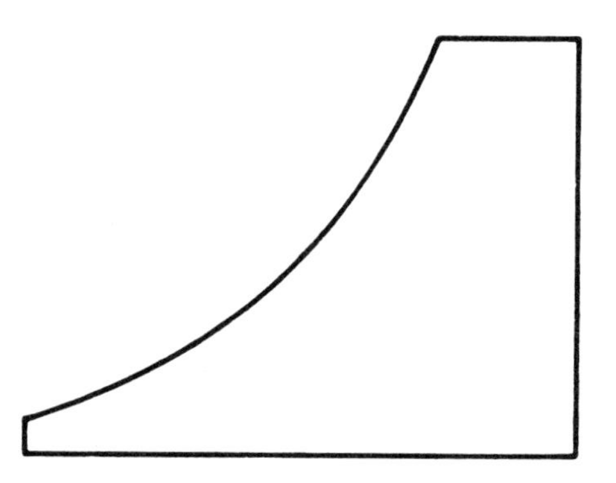

6. Sketch an evaporative cooling process on the psychrometric chart given the following initial and final conditions:

Initial	*Final*
95°F DB	75°F DB
29% RH	80% RH

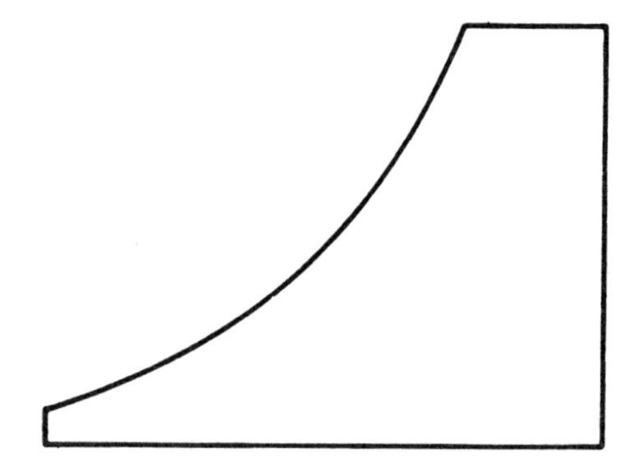

Supply the word or phrase which best completes statements 7-20.

7. ______________________ is the heat required to change a liquid to a vapor.

8. ______________________ is the heat that must be removed to change a liquid to a solid.

9. When applied to the air-conditioning process, the latent heat principle reflects changes in ______________________ .

10. Latent heat increases or decreases do not affect ______________________ .

11. Sensible heating or sensible cooling does not affect ______________ of the air.

12. When sensible heat is added to air:
    - the dry-bulb temperature ______________________
    - the wet-bulb temperature ______________________
    - the dewpoint ______________________
    - the relative humidity ______________________

13. When sensible cooling is added to air:
    - the dry-bulb temperature ______________________
    - the wet-bulb temperature ______________________
    - the dewpoint ______________________
    - the relative humidity ______________________

14. The simultaneous addition of sensible and latent heat is known as a(an) ________ process.

15. The cooling and dehumidifying line is also called the ______________ line.

16. A typical residence may have a sensible heat factor of 0.8. This means that ______________ of the total heat change is sensible heat and ______________ is latent heat.

17. The evaporative cooling process ______________ sensible heat and ________ latent heat.

18. A low bypass factor can mean the following changes in system conditions: ____________ air, ____________ ducts, __________ and ______________ fans and motors.

19. The coil surface temperature is the ______________________ .

20. The average bypass factor for comfort air conditioning varies but usually is in the range from ________ to ____________ .

Solve the following psychrometric problems.

21. Given:

Total air quantity	10,000 cfm
Return air quantity	9,000 cfm
Return air temperature	80°F DB and 62°F WB
Outdoor air quantity	1,000 cfm
Outdoor air temperature	90°F DB and 75°F WB

Find:
a. The dry-bulb temperature of the air mixture. ______________
b. The wet-bulb temperature of the air mixture. ______________

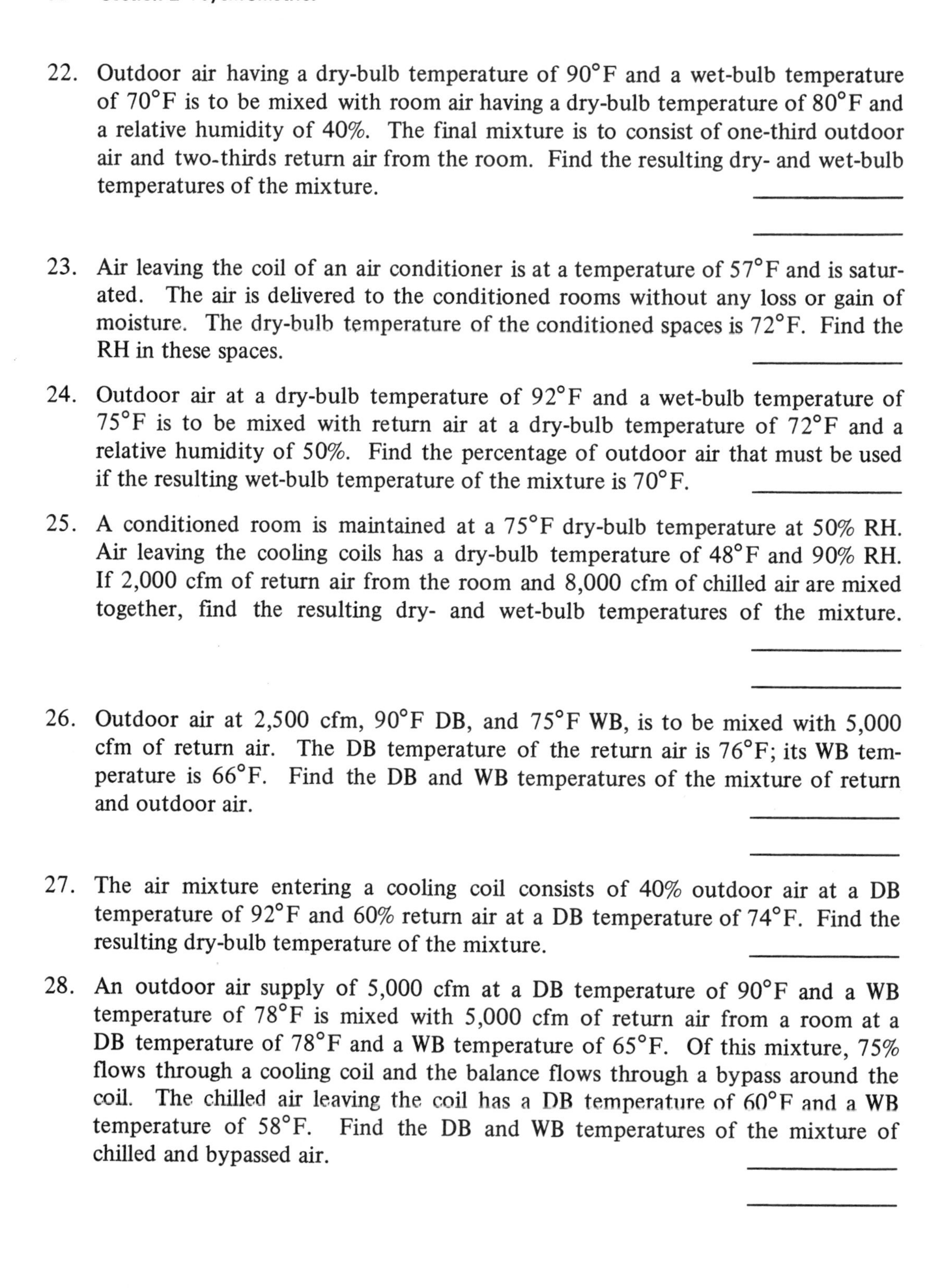

22. Outdoor air having a dry-bulb temperature of 90°F and a wet-bulb temperature of 70°F is to be mixed with room air having a dry-bulb temperature of 80°F and a relative humidity of 40%. The final mixture is to consist of one-third outdoor air and two-thirds return air from the room. Find the resulting dry- and wet-bulb temperatures of the mixture. __________ __________

23. Air leaving the coil of an air conditioner is at a temperature of 57°F and is saturated. The air is delivered to the conditioned rooms without any loss or gain of moisture. The dry-bulb temperature of the conditioned spaces is 72°F. Find the RH in these spaces. __________

24. Outdoor air at a dry-bulb temperature of 92°F and a wet-bulb temperature of 75°F is to be mixed with return air at a dry-bulb temperature of 72°F and a relative humidity of 50%. Find the percentage of outdoor air that must be used if the resulting wet-bulb temperature of the mixture is 70°F. __________

25. A conditioned room is maintained at a 75°F dry-bulb temperature at 50% RH. Air leaving the cooling coils has a dry-bulb temperature of 48°F and 90% RH. If 2,000 cfm of return air from the room and 8,000 cfm of chilled air are mixed together, find the resulting dry- and wet-bulb temperatures of the mixture. __________ __________

26. Outdoor air at 2,500 cfm, 90°F DB, and 75°F WB, is to be mixed with 5,000 cfm of return air. The DB temperature of the return air is 76°F; its WB temperature is 66°F. Find the DB and WB temperatures of the mixture of return and outdoor air. __________ __________

27. The air mixture entering a cooling coil consists of 40% outdoor air at a DB temperature of 92°F and 60% return air at a DB temperature of 74°F. Find the resulting dry-bulb temperature of the mixture. __________

28. An outdoor air supply of 5,000 cfm at a DB temperature of 90°F and a WB temperature of 78°F is mixed with 5,000 cfm of return air from a room at a DB temperature of 78°F and a WB temperature of 65°F. Of this mixture, 75% flows through a cooling coil and the balance flows through a bypass around the coil. The chilled air leaving the coil has a DB temperature of 60°F and a WB temperature of 58°F. Find the DB and WB temperatures of the mixture of chilled and bypassed air. __________ __________

UNIT 9
ADVANCED PSYCHROMETRIC PROCESSES

OBJECTIVES

After completing the study of this unit, the student will be able to

- define specific volume.
- state the relationship of specific volume to air density and how this affects fan and fan motor sizing.
- define enthalpy.
- state how enthalpy is related to the measurement of both latent heat and sensible heat changes.
- state which psychrometric processes use water spray coils; for each of these processes state what changes occur in the wet-bulb, dry-bulb, and dewpoint temperatures.

The discussion in Unit 8 of the fundamental psychrometric processes forms the basis for the study of two additional psychrometric concepts: specific volume and enthalpy. Specific volume is concerned with the space occupied by air; enthalpy refers to the total heat content of the air. A third topic covered in this unit is the spray coil principle involving the operation of the spray coil and the relationship of spray coils to the psychrometric processes.

SPECIFIC VOLUME

Specific volume describes the value in cubic feet of the space occupied by one pound of air at various conditions of temperature and pressure. This subject was covered in a previous unit in terms of the moisture content of the air. The unit showed how to determine the number of grains of moisture in a cubic foot of air.

Specific volume is also a factor in system requirements where air density is a consideration. Since the air density affects the air handling equipment such as fans and fan motors, the specific volume is related to fan performance and fan motor sizes. That is, a high specific volume requires less energy to drive the fan; and, a low specific volume requires more energy to drive the fan. These statements are true if all other conditions such as pressure and fan size are equal.

Specific Volume and the Psychrometric Chart

As shown in figure 9-1, page 80, the specific volume scale extends along the sole of the psychrometric chart from a value of 12.5 cu. ft. per pound of air to 14.5 cu. ft. per pound of air. The specific volume lines, figure 9-1B, are parallel diagonal lines drawn from the instep to the sole of the chart.

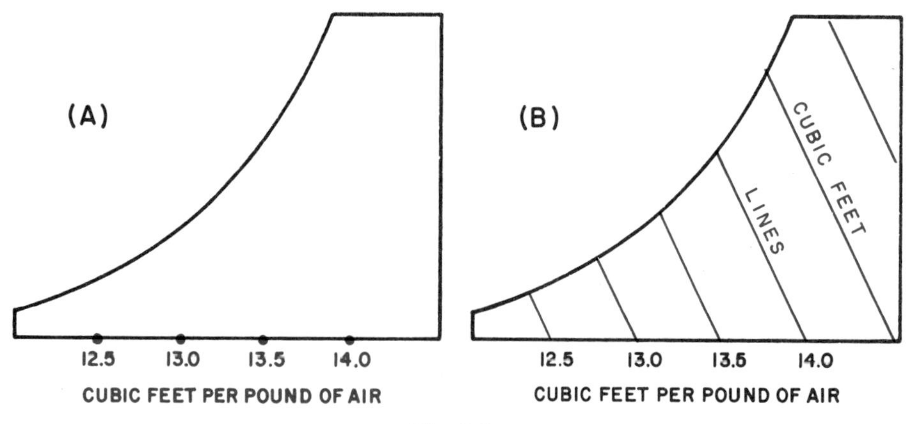

Fig. 9-1

In the following specific volume problems (Problems 1-3), it is assumed that the atmospheric pressure is at sea level and that the air moisture content is zero (the air is dry).

PROBLEM 1 SPECIFIC VOLUME

Given

Air temperature 56°F DB

Find

Specific volume

Solution

1. Find the 56°F DB temperature on the dry-bulb scale at the bottom of the psychrometric chart, figure 9-2.
2. Read the value of 13.0 cu. ft. per pound of air on the specific volume scale.

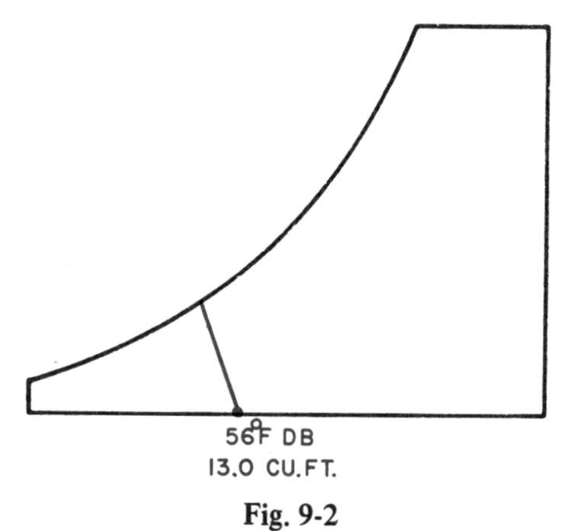

Fig. 9-2

PROBLEM 2 SPECIFIC VOLUME

Given

Air temperature 76°F DB

Find

Specific volume

Solution

1. Find the 76°F DB temperature on the psychrometric chart, figure 9-3.
2. Read the value of 13.5 cu. ft. per pound of air on the specific volume scale.

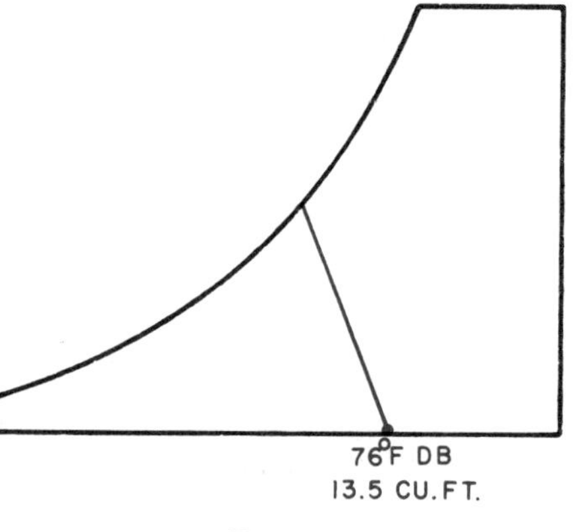

Fig. 9-3

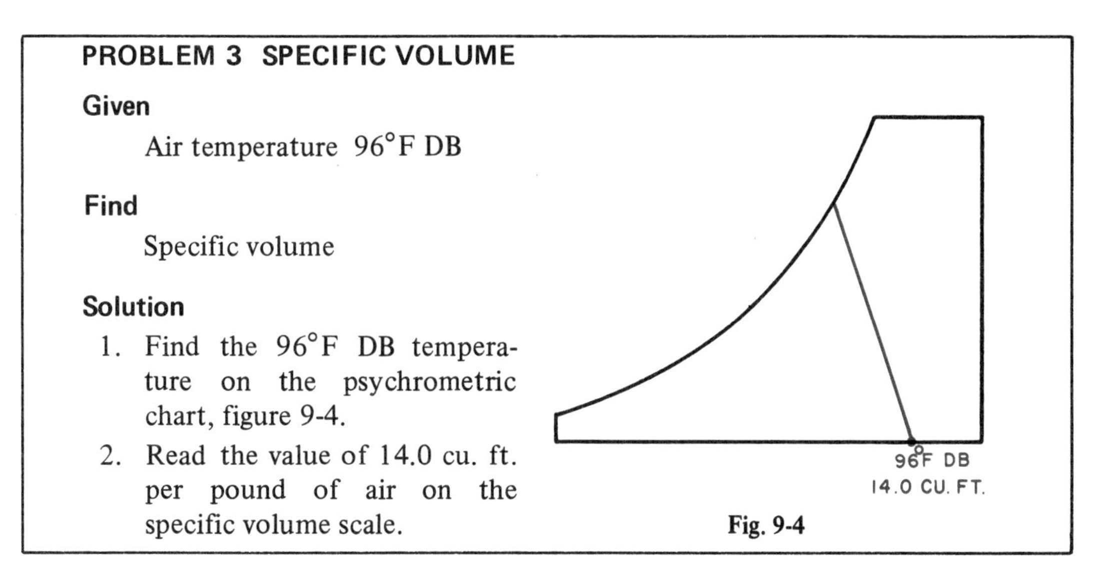

PROBLEM 3 SPECIFIC VOLUME

Given

Air temperature 96°F DB

Find

Specific volume

Solution

1. Find the 96°F DB temperature on the psychrometric chart, figure 9-4.
2. Read the value of 14.0 cu. ft. per pound of air on the specific volume scale.

Fig. 9-4

Problems 1-3 show that as the temperature increases, the volume of the air increases. These problems prove the basic principle that air expands as it is heated; that is, the air is less dense at higher temperatures.

With regard to fans and fan motors, the fact that air expands when it is heated means that less horsepower (or a smaller motor) is required when the specific volume is high as compared to when it is low. This condition results because less energy is needed to move air having a low density (high specific volume).

ENTHALPY

Enthalpy is the total heat content of air. In psychrometric terms, enthalpy defines both the heat in the air and the moisture in the air. Enthalpy is measured in Btu per pound of air.

The enthalpy lines on the psychrometric chart are the same as the wet-bulb temperature lines, figure 9-5A. The determination of enthalpy depends almost entirely on the wet-bulb temperature. The enthalpy scale is located in sections along the instep of the chart, figure 9-5B. The range of enthalpy values extends from 7 to 48 Btu on a normal temperature chart.

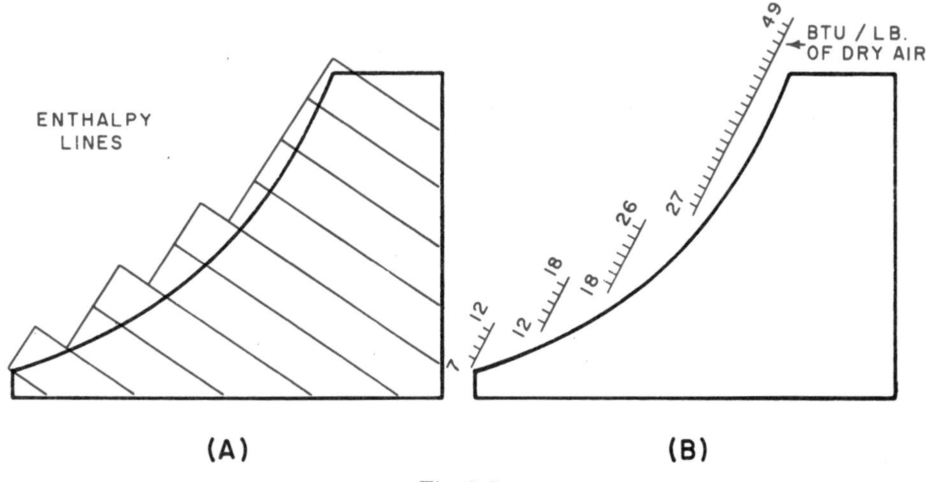

Fig. 9-5

Since enthalpy is the total heat content, the use of the enthalpy value and the psychrometric chart allows one to measure the heat change that takes place in a given psychrometric process. Both the latent heat and sensible heat changes can be measured. Therefore, the use of enthalpy provides a quick means of determining the sensible heat factor.

Problem 4 shows the application of enthalpy to several basic psychrometric processes.

PROBLEM 4 ENTHALPY – COOLING AND DEHUMIDIFYING

Given

Initial dry-bulb temperature	78°F
Initial wet-bulb temperature	65°F
Final dry-bulb temperature	55°F
Final wet-bulb temperature	50°F

Find

A. Total heat removed.
B. Latent heat and moisture removed.
C. Sensible heat removed.

Solution A Total Heat

1. Plot the initial and final temperatures on the psychrometric chart, figure 9-6A.

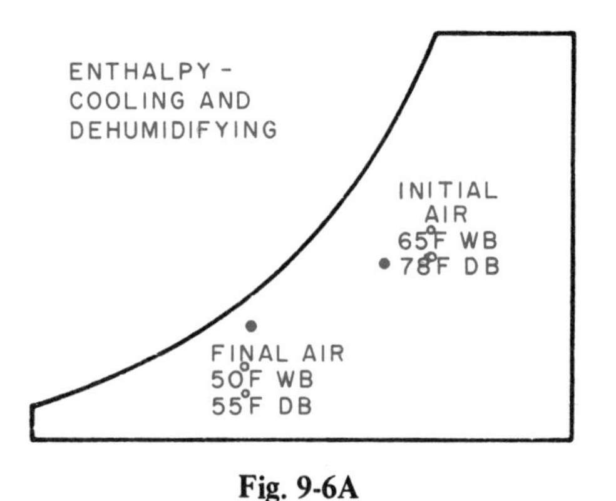

Fig. 9-6A

2. Follow the wet-bulb line from the initial temperature condition to the enthalpy scale. Read a value of 30 Btu, figure 9-6B.

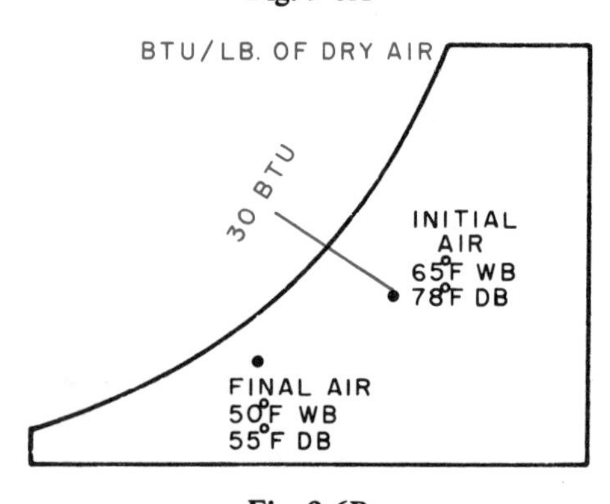

Fig. 9-6B

3. Follow the wet-bulb line from the final temperature condition to the enthalpy scale. Read a value of 20.3 Btu, figure 9-6C.

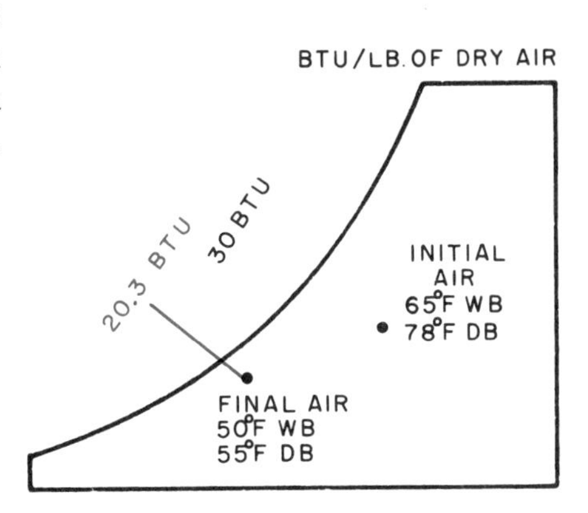

Fig. 9-6C

4. The total heat removed in the cooling process is 30 - 20.3 = 9.7 Btu, figure 9-6D.

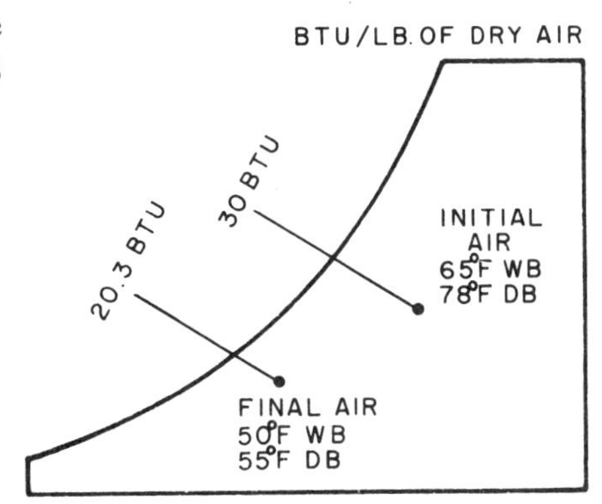

TOTAL HEAT = 30 – 20.3 = 9.7 BTU

Fig. 9-6D

Solution B Latent Heat and Moisture

1. Trace a line horizontally from the final temperature condition (55°F DB and 50°F WB) to the initial dry-bulb temperature line (78°F).

   Follow the dry-bulb line to the initial wet-bulb condition.

   Trace diagonally from the initial condition to the final condition. Note that a triangle is formed, figure 9-6E.

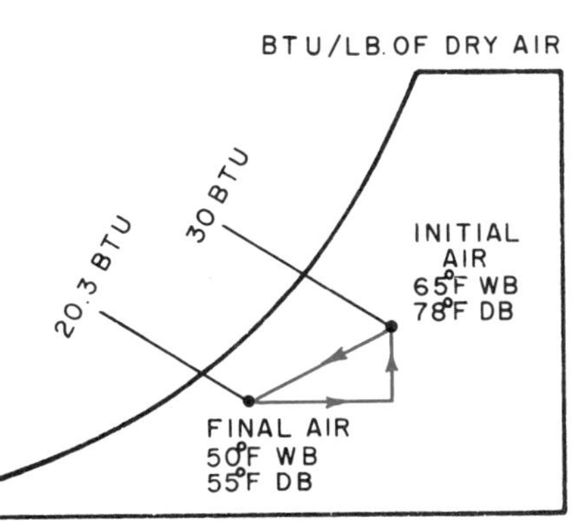

TOTAL HEAT = 30 – 20.3 = 29.7 BTU

Fig. 9-6E

2. Beginning at the intersection of the horizontal and vertical lines of the triangle, trace the wet-bulb line up to the enthalpy scale and read a value of 26 Btu, figure 9-6F.

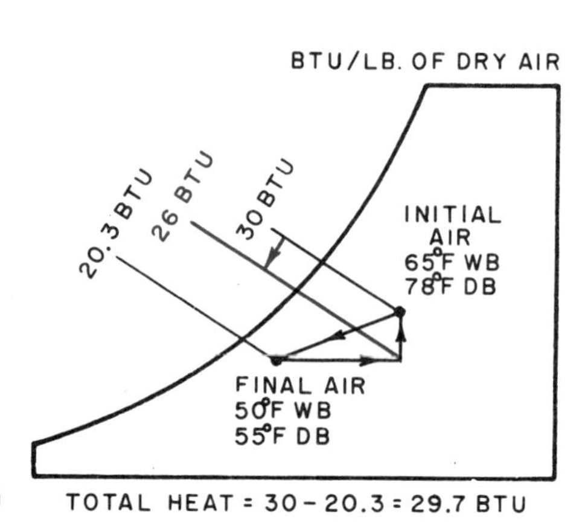

LATENT HEAT = 30-26 = 4 BTU

Fig. 9-6F

3. The vertical leg of the triangle represents the latent heat and the amount of moisture. Thus, the amount of latent heat removed is equal to the heat value at the initial condition (30 Btu) minus the value at the intersection (26 Btu) or 4 Btu per pound of air.

4. The amount of moisture removed in dehumidifying is 71 (grains of moisture at the initial condition) minus 48.7 (grains of moisture at the intersection) or 22.3 grains of moisture per lb. of air, figure 9-6G.

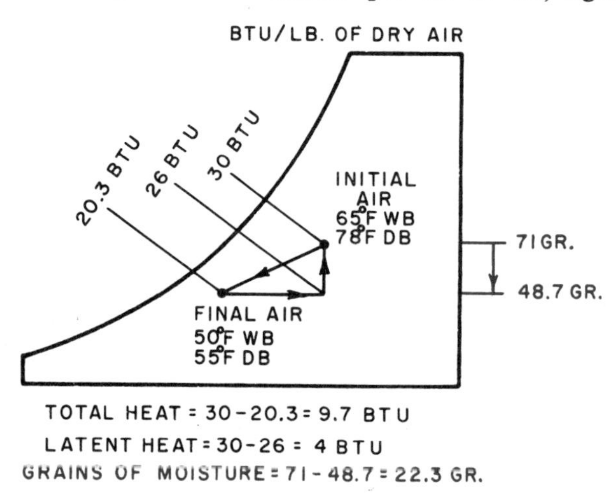

Fig. 9-6G

Solution C Sensible Heat

1. Read the enthalpy value at the intersection of the horizontal and vertical lines of the triangle. This value is 26 Btu, figure 9-6H.

2. Read the enthalpy value at the final condition of the air: enthalpy = 20.3 Btu.

3. The sensible heat removed is 26 – 20.3 = 5.7 Btu per lb. of air.

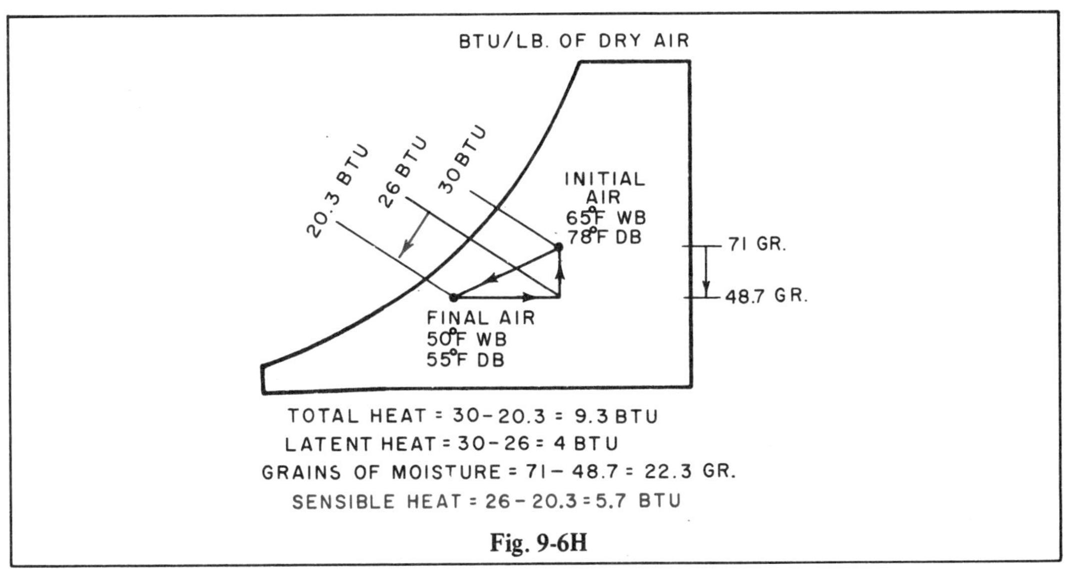

Fig. 9-6H

The use of enthalpy values is another way of determining the sensible heat factor. For Problem 4, the sensible heat factor is obtained by dividing the sensible heat removed (5.7 Btu) by the total heat removed (9.7 Btu): 5.7 Btu/9.7 Btu = 0.59.

SPRAY COIL OPERATIONS

The psychrometric processes described in Unit 8 include cooling and humidifying, cooling and dehumidifying, and heating and humidifying. All three of these processes involve the use of water spray coils inside a spray chamber, figure 9-7.

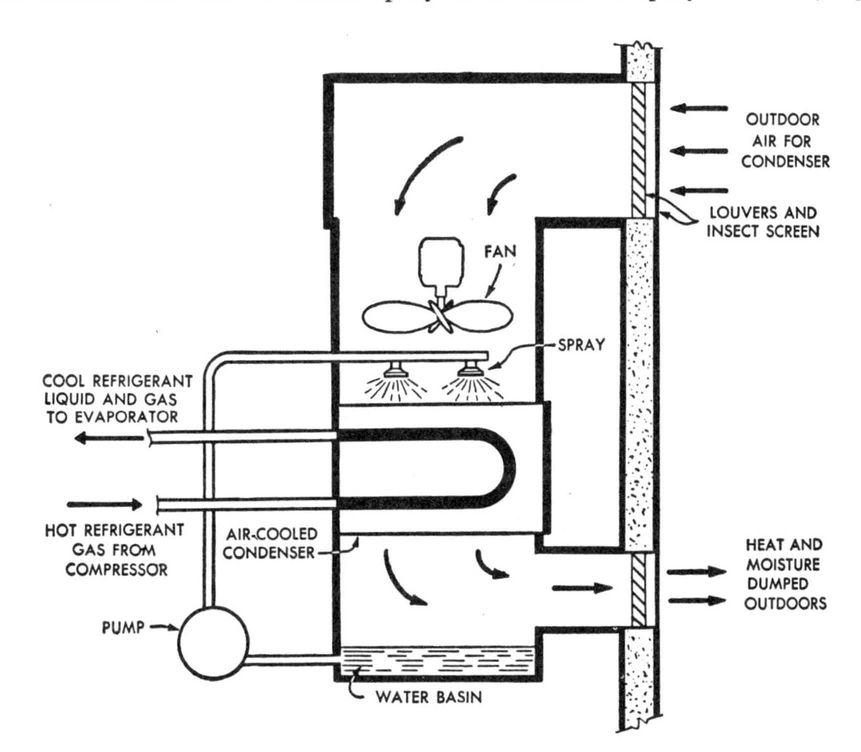

Fig. 9-7 An evaporative condenser

Cooling and Humidifying

The combined process of cooling and humidifying is also known as the evaporative cooling process. When plotted on the psychrometric chart, figure 9-8, this process appears as a line that follows the wet-bulb line. A system in which the evaporative cooling process can take place is shown in figure 9-9. The spray water lowers the dry-bulb temperature of the air and increases the dewpoint. The end result is that the dry-bulb, dew-point, and wet-bulb temperatures of the air leaving the coil are identical.

Figure 9-9 also shows that the water for the sprays is continuously recirculated. The water removes sensible heat and gives up latent heat in equal amounts. If the spray equipment is 100% efficient, the air becomes saturated as it passes through the spray. However, since the spray equipment is not 100% efficient, the condition of the air leaving the spray chamber is below the saturation point.

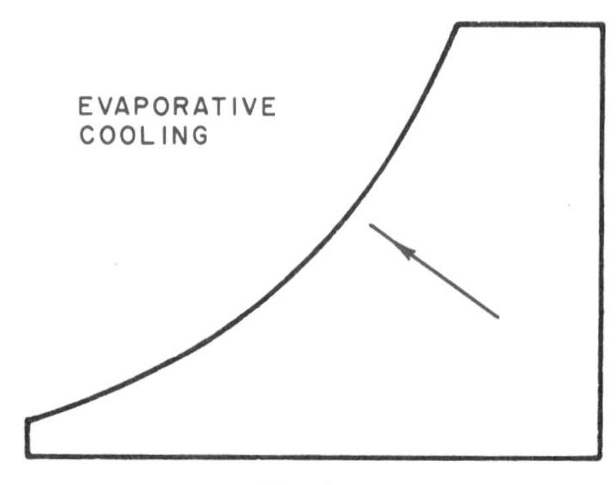

Fig. 9-8

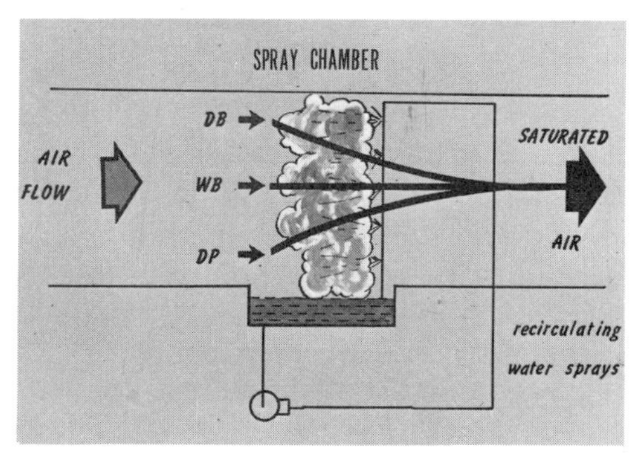

Fig. 9-9

Cooling and Dehumidifying

An arrangement for cooling and dehumidifying the air is shown in figure 9-10. If the temperature of the spray water is below the dewpoint temperature of the entering air, the water removes both heat and moisture from the air. Recirculating water is also used in this process, but a chiller is added to the system to maintain the water temperature below the dewpoint of the air. As the air passes through the chilled water sprays, the dry-bulb, wet-bulb, and dewpoint temperatures of the air are lowered and sensible heat and latent heat are removed.

Another method of cooling and dehumidifying the air uses a cooling coil in the spray chamber, figure 9-11. After the air passes through the water spray, it flows

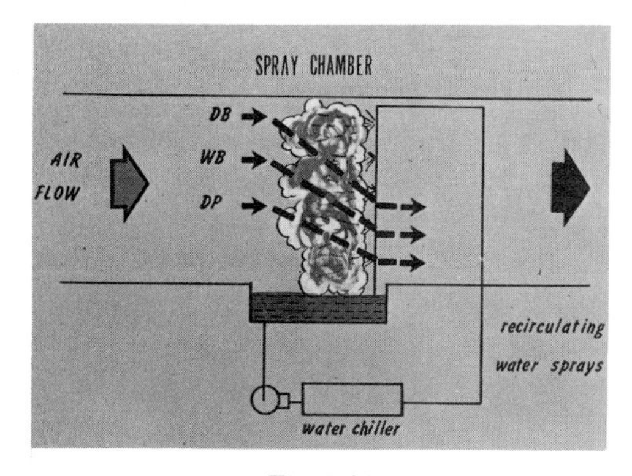

Fig. 9-10

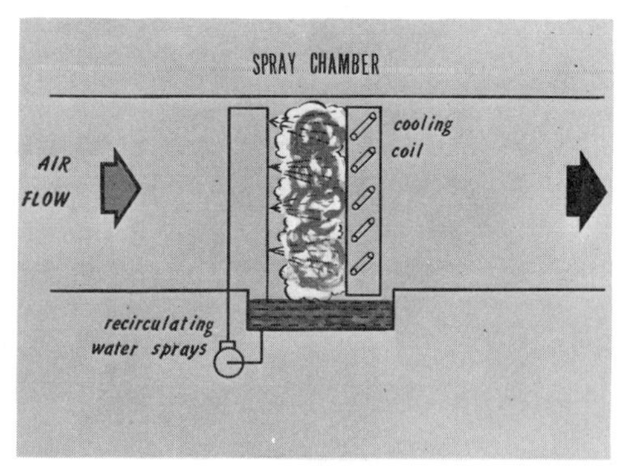

Fig. 9-11

through the cooling coil. The sprays remove heat from the air; the coil removes heat and moisture. This method offers closer control of the condition of the air leaving the system.

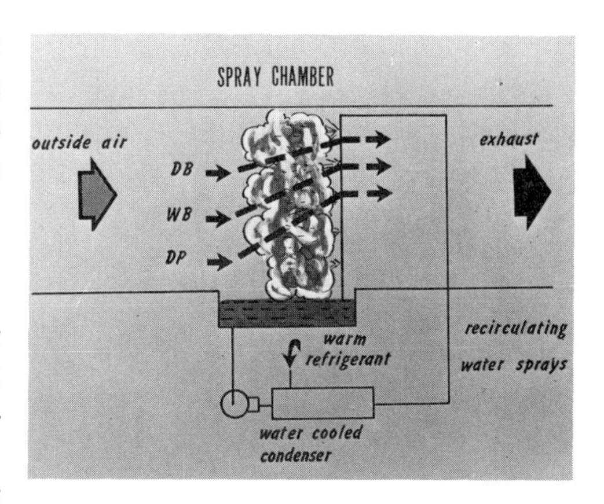

Fig. 9-12

Cooling Tower

The main purpose of a cooling tower is to dispose of the heat removed from the refrigerant by the water in the condenser, figure 9-12.

As shown in figure 9-13, the water in the condenser picks up heat from the refrigerant. The water then flows to the cooling tower where it is sprayed into the air flowing through the tower. This air picks up heat from the water and carries it into the atmosphere. Some of the water evaporates and thus cools the remaining water. The cool water returns to the condenser where it picks up more heat from the refrigerant and the process continues.

Heating and Humidifying

The combined process of heating and humidifying, figure 9-14, uses recirculating water sprays and a steam water heater.

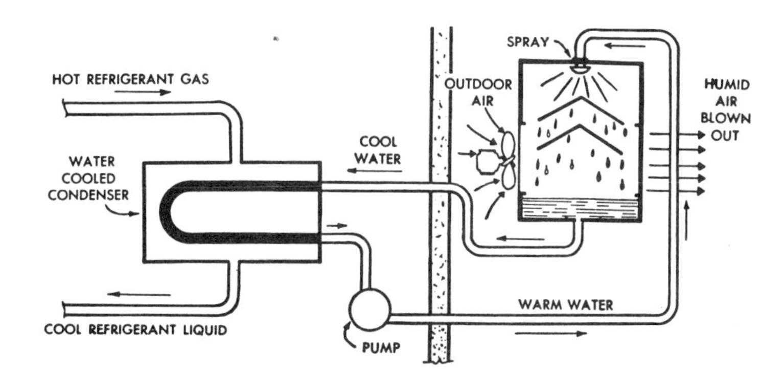

Fig. 9-13 Cooling tower operation

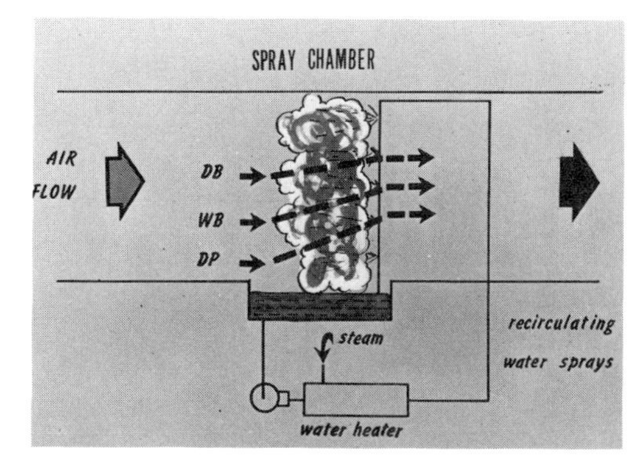

Fig. 9-14

As spray water passes through the heater, it picks up heat. The heated water is then sprayed into the cool airstream where it gives up heat. As a result, part of the spray water evaporates and adds moisture to the air.

The wet-bulb, dry-bulb, and dewpoint temperatures of the air increase in this process.

SUMMARY

- Specific volume refers to the space occupied by air. Enthalpy refers to the total heat content of the air.
- The spray coil concept has a definite relationship to psychrometric processes. Spray coils are used in the processes of cooling and humidifying, cooling and dehumidifying, and heating and humidifying.
- The specific volume is the number of cubic feet occupied by one pound of dry air.
- Specific volume lines on the psychrometric chart are parallel, diagonal lines extending from the instep to the sole of the chart.
- The psychrometric chart shows that as the temperature increases, the specific volume increases.
- A high specific volume requires low horsepower.
- Enthalpy is measured in Btu per pound of air.
- Enthalpy lines on the psychrometric chart are the same as the wet-bulb lines.
- Enthalpy depends almost entirely on the wet-bulb temperature.
- Enthalpy is used as a measure of both sensible heat and latent heat changes on the psychrometric chart.
- Enthalpy can be used to determine the sensible heat factor.
- The combined process of cooling and humidifying is an evaporative cooling process.
- In the cooling and humidifying process, the spray water lowers the dry-bulb temperature and raises the dewpoint temperature.
- In the cooling and dehumidifying process, the chilled spray water lowers the dry-bulb, wet-bulb, and dewpoint temperatures.
- A cooling tower disposes of heat removed from the refrigerant by the water in the condenser.
- The heating and humidifying process uses the spray water to add moisture to the air. In this process, the wet-bulb, dry-bulb, and dewpoint temperatures of the air increase.

REVIEW

Supply the word or phrase which best completes statements 1-7.

1. Enthalpy refers to ______________________ of the air.
2. ______________________ is the number of cubic feet occupied by one pound of dry air.
3. The psychrometric chart shows that as the temperature increases, the specific volume ______________________ .
4. ______________________ specific volume requires low horsepower.
5. Enthalpy is measured in ______________________ .
6. Enthalpy depends almost entirely upon ______________________ temperature.
7. Cooling and humidifying is a(an) ______________________ process.
8. In the cooling and humidifying process, what effect does the spray water have on the dry-bulb temperature and the dewpoint temperature?
9. In the cooling and dehumidifying process, what effect does the chilled spray water have on the dry-bulb temperature, wet-bulb temperature, and dewpoint temperature?
10. In the heating and humidifying process, what is the effect on wet-bulb temperature, dry-bulb temperature, and dewpoint temperature?

Solve each problem and prepare a psychrometric chart for each.

11. Outside air at a 40°F DB temperature and 60% RH is heated to a 90°F DB temperature. No humidification equipment is used. Determine:
    a. the RH of the treated air, and ____________
    b. the heat added per lb. of air. ____________
12. Use the same conditions as given in Problem 11. In this case, however, a humidifier is operated to maintain 50% RH. Determine:
    a. the heat added per lb. of air, and ____________
    b. the moisture added per lb. of air. ____________
13. A direct-expansion air-conditioning coil receives air at 90°F DB and 50% RH. The air leaving the coil is 60°F DB and 90% RH. Find:
    a. the apparatus dewpoint, ____________
    b. the heat removed per pound of air, and ____________
    c. the moisture condensed per pound of air. ____________
14. It is desired to cool a hospital laboratory in the summer. All outside air is used. The condition of the outside air is 85°F DB and 50% RH.
    a. If no dehumidification is to occur, what is the apparatus dewpoint? ____________
    b. The conditioned space is to be maintained at 75°F DB. What is the RH in the space? ____________
15. If an air conditioner is handling air at 98°F DB and 75°F WB, how much sensible heat must be removed to cool the air to 75°F DB and 50% RH? ____________

SECTION 3
PRINCIPLES OF LOAD ESTIMATING

UNIT 10
SOURCES OF HEAT

OBJECTIVES

After completing the study of this unit, the student will be able to

- identify the heat sources that affect the cooling load.
- describe the variations in the solar heat load through glass areas of a building in relation to the time of day.
- define stored cooling capacity.
- describe the use of zoning as an effective method of handling varying load conditions.
- identify the heat losses that affect the heating load.

Sections 1 and 2 of this text have investigated the principles of air conditioning and its effect on human comfort. The psychrometric chart has been introduced and problems given to show its use in predicting the characteristics of air for a given set of conditions. All of this information is now applied in this section to some of the more practical phases of air conditioning such as the air-conditioning load and estimating the cooling and heating load.

COOLING LOAD HEAT SOURCES

The heat sources to be described basically are those which comprise the summer cooling load. Some heat sources, such as people, lights, and small domestic appliances, are so variable that they usually are not considered when determining the load for winter heating in small residences. In large commercial buildings, however, the heat generated by these heat sources is significant and is stable enough to be an important factor.

Outdoor Heat Sources

The major part of the summer cooling load arises from heat sources outside a structure, figure 10-1. The greatest heat source is the sun; this heat is known as *solar heat*. Solar heat enters a structure directly (through glass) and by conduction through the building materials.

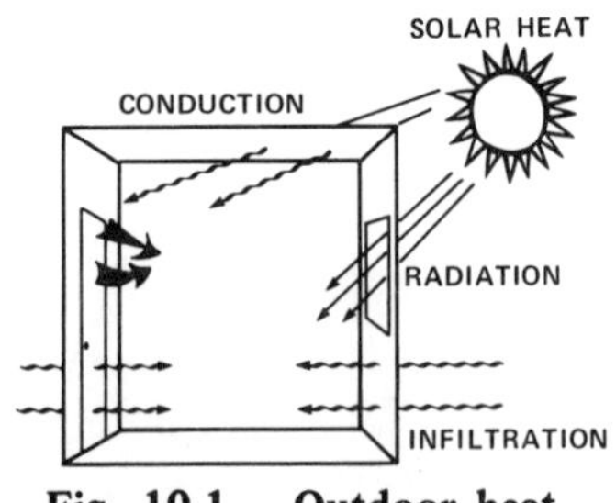

Fig. 10-1 Outdoor heat sources

The solar heat entering a structure through glass is immediately absorbed in the room. Its effect is felt at once. The heat entering by conduction through the walls and roof is not immediately absorbed in the room. Depending on the construction material used, the effect of conducted solar heat may not be felt for several hours. In some instances, the heat may not reach the inside area until after sunset. The amount of heat that enters a structure is measured in terms of a U-factor. Each different type of building material has its own U-factor (Units 11 and 12). The U-factor is a value applied to the quantity of heat that flows through one square foot of building surface.

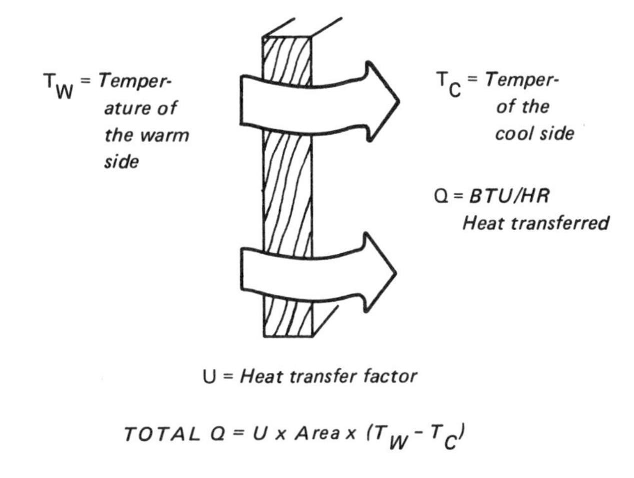

Fig. 10-2 Finding the conducted heat

Heat conduction through materials is the result of differences between the indoor and outdoor air temperatures. The greater the temperature difference, the faster is the flow of heat. The quantity of heat conducted in this manner depends on the size of the wall or roof area and on the resistance offered by the material to the heat flow.

To find the heat that flows through the building surface by conduction, figure 10-2, the U-factor is multiplied by the surface area in square feet. This product is then multiplied by the difference in temperature between the indoor air and the outdoor air. The resulting value is the total heat flow, Q. Q is expressed in Btu per hour of conducted heat flow.

PROBLEM 1 TOTAL SENSIBLE HEAT CONDUCTED (Q) (Through Frame Wall)

Given

Wall surface	8 ft. x 20 ft.
U-factor	0.25
Outdoor temperature	90°F
Indoor temperature	78°F

Find

Total heat conducted (Q)

Solution

Q = Surface area x U-factor x (outdoor temperature - indoor temperature

Q = (8 x 20) x 0.25 x (90 – 78)

Q = 160 x 0.25 x 12

Q = 480 Btu per hour (sensible heat conducted through frame wall)

Building Orientation

The position of the building with relation to the sun is a factor that does not change the total heat load, but which can be put to practical use. For example,

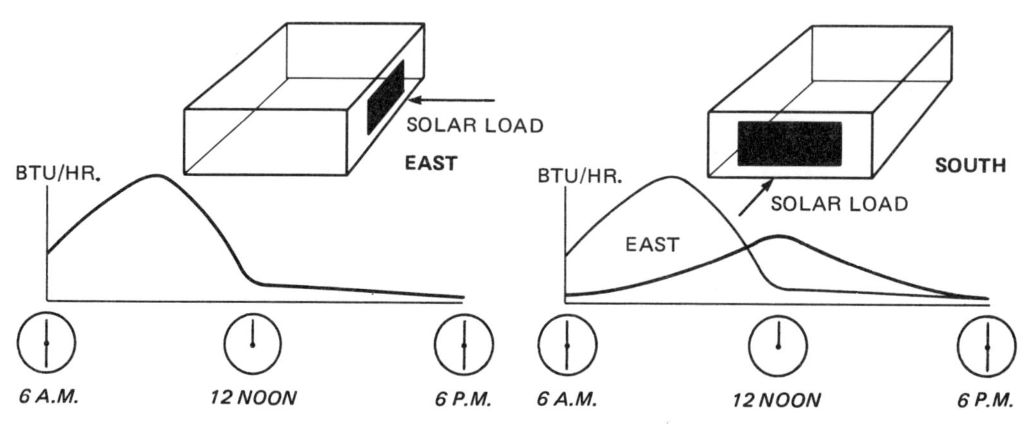

Fig. 10-3 Solar load from the east

Fig. 10-4 Solar load from the south

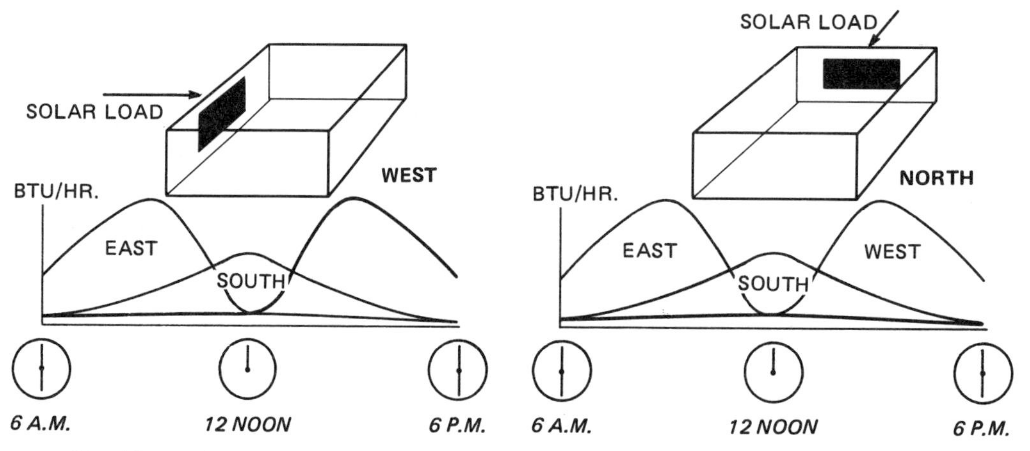

Fig. 10-5 Solar load from the west

Fig. 10-6 Solar load from the north

figures 10-3 through 10-6 illustrate the possible placement of the glass areas of a building with respect to the sun. The resulting solar heat load and the time relationships are also shown. For these figures, it is assumed that there is equal glass area in each wall of the building and the midsummer conditions apply.

Since the sun rises in the east, the solar load through the glass in the east wall is greatest in the morning, figure 10-3. By noon, the solar load in the east is somewhat reduced and its effects are now greater through the glass in the south wall, figure 10-4.

By midafternoon, the solar heat reaches its greatest intensity; it is now felt through the glass in the west wall, figure 10-5. The north wall, of course, is exposed to the less intense late afternoon and early evening sun; thus, the load through the glass on the north side is about 10% of the load on the east or west, figure 10-6.

The information given in figures 10-3 through 10-6 is important in its application to practical situations where the effects of the solar load are a factor. For example, if a building used for a retail business has a large glass surface facing toward the east and the remaining walls have only small glass areas, it is possible to reduce the initial cost of the air-conditioning equipment by 25%.

To reduce the initial cost of the system, the equipment is selected not to handle the highest load (the solar load through the glass on the east side), but rather to handle the next-to-highest load (such as the afternoon solar load on the west wall).

Although the solar heat in the afternoon is even more intense than the morning solar heat, the west wall of the building in this example has a limited glass area. Therefore, it is reasonable to expect a smaller solar load through the west wall. Since the maximum load to be handled is less, smaller equipment can be selected. Thus, the owner must accept slightly warmer building temperatures in the morning hours when customers are likely to be infrequent. The morning temperatures are warmer because the equipment selected can handle only the maximum midafternoon building load. The equipment is not large enough to handle the total morning load.

Walls and Roofs

Heat from the sun enters a building through the walls and roof at a much slower rate than it does through glass, figure 10-7. As the solar heat penetrates the surface skin of the building, some of the heat enters the building material and some is reflected to the atmosphere. The heat absorption process is continuous and the amount of solar heat entering the building material penetrates deeper until it reaches the inside surface. If the sun were stationary so that it could shine continuously in one location with the same intensity, approximately seven hours would be required for the heat to reach to the interior surface of a 12-in. thick brick wall.

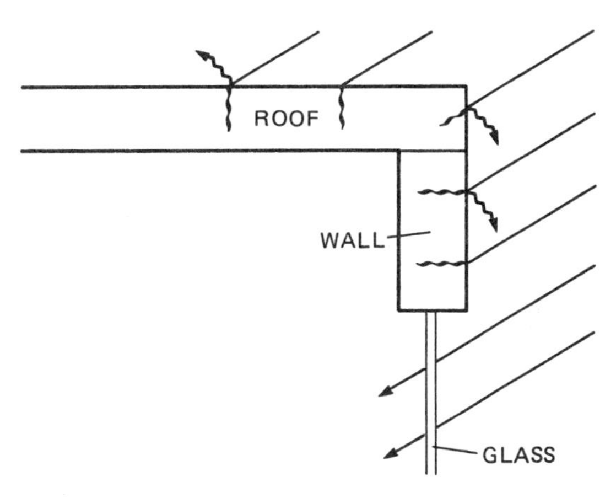

Fig. 10-7 Heat gain through the walls and roof

Figure 10-8 is a comparison of the heat gain through equal areas of roof surface, wall surface, and glass. This graph clearly shows that the solar heat penetrates glass with much greater ease than it does nontransparent materials of greater thickness.

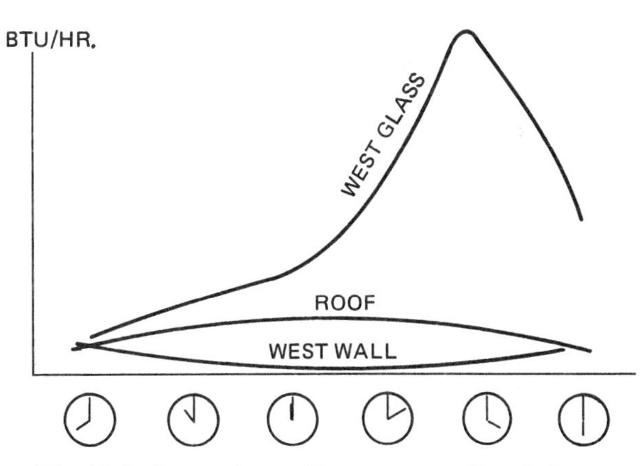

Fig. 10-8 Comparison of heat gain: wall and glass

Infiltration

Another source of heat that must be considered when making a cooling load estimate is infiltration. The heat due to this source is in the air entering the building through cracks around doors and windows and through open doors. This heat load is directly related to the quality of the building construction and to the presence or absence of weatherstripping. If good construction practices were followed, there is less total crack area. Therefore, the cooling load estimate is smaller. The degree of infiltration is also affected by wind velocity, that is, the stronger the wind, the greater is the amount of infiltration.

A second source of heat through infiltration is the heat entering the building through ducts that supply outside air for ventilation. Ventilation standards are governed by the American Society of Heating, Refrigerating, and Air-conditioning Engineers (ASHRAE). These standards define the amount of outside air necessary to dilute unwanted odors such as those resulting from smoking.

To determine the ventilation air load (heat added by ventilation air), multiply the amount of air in cubic feet per minute used for ventilation x a factor of 1.08 x the difference between the outdoor and indoor air temperatures. (Note: the factor of 1.08 applies to sensible heat only.)

PROBLEM 2 VENTILATION AIR SENSIBLE COOLING LOAD (SUMMER)

Given

Ventilation air	500 cfm
Outdoor temperature	90°F
Indoor temperature	78°F
Multiplying factor	1.08

Find

Heat added, Q

Solution

Q = (cfm x 1.08) x (outdoor temperature - indoor temperature)
Q = (500 x 1.08) x (90 - 78)
Q = 540 x 12
Q = 6,480 Btu per hour sensible heat

Problem 3 explains the method of determining the latent heat (moisture) load in ventilation air. In this instance, the moisture difference (in grains per pound of air) rather than the temperature difference is used; the multiplying factor for latent heat is 0.68.

PROBLEM 3 VENTILATION AIR LATENT COOLING LOAD (SUMMER)

Given

Ventilation air	500 cfm
Outdoor temperature	90°F and 50% RH
Indoor temperature	78°F and 45% RH
Multiplying factor	0.68

Find

Heat added, Q

Solution

Q = (cfm x 0.68) x (outdoor - indoor grains per pound of air)
Q = (500 x 0.68) x (107 - 62)
Q = 340 x 45
Q = 15,300 Btu per hour latent heat

The total heat added by ventilation air is the sum of the sensible heat plus the latent heat. Thus, based on the results of Problems 2 and 3, the total heat added for these conditions is:

$$Q = 6{,}480 + 15{,}300 = 21{,}780 \text{ Btu per hour.}$$

The final important source of outside heat is from moisture. Moisture enters the building by infiltration and is called the latent load. The moisture enters through cracks and becomes part of the room load; or, it is carried with the ventilation air and becomes part of the outdoor supply air load.

Indoor Heat Sources

When estimating the heat load, indoor heat sources must be considered also. These sources include people, lights, appliances, and motors.

People are a source of both sensible heat and latent heat, figure 10-9. The heat produced by a person depends upon the energy that is being exerted. A person at rest causes less heat than a person being very active. For example, a person sitting quietly produces about 1/7 as much heat as a person who is bowling.

Fig. 10-9 People: sources of sensible and latent heat

All lights give off heat. The heat emitted by incandescent bulbs is directly related to the wattage of the bulbs. The heat produced by fluorescent lamps is approximately 25% greater than that expected from the rated wattage. This heat increase is due to the additional electricity required by the ballast. The heat load from all types of lights varies according to the usage.

Motors, appliances, and office machines are additional sources of indoor heat. This heat load is a direct function of the energy that is used. Motor heat generally is based on the horsepower rating, but varies according to usage and to the starting and stopping characteristics of the motor. Heat from appliances and office machines is directly related to the fuel or energy consumed. Again, the actual load from these items is affected by usage.

STORAGE

Every structure can absorb and retain heat. As indicated previously in this unit, a long time may be required for heat to penetrate to the inside surface of a wall or roof. In some cases, the heating effect may not be felt until after sunset. Thus, the interior as well as the inside surface of the wall or roof contains heat. In addition, all objects in the building (such as furniture and floors) contain heat. If this heat is present when the air-conditioning equipment is shut down at night, a percentage of the heat is retained. This heat becomes part of the heat load present when the equipment starts again in the morning.

This portion of the start-up load can be decreased by operating the air-conditioning equipment during the night or during the early morning hours before the building is occupied. Since a solar heat load is not a factor during these periods, the total building heat load is considerably below the load that the equipment is capable of handling. Therefore, it is possible to reduce the temperature of the walls, roof, and objects in the building to the point at which they are no longer a heat load. Actually, these surfaces can be cooled until they become a source of stored cooling capacity. This stored capacity can then be used to offset part of the solar heat load during the morning and the load that results when the building is again occupied.

This principle of stored cooling capacity can be applied to smaller commercial and residential installations whenever the total heat load is below the maximum capacity of the cooling equipment. However, care must be taken to prevent the temperature from becoming too cool for comfort if the building is occupied.

The stored cooling capacity can be used to reduce the size of the air-conditioning equipment. Figure 10-10 shows that the equipment capacity can be smaller than that required to handle the maximum load if the stored cooling capacity is used.

The storage principle can be applied to any comfort installation in which the maximum load is present for less than two to three hours. It is not recommended where a change in temperature is critical.

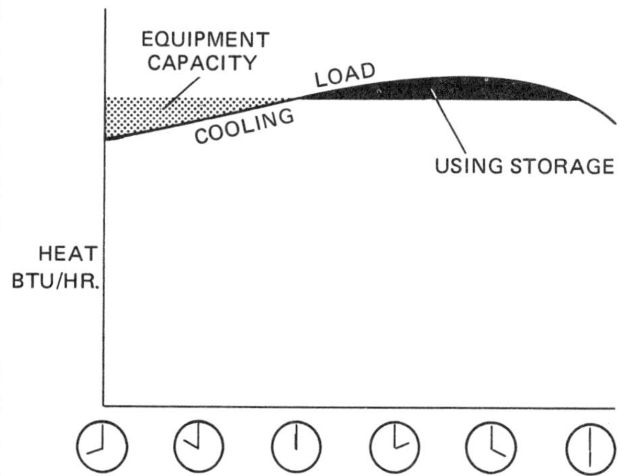

Fig. 10-10 Stored cooling capacity used

ZONING

In air conditioning, a zone is an area in a building that is set apart from other areas, usually by partitioning. Zoning makes it possible to handle load conditions in the various areas on an individual space or zone basis. Zoning usually results in a lower equipment operating cost; nonzoning usually results in a lower initial equipment cost.

If different loads occur in different parts of a building, or if different temperatures are required in different parts of a building, zoning may be required. If zoning is to be used, the load estimate is usually based on the use of one central equipment (heating and cooling) installation rather than separate installations for each zone. Separate installations cost more. In addition, the use of separate installations requires a greater total cooling capacity because each installation must be able to handle the maximum load for its zone.

Figure 10-11 shows one possible arrangement of zones within a building. In this case, the building is divided into four zones: north, south, east, and west.

Assume that the maximum loads are as follows: the north zone, 5 tons; the south zone, 5 tons; the east zone, 10 tons; and the west zone, 12 tons. Thus, the total cooling capacity required when each zone is handled separately is 5 + 5 + 10 + 12 or 32 tons, figure 10-11B. If the same zones are handled by a central system, the maximum cooling capacity required equals the maximum building load at any given time of the day. All of the zones do not experience the full solar load at the same time. As a result, the maximum loads of the zones do not occur at the same time. Therefore, the amount of cooling to be supplied at any given time must be just large enough to handle the zone experiencing the maximum load plus the load

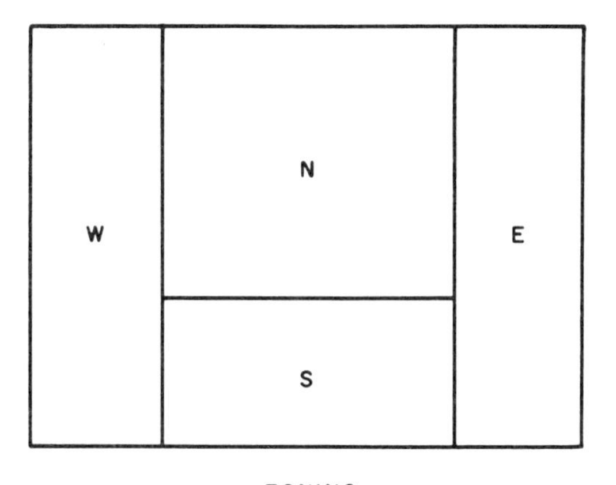

ZONING
(A)

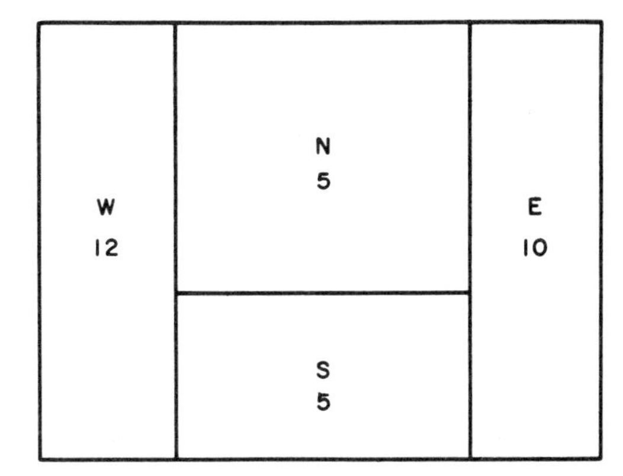

PEAK LOADS PER ZONE
(B)

Fig. 10-11

in the remaining zones. In most instances, at least one of the other zones is not at its peak load.

For the building shown in figure 10-11, the west zone has the highest maximum load. Thus, the maximum building load probably occurs about 4 P.M., figure 10-12. Since the sun strikes only the west wall at this time, the zone loads are as follows: north, 5 tons; south, 5 tons; east, 5 tons; and west, 12 tons. The total maximum cooling capacity required for the building is 5 + 5 + 5 + 12 = 27 tons. This value is 5 tons less than the capacity required when the load is estimated according to the individual maximum loads for each zone. This reduction of 5 tons is due to the reduction in the load of the east zone at 4 P.M.

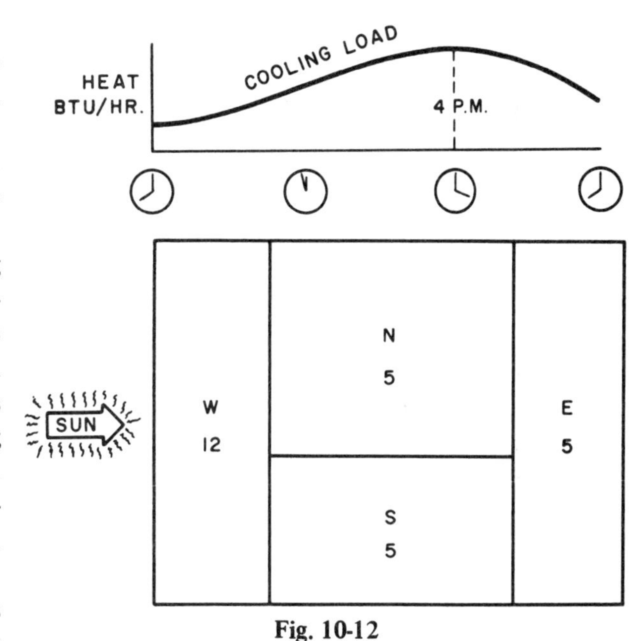

Fig. 10-12

Note that the term *maximum load* refers to the maximum cooling required on any given day of the year.

HEATING LOAD HEAT LOSSES

A portion of the heat within a building is lost through open doors, by conduction, and by filtration through cracks around windows and doors.

Conduction

During the heating season, heat is lost in a manner similar to the way in which some heat is gained during the cooling season. That is, heat is lost by conduction

Fig. 10-13 Heat lost through conduction

Fig. 10-14 Conduction takes place through glass, floors, and foundations.

through walls and roofs, glass areas, basement floors, and foundations, figures 10-13 and 10-14. The total heat lost, therefore, consists of losses by conduction through materials and by filtration through cracks and open doors.

The amount of heat lost is affected by the materials used in construction and by the difference in temperature between the indoor air and the outdoor air. The U-factor rate of heat flow is used to determine the total heat quantity Q that is lost through the materials. In a heating operation, however, the temperature difference between the inside and outside air is usually quite large. As a result, the transfer of heat by conduction is usually greater than that of a cooling operation. For example, the average temperature difference for cooling may be 10 to 20 degrees; for heating, this difference is in the range of 30 to 70 degrees.

PROBLEM 4 TOTAL HEAT LOST BY CONDUCTION (Q) (FRAME WALL)

Given

Wall surface	8 ft. x 20 ft.
U-factor	0.25
Outdoor temperature	20°F
Indoor temperature	75°F

Find

Total heat lost through wall, Q

Solution

Q = surface area x U-factor x temperature difference
Q = (8 x 20) x 0.25 x (75 – 20)
Q = 160 x 0.25 x 55
Q = 2,200 Btu per hour

Heat is also lost through floors on ground level (slab construction) and through floors and walls below the ground level (basement area), figure 10-15. There is usually a greater heat loss from the outer areas of slab floors on ground level than through basement floors and walls that are below the ground level. This difference in heat loss is due to the difference in temperature between surface ground and the ground below the surface. The surface ground temperature varies with the air temperature. This temperature variation decreases almost uniformly as the depth below the ground surface increases. At approximately eight feet, the ground temperature remains relatively constant since the surface temperature has little effect. At this depth, the temperature may be a minimum of 45°F. U-factors for heat loss through floors at the ground level and below ground level are given in Unit 11.

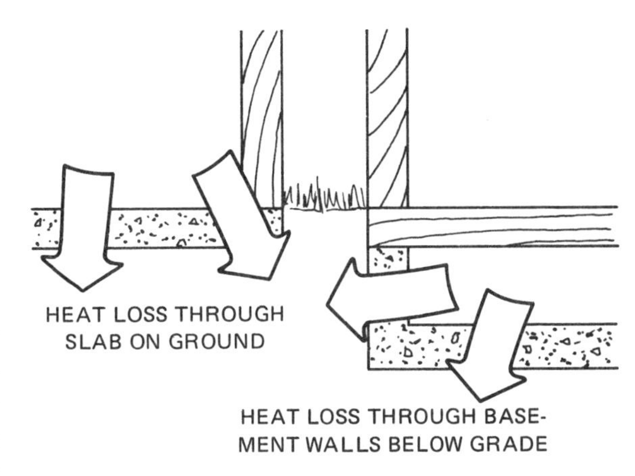

Fig. 10-15 Heat lost on and below ground level

Ventilation

Heat loss also occurs because heat is required to increase the temperature of the air used for ventilation. A better term for this heat is perhaps the *source of the heat load*, rather than *heat loss*. The heat used to increase the ventilation air temperature is not actually lost since it enters the heated space. However, this heat is an additional requirement (heat load) since the cold ventilation air must be heated.

The ventilation air heat load is determined in the same manner as the cooling load: multiply a factor of 1.08 by the cubic feet per minute of air and the indoor temperature minus the outdoor temperature. (The factor of 1.08 applies to sensible heat only.)

PROBLEM 5 VENTILATION AIR SENSIBLE HEAT LOAD (WINTER)

Given

Ventilation air	500 cfm
Indoor temperature	75°F
Outdoor temperature	20°F
Multiplying factor	1.08

Find

Heat added, Q

Solution

Q = cfm x 1.08 x (indoor temperature – outdoor temperature)
Q = 500 x 1.08 x (75 – 20)
Q = 540 x 55
Q = 29,700 Btu per hour, sensible heat

If humidity conditions indicate that a latent heat load is to be added, the latent heat required is determined as follows:

Q = cfm x 0.68 x (indoor humidity in grains per pound minus outdoor humidity in grains per pound)

PROBLEM 6 VENTILATION AIR LATENT HEAT LOAD

Given

Indoor conditions	75°F DB and 35% RH
Outdoor conditions	20°F DB and 20% RH
Ventilation air	500 cfm
Multiplying factor	0.68

Find

Heat added, Q

Solution

Q = cfm x 0.68 x (indoor-outdoor grains per pound)
Q = 500 x 0.68 x (46 gr - 4 gr)
Q = 340 x 42
Q = 14,280 Btu per hour, latent heat

Heat Load Reduction

Several methods can be used to reduce the heat load. For example, the walls, roofs, ceilings, and floors can be insulated; storm windows or double-pane glass windows and weatherstripping around windows and doors can be installed.

An insulation thickness of four to six inches is recommended for maximum reduction of the heat load. Actually, the first four inches of insulation are the most effective.

The addition of storm or double-pane glass windows reduces the conducted heat flow through glass by 40% to 60%.

SUMMARY

Load Description

- The air-conditioning load is the amount of heat that must be added to or removed from a structure to maintain the desired conditions.
- The determination of the amount of heat to be handled to establish and maintain the required comfort conditions is called load estimating.
- Knowledge, experience, and common sense are required to produce a good load estimate.

Cooling Load (Outdoor Heat Sources)

- The majority of the summer cooling load consists of outdoor heat sources such as direct solar heat through glass, conducted heat, and heat in ventilation and infiltration air.
- Solar heat enters the building directly through glass by radiation and through building materials by conduction.
- The effects of direct solar heat are felt immediately. Conducted heat may not be felt for several hours.

- The amount of conducted solar heat entering a building depends upon the construction materials.
- The U-factor is a value applied to the quantity of heat flowing through one square foot of building material.
- Heat conduction through the construction materials takes place as a result of the temperature differences between the indoor air and the outdoor air.
- The amount of conducted heat entering the building depends upon the size of the surface area and the resistance to heat flow of the material used in the area.
- Q is the total heat load and is measured in Btu per hour.
- Building orientation can reduce the cost of air-conditioning equipment up to 25%.
- Approximately seven hours of continuous sun are required if the heat of the sun is to reach the inside of a 12-inch thick brick wall.
- Solar heat penetrates glass in a shorter time than is required for other materials such as wood or brick.
- The infiltration heat load enters the building through cracks around windows and doors and through open doors.
- Ventilation air is a source of heat.
- Moisture entering a building by infiltration or ventilation is a heat load.
- The sensible heat load resulting from infiltration or ventilation air is determined as follows: Q = (cfm x 1.08) x (outdoor temperature – indoor temperature)
- The latent (moisture) heat load resulting from infiltration or ventilation air is determined as follows: Q = (cfm x 0.68) x (moisture difference between the indoor air and the outdoor air)
- The total Q is obtained by adding the sensible heat and the latent heat.

Cooling Load (Indoor Heat Sources)

- Indoor heat sources include people, lights, appliances, motors, and machines.
- Active people produce more heat than do people at rest.
- Lights produce heat according to the wattage value of the lamps.
- Motors, appliances, and machines produce heat according to the energy or fuel consumed.

Cooling Storage

- Every structure absorbs and retains heat.
- Stored heat can become an additional heat source for the morning startup of the conditioning equipment.

- The stored cooling capacity can be used to offset part of the heat load during maximum load periods.
- The use of the cooling storage principle means that smaller (less costly) equipment can be selected.

Zoning

- A zone is an area of a building that is set apart from other areas, usually by partitions.
- Zoning makes it possible to maintain different temperatures in any part in the building.
- Zoning usually results in a lower equipment operating cost (although the initial equipment cost may be higher).
- If zoning is required, the load estimate should be based on the use of one central air-conditioning equipment station.
- A zoning estimate should be based on the zone with the maximum load plus the simultaneous load in the other zones.
- The maximum load refers to the maximum cooling required on any given day of the year.

Heating Load Heat Losses

- The heating load consists of heat escaping from a building by conduction and by filtration through cracks around windows and doors.
- The rate of conducted heat flow depends upon the type of construction material used and the temperature difference between the inside air and the outside air.
- The amount of heat loss through cracks depends upon the quality of construction and the wind velocity.
- Heat loss due to conduction is determined as follows:

  Q = surface area x U-factor x temperature difference
- Cold ventilation air is a source of heat load. Ventilation air sensible heat load is determined as follows:

  Q = cfm x 1.08 x temperature difference
- Ventilation air latent heat load is determined as follows:

  Q = cfm x 0.68 x moisture difference in grains per lb.

Heat Load Reduction

- Heat load can be reduced by the addition of insulation, weatherstripping, storm windows, and double-pane glass windows.
- Two or four inches of insulation are recommended for walls; four to six inches are most effective for roofs.

REVIEW

1. Name the chief sources of outdoor heat that make up the major part of the summer cooling load.
2. What is a U-factor?
3. Define Q.
4. What equation is used to determine the ventilation air sensible heat load?
5. What equation is used to determine the ventilation air latent heat load?
6. What are the chief sources of indoor heat?
7. How is the amount of heat from lights and motors determined?
8. Define air-conditioning load.
9. Total heat includes two kinds of heat. Describe each type.
10. What is load estimating?

UNIT 11

COOLING AND HEATING LOAD ESTIMATING GUIDES

OBJECTIVES

After completing the study of this unit, the student will be able to

- describe what effects the following factors have on the estimation of the cooling load and the heating load –

  construction materials
  water sprays
  ventilation
  infiltration
  people
  lighting

- state the procedure for calculating heat gains.

The following factors in estimating cooling and heating loads are covered in this unit: U-factor values, percent heat flow reduction, ventilation recommendations, and the effects of materials, lights, people, and equipment. While the values given in this unit are to serve as guides only, they are useful in making rough preliminary estimates of the cooling and heating loads.

	U-FACTORS	SQUARE FEET COMPARED TO GLASS
WINDOW ORDINARY GLASS	1.13	
RESIDENTIAL WALL	.25	4½
RESIDENTIAL ROOF and CEILING	.31	4
COMMERCIAL WALL	.33	3½
COMMERCIAL ROOF and CEILING	.40	3

Fig. 11-1 U-factors for glass and construction materials

COOLING LOAD

Glass

Figure 11-1 gives the U-factors for glass and for typical construction materials. The figure also compares the heat flow through glass with the heat flow through the various materials. For a given set of air conditions, the U-factor for glass is more than twice that of the other materials. In other words, glass allows as much heat to pass through as approximately four sq. ft. of residential wall or roof area, or three sq. ft. of commercial wall or roof area.

U-FACTOR	% REDUCTION IN HEAT LOSS
WINDOW 1.13	—
STORM WINDOW .45	60
DOUBLE WINDOW .65	43

Fig. 11-2 Effects of types of glass and glass construction on conducted heat quantity

Reduction of Conducted Heat Through Glass

Different types of glass and glass construction have varying effects on the quantity of conducted heat, figure 11-2. For example, a storm window that fits tightly reduces the heat flow by more than 50%. A sealed double-pane window with a 1/4-in. air space between the panes reduces the conducted heat flow by approximately 40%. The double-pane window and the storm window are effective for both summer cooling and winter heating conditions. For an air-conditioned structure, the storm windows remain in place year-round, except when cleaning is required.

Reduction of Solar Heat Through Glass

The heat load caused by direct solar heat through glass can be reduced by using different types of glass and glass construction, figure 11-3. Certain shading devices are also effective in combating direct solar heat, figure 11-4.

	SOLAR HEAT REDUCTION
HEAT ABSORBING GLASS	25%
DOUBLE PANE	10-20%
STAINED GLASS	30-65%

- Special heat absorbing glass reduces the solar heat load by approximately 25%.
- Double-pane glass not only reduces the conducted load but also is 10% to 20% effective in reducing the solar load.
- Stained glass can be up to 65% effective in reducing the solar load depending upon the color used.

Fig. 11-3 Solar heat reduction through glass

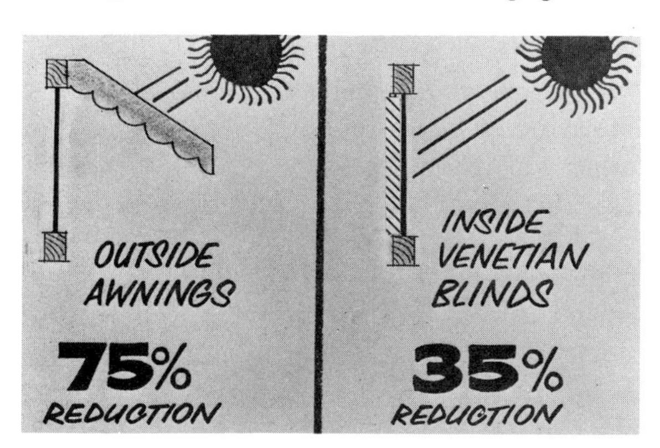

- Awnings or other types of shading devices installed on the outside of the window can be up to 75% effective in reducing the solar load.
- Venetian blinds or curtain shades on the inside of the window reduce the solar load by approximately 35%.

Fig. 11-4 Solar heat reduction: shading devices

	U-FACTORS	SQUARE FEET COMPARED TO GLASS
WINDOW ORDINARY GLASS	1.13	
RESIDENTIAL WALL	.25	4½
RESIDENTIAL ROOF and CEILING	.31	4
COMMERCIAL WALL	.33	3½
COMMERCIAL ROOF and CEILING	.40	3

Fig. 11-5 U-factors for walls and roofs

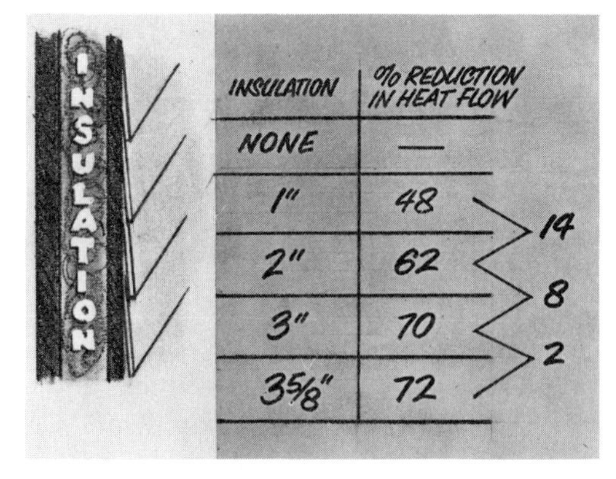

Fig. 11-6 Heat flow reduced by insulation

Walls and Roofs

The heat transfer U-factor for walls and roofs ranges from 0.25 for typical residential walls to 0.40 for typical commercial roofs, figure 11-5.

Reduction of Conducted Heat Through Walls and Roofs

The flow of heat through walls can be reduced by insulation, figure 11-6. One inch of insulation reduces the heat flow by 48%. The second inch adds a 14% reduction in heat flow for a total of 62%. The third inch adds 8% for a total reduction of 70%. The fourth inch adds a further reduction of 2%.

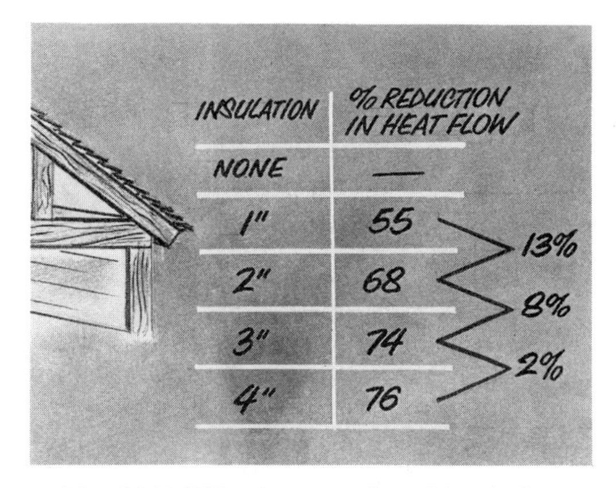

Fig. 11-7 Effectiveness of roof insulation

Insulation in the roof or ceiling is even more effective in reducing the conducted heat, figure 11-7. One inch of insulation reduces the heat flow by approximately 55%. The second inch reduces the heat flow by another 13% for a total of 68%. The third inch adds an 8% reduction and the fourth inch another 2% reduction for a total reduction of 76%. Although additional inches of insulation reduce the heat conducted through ceilings and roofs by only a small percentage, a depth of six in. of insulation is recommended for modern residential construction.

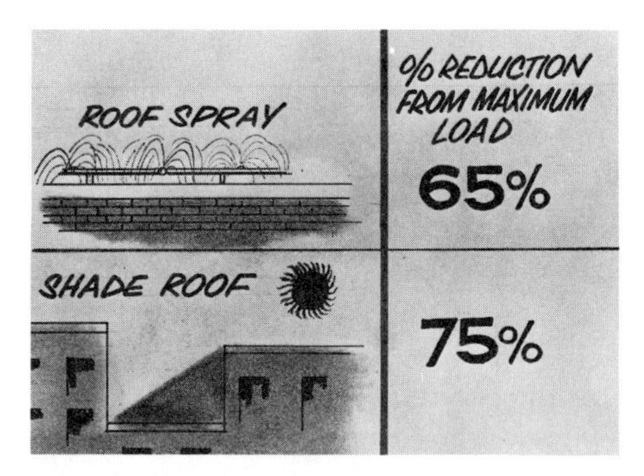

Fig. 11-8 Methods of reducing roof heat load

Figure 11-8 shows two methods of reducing the roof heat load. One method sprays water on the roof at a controlled volume. As the moisture evaporates, the surface of the roof is cooled. This method reduces the roof heat load by approximately 65%. When estimating the cooling load for summer operation, the effect of shading must also be considered. Any part of a roof that is protected from the sun by shade from adjoining structures shows a 75% reduction of the solar load in that area.

Roof Materials

The types of construction materials used and their resistance to heat flow also affect the roof heat load. A wooden frame roof is considered to be light construction. Such a roof allows approximately 18 Btu per hour to pass through each square foot of surface, under maximum load conditions, figure 11-9.

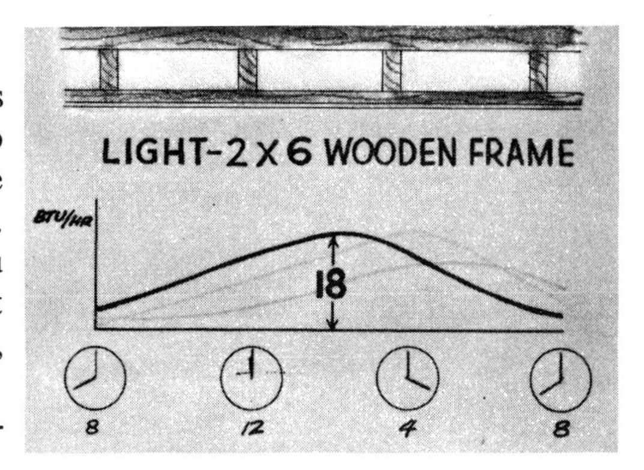

Fig. 11-9 Btu/hr. for light construction

Four inches of concrete are considered to be medium roof construction. This type of roof passes approximately the same amount of heat (19 Btu per hour) under maximum load conditions as the wood frame roof.

In general, heavy roof construction is considered to be six inches of concrete. This construction allows only 15 Btu per hour to pass through each square foot of roof surface.

If the designer is aware of the types of roof materials and the effectiveness of insulation, the estimate of the solar load on the roof can be reduced considerably. For example, it was shown that a medium concrete roof has a maximum heat load of 19 Btu per hour per square foot. However, a heavy concrete roof with four inches of insulation has a heat load of approximately four Btu per hour per square foot at maximum conditions.

For a roof that is 50 ft. by 100 ft., the heat load difference between the medium roof and the heavy roof with insulation is:

(100 ft. x 50 ft.) = 5,000 sq. ft. x (19 - 4) = 75,000 Btu per hr. (approximately)

Ventilation

Acceptable ventilation standards range from 5 to 7 1/2 cubic feet of air per minute (cfm) per person. If a moderate amount of smoking occurs in an area, a minimum of 15 to 30 cfm of ventilation air is recommended. For special areas where heavy smoking is expected, the ventilation air should be increased to 30 to 50 cfm per person.

Figure 11-10 shows typical ventilation air quantities per person for office, shop, and department store conditions. These air quantities are also shown in terms of the amount of sensible heat per person in Btu/hr. based on 95°F outside and 80°F inside temperatures. As explained in Unit 10, the total ventilation air cooling load includes both sensible heat and latent heat. The latent heat must be added to the Btu values shown in the figure to obtain the total ventilation air load.

VENTILATION AIR	SENSIBLE HEAT PER OCCUPANT SUMMER
OFFICE (15 CFM)	240 BTU/HR.
RETAIL SHOP (10 CFM)	160 BTU/HR.
DEPARTMENT STORE (7½ CFM)	120 BTU/HR.

Fig. 11-10 Typical ventilation quantities

SUMMER COOLING

	SENSIBLE HEAT
WINDOW 3'X5'	350 BTU/HR.
RETAIL SHOP 3'X7'	65 BTU/HR. PER PERSON IN STORE

Fig. 11-11 Infiltration load

HEAT FROM PEOPLE

	SENSIBLE BTU/HR.	LATENT BTU/HR.	TOTAL BTU/HR.
THEATRE	195	155	350
OFFICE	200	250	450
DANCING	245	605	850
BOWLING	465	985	1450

Fig. 11-12 Latent and sensible heat in occupied areas

Infiltration

Infiltration of heat through doors and windows depends on the quality of construction. A casement window with a 1/64-in. crack has a cfm factor that is half as large as the same window with a 1/32-in. crack. In other words, approximately twice as much air infiltrates through the 1/32-in. opening in the window.

An average infiltration load for summer cooling is 350 Btu per hour for a 3 x 5-ft. window and 65 Btu per hour per person using a 3 x 7-ft. door in a retail shop, figure 11-11. To determine the total infiltration in Btu for the door of a retail shop, multiply 65 Btu per hour by the number of customers per hour.

People

People add both sensible heat and latent heat. Typical values for the total heat per person for theaters, offices, dance areas, or bowling centers range from 350 Btu (theater) to 1,450 Btu (bowling), figure 11-12. These values are equivalent to approximately 3 to 12 tons of cooling per 100 people.

Figure 11-12 also shows the latent and sensible heat breakdown of the total heat load per person. The cooling load for the office conditions is approximately four tons per 100 people. For dance areas, the cooling load is approximately seven tons per 100 people.

In general, the sensible and latent load for theaters, offices, retail shops, and similar types of occupancies ranges from 35,000 Btu to 45,000 Btu per 100 people. This load requires roughly three to four tons of cooling capacity.

The combined ventilation and people load at frequently encountered conditions (95°F DB temperature and 40% outdoor RH and 80°F DB temperature and 50% indoor RH) is approximately 70,000 to 90,000 Btu per hour per 100 people. Such a load requires 6 to 8 tons of cooling capacity. The ventilation load is roughly equivalent to the people load.

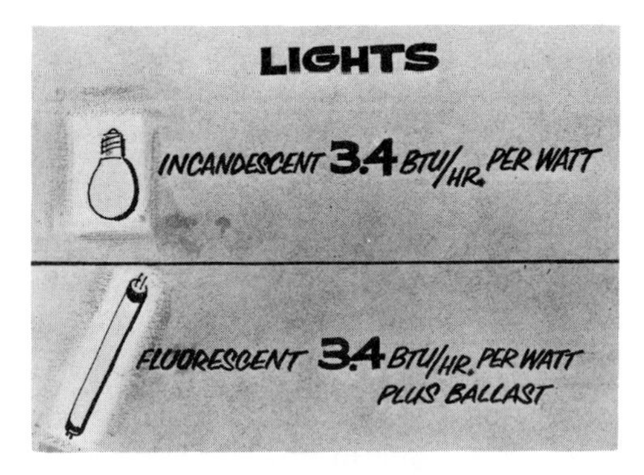

Fig. 11-13 Heat load from lights

Lights

The load from incandescent lights is 3.4 Btu per hour per watt, figure 11-13.

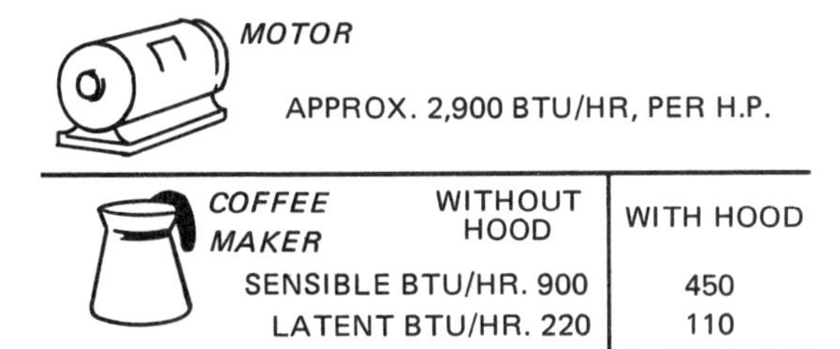

Fig. 11-14 Heat load from motors, appliances, and office machines

For a bulb rated at 100 watts, the load is 340 Btu; for a 1,000-watt bulb (1 kilowatt), the load is 3,400 Btu per hour. A value of ten kilowatts equals a load of 34,000 Btu. This load requires approximately 3 tons of cooling. Fluorescent lights require an addition of about 25% to these values to account for the ballast.

Motors, Appliances, and Office Machines

Figure 11-14 shows the heat given off by various types and sizes of devices. The nameplate rating of each of these items is not always an accurate indication of the resulting heat load. The actual load depends on the usage and the starting and stopping characteristics of each device. It is apparent from figure 11-14 that some office machines are a source of considerable heat and can require 2 1/2 tons or more of cooling.

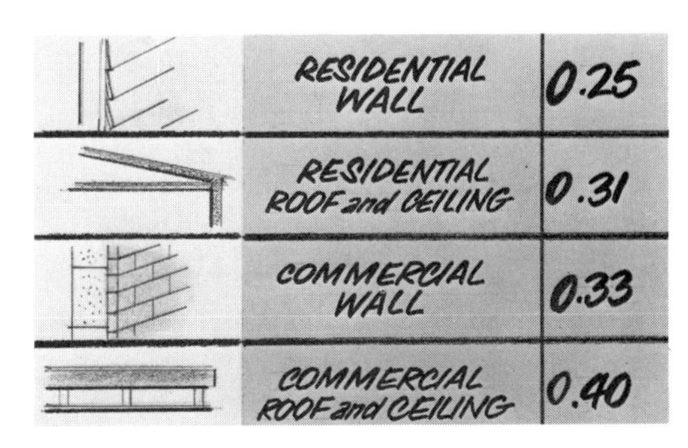

Fig. 11-15 U-factors for walls and roofs

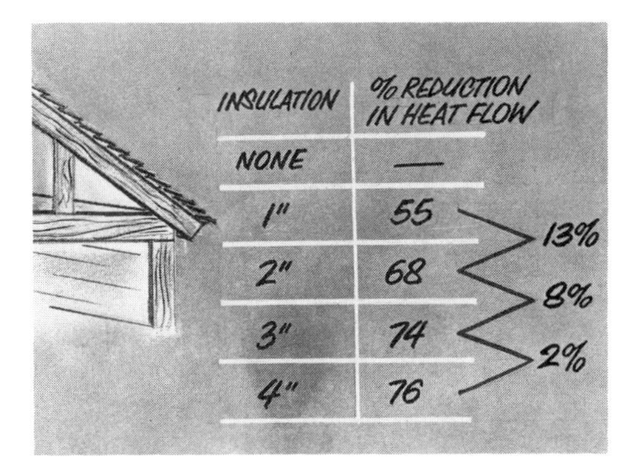

Fig. 11-16 Effectiveness of roof insulation

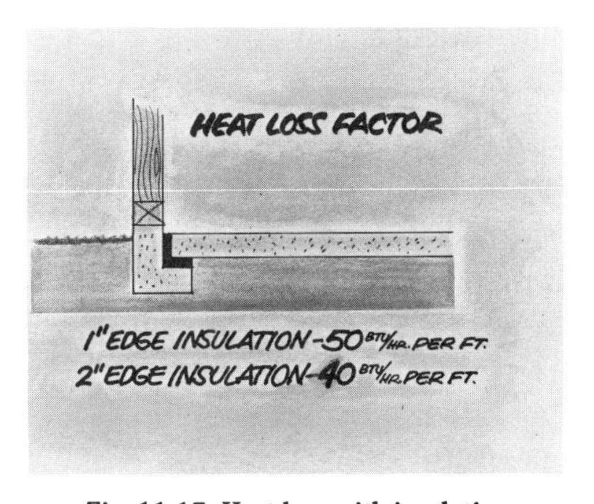

Fig. 11-17 Heat loss with insulation

HEATING LOAD ESTIMATE

Glass, Walls, Roofs, Floors

The heat flow through materials is the same whether the flow is toward the outside or toward the inside. Thus, the same U-factors apply for heating and for cooling. Figures 11-15 and 11-16 give the U-factors already presented in the cooling estimate section of this unit. The comparison values are also the same. Insulation reduces the heat flow through the roof by the same percentages given for summer cooling.

For slab floor construction, insulation around the edge of the slab is probably more effective in winter than in summer, figure 11-17. This situation is reasonable since the heating loss is likely to be greater in winter than the heat gain in summer.

For example, at 0°F the heat loss around the perimeter of a slab is 50 Btu per linear foot with 1 in. of insulation. For a slab having dimensions of 30 ft. x 40 ft., the heat loss is 50 x 140 = 7,000 Btu per hour, figure 11-18, page 110.

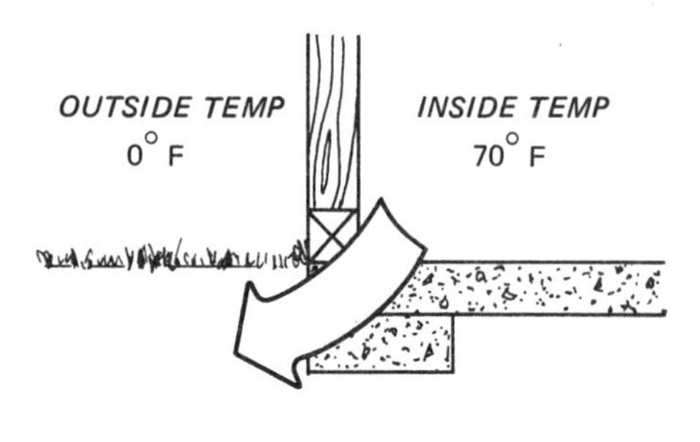

HEAT LOSS Q = F xP
= 50 x (140)
= 7,000 BTU/HR.

Fig. 11-18 Heat loss around slab floor

Infiltration

The infiltration heat losses shown in figure 11-19A are based on the following conditions: outdoor temperature of 0°F, indoor temperature of 70°F, and a 15 mph wind. The closed door and window losses are relatively small. However, the losses through an open door in a retail shop are nearly 6 to 7 times greater than the closed door and window losses. Weatherstripping will reduce crack infiltration by a value that can be as high as 50%, figure 11-19B.

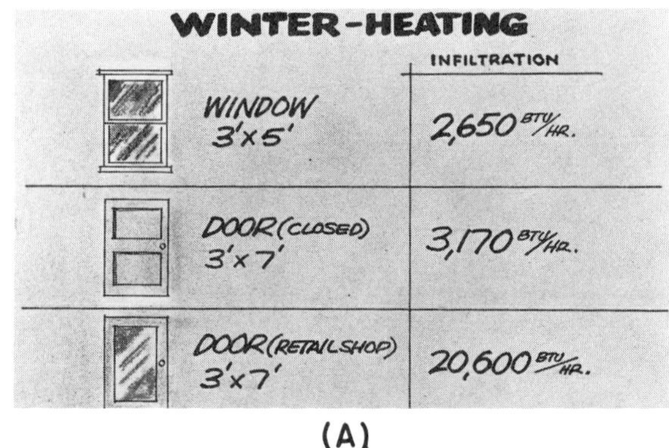

(A)

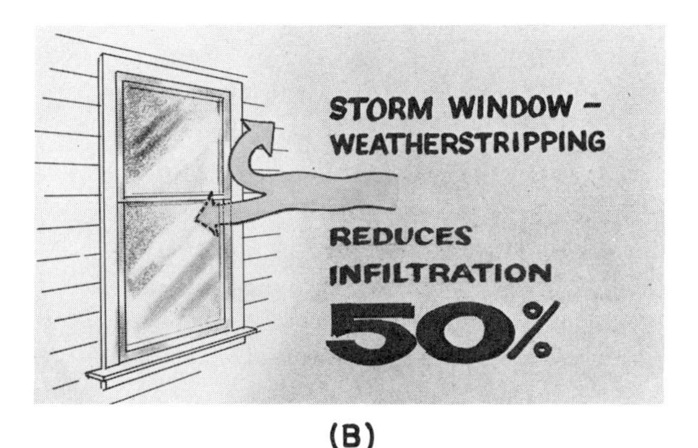

(B)

Fig. 11-19 Infiltration losses

Ventilation Air Heating Load

The ventilation air sensible heat load values are based on an outdoor temperature of 0°F and an indoor temperature of 70°F, figure 11-20.

The ventilation air latent heat load is based on the temperature and humidity difference.

VENTILATION AIR	SENSIBLE HEAT PER OCCUPANT WINTER
OFFICE (15 CFM)	1140 BTU/HR
RETAIL SHOP (10 CFM)	760 BTU/HR
DEPARTMENT STORE (7½ CFM)	570 BTU/HR

Fig. 11-20 Ventilation air sensible heat load

PROBLEM 1 VENTILATION AIR LATENT HEAT LOAD

Given

Outdoor temperature and humidity	0°F and 100% RH
Indoor temperature and humidity	70°F and 30% RH
Ventilation air quantity (office)	500 cfm
Multiplying factor	0.68

Find

Latent heat load, Q

Solution

Q (latent) = cfm x 0.68 x (indoor grains/lb. - outdoor grains/lb.)
Q (latent) = 500 x 0.68 x (33-11)
Q (latent) = 340.00 x 22 = 7,480 Btu per hour

Since the office worker requires 15 cfm of ventilation air, the air quantity supplied (500 cfm) is sufficient for 34 people. The ventilation latent load per person then is 7,480 ÷ 34 = 220 Btu per hour.

CALCULATIONS OF HEAT GAINS

The calculation of the heat gain through the walls, roof, ceilings, windows, and floors of a structure requires three simple steps:

1. determine the net area in square feet;
2. find the proper heat gain factor from a table; and
3. multiply the area by the factor to find the product expressed in Btu/hr.

The term net area is used in the calculation because the total wall area cannot be used. That is, the total wall is found and then any window and door areas are subtracted to obtain a net wall area. This area is then combined with the areas of the roof, ceilings, and floor to obtain the total net area.

The phrase *proper heat gain factor* is used to point out that the factors may differ, depending on the tables used. The text to this point has presented some U-factors. Unit 12 gives tables of different factors known as combination factors. In addition, manufacturers of air-conditioning equipment compile tables for use with their estimating forms. These tables again may vary with each manufacturer depending on the particular information used to develop the tables. Regardless of the source of the factors used, the following steps are basic to the determination of the heat load.

Step 1: Determining the Net Area

Find the net area of a wall 24 ft. 6 in. long and 8 ft. high containing three 3 ft. x 4 ft. windows.

a. Total wall area = length x height
= 24.5 ft. x 8 ft. = 196 sq. ft.

(Note: the dimensions in feet and inches are converted to decimal feet)

b. Window area = (3 ft. x 4 ft.) x 3
= 12 sq. ft. x 3 = 36 sq. ft.

c. Net area = 196 - 36 = 160 sq. ft.

Step 2: Finding the Heat Gain Factor

The wall in this example is an 8-in. brick exterior wall with no interior finish. Table A of the Appendix indicates that the heat gain factor is 0.50.

Diagram of wall and type of masonry	Thick-ness of masonry, in.	Plain wall, no interior finish
SOLID BRICK	8 12 16	0.50 0.35 0.28

Fig. 11-21 Excerpt from Table A of Appendix

Step 3: Multiplying the Area by the Factor

Net area x factor = Btu/hr.
160 x 0.50 = 80 Btu/hr. for the given wall

This example is a simple one but it illustrates the essential steps that must be performed to determine the heat gain. Factors such as the building orientation, latitude of the structure, amount of shade and sunlight, and similar conditions which affect the true condition are omitted from the calculation although normally they are considered.

Succeeding units repeatedly use the formula, heat = area x factor = Btu/hr., in residential and commercial estimating.

SUMMARY

- A U-factor is a coefficient of heat transfer. Since materials differ in their heat transmission properties, each material has a different U-factor. Complete U-factor tables are included in the Appendix.
- Heat gains through glass can be reduced by:
  1. storm windows
  2. double-pane (Thermopane®) windows
  3. stained or tinted glass
  4. awnings, blinds, or screens
  5. shading due to roof overhang
- Heat gains through walls can be reduced by:
  1. insulation
  2. light exterior color
  3. shading (roof overhang, other buildings)
- Heat gains through the roof can be reduced by:
  1. natural or forced ventilation of the attic
  2. insulation of ceiling or rafters
  3. roof spray or pond on flat roof
  4. shading from another building
  5. type of construction (light, medium, heavy)

- Ventilation air contains both sensible and latent heat.
- The ventilation air sensible load is based on the temperature difference between the inside air and the outside air. Q (sensible) = cfm x 1.08 x (temperature difference)
- The ventilation air latent load is based on the humidity difference between the inside air and the outside air. Q (latent) = cfm x 0.68 x (humidity difference in gr./lb.)
- Heat losses are calculated using the same U-factors as for heat gains since heat flow is the same whether toward the outside or the inside.
- People add both sensible heat and latent heat. The amount of heat of both types added depends on the activity of the people. The heat ranges from 350 Btu/hr. per person at rest to 1,450 Btu/hr. per person engaging in strenuous exercise.
- Lights add sensible heat: 3.4 Btu/hr. per watt for incandescent lamps and 3.4 Btu/hr. per watt plus 25% for fluorescent lamps.
- Motors add sensible heat: 2,545 Btu/hr. per horsepower. The efficiency of the motor, however, reduces this value to 80% to 90%.

REVIEW

Solve the following problems.

1. A wall 30 ft. long by 8 ft. high has three windows. Each window measures 3 ft. 6 in. by 5 ft. The outside DB temperature is 90°F and the inside DB temperature is 75°F.
   a. What is the net area of the wall?
   b. If single-pane glass is used, what is the total heat gain through the glass?
   c. How much is the heat gain through glass reduced if double-pane glass is used?
   d. What is the heat gain through the wall if the construction shown in figure 11-22A is used?
2. If a wall whose net area is 800 sq. ft. is constructed as shown in figure 11-22B, what is the heat gain through the wall?

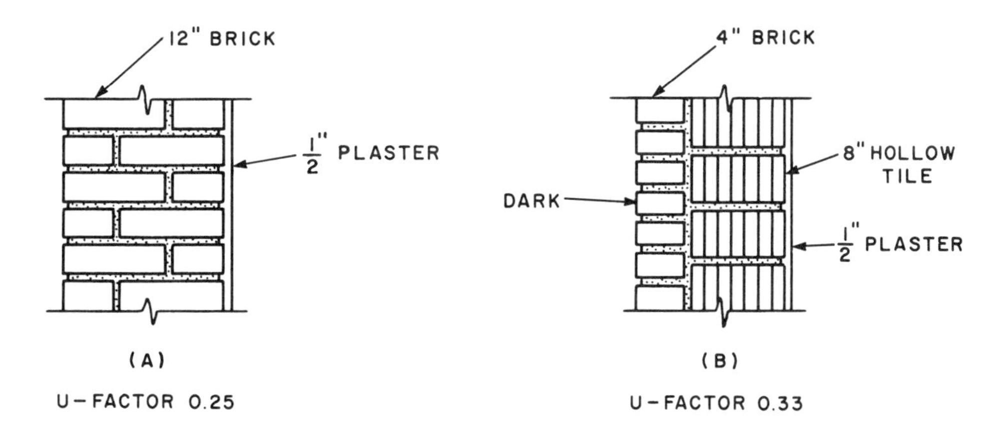

Fig. 11-22

3. An outside wall has three windows and one door. The windows measure 4 ft. 4 in. x 6 ft. each and have single-pane glass. The door is 4 ft. x 7 ft. The gross wall area is 28 ft. x 9 ft. The outside DB temperature is 90°F; the inside DB temperature is 75°F.
   a. What is the heat gain through the glass?
   b. What is the heat gain through the wall if it is constructed as in figure 11-22A?
   c. What is the total heat gain (wall and glass) if double-pane glass and the construction shown in figure 11-22B are used?
4. a. A room has an incandescent lighting load of 3,000 watts. What is the heat gain from the lights?
   b. What is the heat gain if fluorescent lamps of the same wattage are used?
5. A theater has a capacity of 500 people. How much outdoor air should be introduced for ventilation?
6. If moderate smoking is the usual condition in a room seating 200 people, how much outdoor air is needed for ventilation?
7. A retail shop serves a daily average of 150 customers. The entrance door is 3 ft. by 7 ft. What is the total heat gain due to infiltration?
8. A slab floor measures 26 ft. x 40 ft. and has 1 in. of edge insulation. What is the heat loss of this floor at 0°F?
9. An office seats 10 people. What is the total heat load from the people (sensible and latent heat)?
10. State the basic formula for determining heat transfer.
11. List, in order of effectiveness, four methods of reducing the heat gain through glass.
12. List three ways of reducing the heat gain through walls.
13. List four ways of reducing the heat gain through roofs.
14. What factor is used to calculate the sensible heat added by (a) lights and (b) motors?
15. State the formulas used to calculate the sensible heat and latent heat added by ventilation air.
16. Why are the same U-factors used whether the heat transfer results in a gain (summer cooling) or a loss (winter heating)?

UNIT 12
ESTIMATING THE AIR-CONDITIONING LOAD

OBJECTIVES

After completing the study of this unit, the student will be able to

- use forms and associated tables supplied by the Air Conditioning and Refrigeration Institute (ARI) to estimate the air-conditioning load (both heating and cooling loads).
- Explain how the following items are important to a load estimate

survey pattern	people
building orientation	motors, lights, and appliances
building size	ventilation
building shape	equipment location
materials of construction	water, gas, and electrical services
glass areas	local and state building codes
infiltration	

- state the purpose of a load estimate.
- use appropriate tables of factors in determining sensible and latent heat gain.
- determine ventilation using either the cubic feet of air per minute (cfm) per person in a building or the cubic feet of air per minute (cfm) required per square foot of floor space in a building.
- determine infiltration using the air change approach.

Units 10 and 11 presented the sources of heat gains and losses which must be considered in determining the total air-conditioning load. The basic heat transfer formula used to obtain U-factors was described and some U-factors were illustrated. (A complete tabulation of U-factors appears in the Appendix.) The forms used in this unit are published by the Air Conditioning and Refrigeration Institute (ARI). These forms are samples of the numerous types of forms and tables that are available. Subsequent units of this text dealing with residential and commercial estimating for specific structures will introduce forms from various manufacturers.

SURVEY PATTERN

A survey is a review of a particular structure to determine the best possible air-conditioning system for that structure and the best method of installing the system. The review is most effective when it follows a specific pattern. Surveys for larger buildings (such as office buildings) are usually more comprehensive than surveys for smaller commercial or residential structures.

Building Orientation

The location of the structure must be identified in relation to the compass points, the sun, wind direction, and the surrounding buildings. It is necessary to determine the direction in which the building will face. It is also important to know on which sides of the building the sun will shine and at what times of the day, figure 12-1. This information is related to the solar load on the building and can also be used to determine the layout of the duct system. For example, if the front of the building is glass surfaced and faces east, the major solar load probably occurs through the glass surface early in the morning. If the front surface is glass and faces south, the major solar load probably occurs at noon or during the early afternoon.

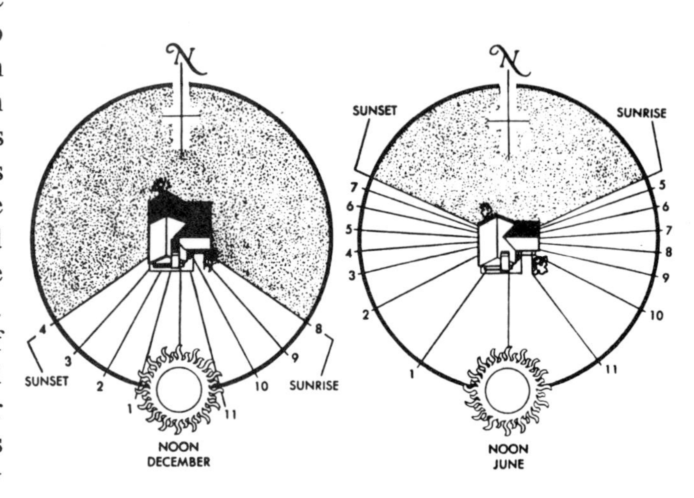

Fig. 12-1 Hourly direction of the sun in winter and summer

The prevailing wind can affect the infiltration load, figure 12-2. For example, the infiltration load on the windward side is greater if the building has many windows as compared to a building with a solid wall on that side. The amount of infiltration through any cracks around the windows depends, to a degree, on the type of construction. The location of the building in relation to other buildings also has an effect on both the solar load and the infiltration load. For example, if the building is partially surrounded by taller buildings, the solar load is reduced because of the shade offered by the taller buildings. In addition, the effect of the prevailing wind is reduced as a factor in the infiltration load.

Thus, the load estimater must be familiar with the location or orientation of a building so that the effects of the weather on the building air-conditioning load can be anticipated and included in the estimate.

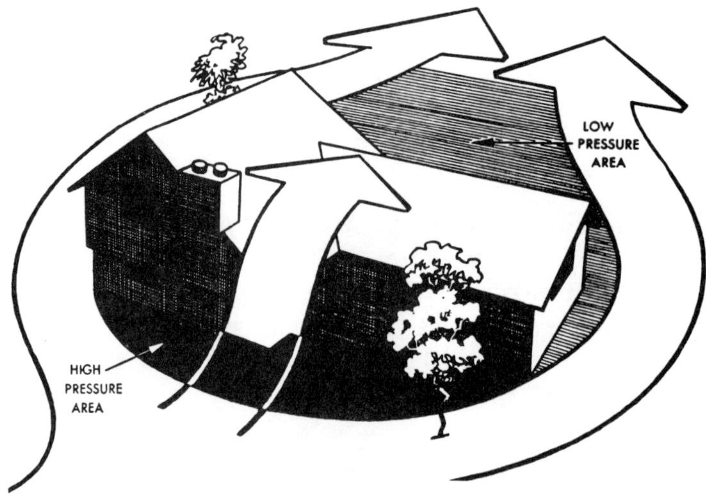

Fig. 12-2 High- and low-pressure areas from wind effect

Building Size

Important factors in estimating the air-conditioning load are the areas of the walls, roof, and exposed floor surface. These areas are used to determine the conduction and convection cooling or heating load due to the sun, wind, and other outdoor weather conditions. The total wall, roof, and floor area estimates are modified by considering the percentage of glass area and the type of building construction.

Building Shape

The shape of the building is also an important factor since it affects the layout of the duct system and is a consideration when locating the air-conditioning and refrigeration equipment. For example, the duct system installed in an L- or U-shaped building is different from the duct system installed in a circular building.

Materials of Construction

The materials used to construct the building must be included in a complete survey. The cooling or heating load of the building varies according to the type of material used, such as glass, wooden frame, or heavy or light concrete. Quality or well-fitted construction usually requires less cooling and heating than does loose construction – particularly with regard to window and door framing. The use of insulation and the amount used also affect the cooling and heating load.

Glass Areas

The percentage of glass area, the type of glass used, and the direction of the wall in which the glass is installed are all major factors in determining the air-conditioning load. A determination of the time at which the maximum solar load occurs is based in part on the percentage of glass on a given surface of a building and the length of time that the glass area is exposed to direct solar radiation. The type of glass used (especially in those structures where large glass surfaces are installed) is a significant factor in load estimating. Double-pane sealed windows or windows having storm sashes reduce the conduction load considerably. If the glass area is shaded indoors by blinds or outdoors by awnings or building overhangs, the radiation load is reduced considerably.

In summary, the survey should indicate the percentage of glass area, the direction in which the glass areas face, and any devices or details of building construction that shade all or a portion of the glass area.

Infiltration

The major portion of infiltration takes place through doors and cracks around windows. It is important, therefore, to note the locations of doors and window areas in relation to the prevailing wind, figure 12-3, page 118. An important consideration is whether the doors are located on one, two, three, or even four or more sides of the building. If doors are open simultaneously on opposite sides of the building, the amount of infiltration that takes place is considerably greater than if just one door in the building is opened. If weatherstripping is applied to all windows and doors, the infiltration loss can be reduced by as much as 50%. This is especially true if the window construction results in wide cracks over several linear feet.

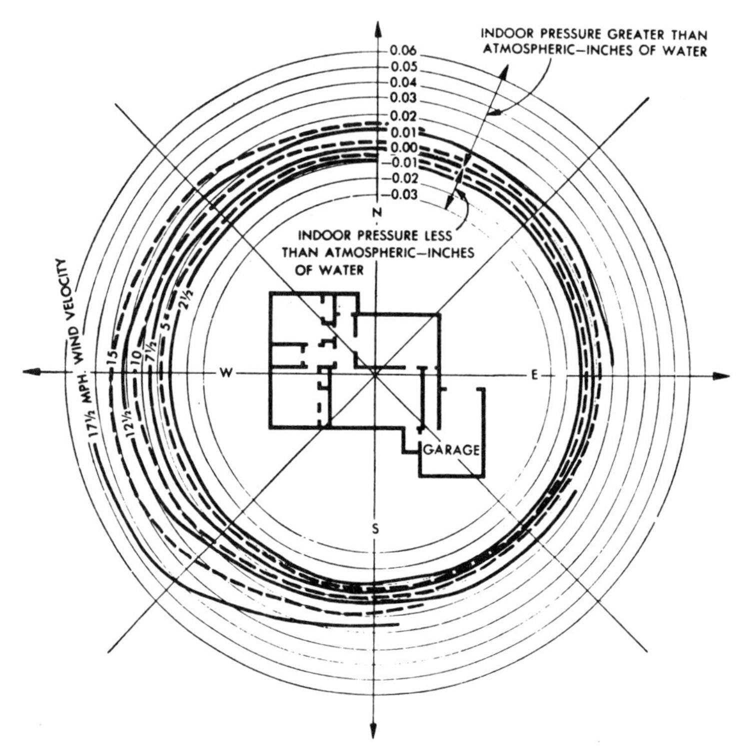

Fig. 12-3 Indoor pressure is greater than the outdoor pressure because of wind action

People Occupancy

A commercial building survey indicates the estimated number of people who are to occupy the building at a given time. This information includes an estimate of the maximum number of people who may be present in the building. The load estimator should determine if the maximum people load occurs at the same time as the maximum solar load. The answer depends to a large extent on the locations of the glass areas and the effective solar load on the glass areas in relation to the time of the maximum people load. Thus, the maximum load on the building is likely to occur at a time when the maximum number of people are in the building and the solar load through the major glass area is the largest. For example, if the people load in the morning is light and the east side of the building is of predominantly solid construction (little or no glass areas), the morning load is probably light. If the people load is concentrated in midafternoon and the south side of the building is predominantly glass area, then the midafternoon load is probably the maximum load for the building. If the people load is large, such as in theaters, department stores, and supermarkets, this fact must be entered on the survey. The specific time at which the largest people concentration is likely to occur must also be recorded.

Motors, Lights, and Appliances

The estimator must determine the load from motors, lights, and appliances according to the wattage or power used and the fuel consumed. The length of time that the lights, motors, and appliances are used and the frequency of use of these devices should be determined. The values for the total wattage of the lights, the total motor horsepower, and the appliance power or fuel consumed should be recorded.

These values can vary and are established by their use in a building. For example, a theater may have a lower lighting load as compared to the people load, especially in the auditorium section. Similarly, a restaurant may have a greater people and appliance load than a lighting load.

Ventilation

As stated in previous units, ventilation air in large buildings enters through the ductwork. The amount of ventilation air necessary is determined by the number of people who are to occupy the building and by standard building codes. Ventilation air requirements form a considerable portion of the load, especially during extremely cold days (heating cycle) or extremely hot days (cooling cycle). In the heating cycle, the cold outdoor air must be heated before it can be introduced to the conditioned space. In the cooling cycle, the hot outdoor air must be cooled before it can be introduced to an air-conditioned space.

In residential and small commercial structures, ventilation air is not necessarily introduced through the ductwork. Cracks around windows and doors usually supply the required air change in the structure.

Equipment Location

The location of the air-conditioning and refrigeration equipment must be considered with regard to the available space and possible building construction obstructions. In smaller stores and restaurants (where self-contained equipment is used), the location of this equipment can be a critical factor in combating draft conditions. The proper location of the equipment can insure that generally good air conditioning results. In many installations, the equipment location can be a factor in minimizing the amount of ductwork required. It is preferred to place the equipment in a central location rather than at an extreme end of a building.

Water, Gas, and Electrical Services

The placement of the water, gas, and electrical services should be determined and noted by the estimator. Lower installation costs can be realized if the air-conditioning and refrigeration equipment can be located near these services. In some instances, the location of the services may dictate the site of the air-conditioning and refrigeration equipment.

Local and State Building Codes

State building codes do not vary from area to area, but are consistent throughout the state. On the other hand, each locality may have a different set of building codes. One city may have more stringent construction specifications than another city. Similarly, the local gas, electrical, and piping codes may be more or less stringent in one area than another. In all instances, however, the state code applies.

These codes were developed and are enforced principally to maintain safety standards for all types of construction.

In later units of this text, typical surveys and check lists are used to estimate residential and commercial buildings from plans.

THE LOAD ESTIMATE

The load estimate is based on the design conditions inside the building to be air conditioned and the outside atmospheric conditions surrounding the building. The outside design conditions are the maximum extremes of temperature occurring in a specific locality. The inside design condition is the temperature and humidity to be maintained for optimum comfort.

The purpose of a load estimate is to determine the size of the air-conditioning and refrigeration equipment required to maintain the inside design conditions during periods of maximum extremes in the outside temperatures. The design load is, in effect, the air-conditioning and refrigeration equipment capacity required to produce and maintain satisfactory indoor conditions. For example, if the city of Minneapolis, Minnesota has a design outside summer temperature of 98°F and an outside winter

State	City	Winter (heating) dry-bulb, °F	Summer (cooling)	
			Dry-bulb, °F	Wet-bulb, °F
Alabama	Birmingham	10	95	78
	Mobile	15	95	80
Arizona	Flagstaff	-10	90	65
	Phoenix	25	105	76
	Yuma	30	110	78
Arkansas	Little Rock	5	95	78
California	Bakersfield	25	105	70
	El Centro	25	110	78
	Fresno	25	105	74
	Los Angeles	35	90	70
	San Diego	35	95	68
	San Francisco	35	85	65
Colorado	Denver	-10	95	64
	Pueblo	-20	95	65
Connecticut	Bridgeport	0	95	75
Delaware	Wilmington	0	95	78
District of Columbia	Washington	0	95	78
Florida	Jacksonville	25	95	78
	Miami	35	91	79
Georgia	Atlanta	10	95	76
	Savannah	20	95	78
Idaho	Boise	-10	95	65
Illinois	Chicago	-10	95	75
	Springfield	-10	98	77
Indiana	Indianapolis	-10	95	76
	Terre Haute	0	95	78
Iowa	Des Moines	-15	95	78
	Sioux City	-20	95	78
Kansas	Topeka	-10	100	78
Kentucky	Louisville	0	95	78
Louisiana	New Orleans	20	95	80
Maine	Augusta	-15	90	73
Maryland	Baltimore	0	95	78
Massachusetts	Boston	0	92	75
	Worcester	-5	93	75
Michigan	Detroit	-10	95	75
	Lansing	-10	95	75
Minnesota	Duluth	-25	93	73
	Minneapolis	-20	95	75
Mississippi	Vicksburg	10	95	78
Missouri	Kansas City	-10	100	76
	St. Louis	0	95	78
Montana	Butte	-20		
	Miles City	-35		
Nebraska	Omaha	-10	95	78
Nevada	Reno	-5	95	65
New Hampshire	Concord	-15	90	73
New Jersey	Newark	0	95	75
New Mexico	Albuquerque	0	95	70
New York	Buffalo	-5	93	73
	New York	0	95	75
	Syracuse	-10	93	75
North Carolina	Asheville	0	93	75
	Raleigh	10	95	78
North Dakota	Bismarck	-30	95	73
Ohio	Akron	-5	95	75
	Dayton	0	95	78
	Toledo	-10	95	75
Oklahoma	Tulsa	0	101	77
Oregon	Portland	10	90	68
Pennsylvania	Philadelphia	0	95	78
	Pittsburgh	0	95	75
Rhode Island	Providence	0	93	75
South Carolina	Charleston	15	95	78
South Dakota	Sioux Falls	-20	95	75
Tennessee	Nashville	0	95	78
Texas	Austin	20	100	78
	Dallas	0	100	78
	El Paso	10	100	69
	Houston	20	95	78
Utah	Salt Lake City	-10	95	65
Vermont	Burlington	-10	90	73
Virginia	Richmond	15	95	78
Washington	Seattle	15	85	65
	Spokane	-15	93	65
West Virginia	Charleston	0	95	75
	Wheeling	-5	95	75
Wisconsin	Milwaukee	-15	95	75
Wyoming	Cheyenne	-15	95	65

Table 12-1 Recommended outside design conditions for localities in the continental United States

Place	Winter, dry-bulb, °F	Summer Dry-bulb, °F	Summer Wet-bulb, °F	Place	Winter, dry-bulb, °F	Summer Dry-bulb, °F	Summer Wet-bulb, °F
Athens, Greece	30	95	72	Mexico City, Mexico	35	79	61
Berlin, Germany	5	91	70	Montreal, Canada	-10	86	74
Bogota, Columbia	30	70	65	Moscow, Russia	-40	85	72
Bombay, India		93	82	Osaka, Japan	20	93	81
Buenos Aires, Argentina	30	90	75	Panama, Panama		88	78
Cairo, Egypt	40	100	80	Paris, France	15	90	75
Calcutta, India	60	100	82	Rio de Janeiro, Brazil	55	90	78
Capetown, South Africa	45	88	72	Rome, Italy	30	92	73
Havana, Cuba	60	90	78	Shanghai, China	30	95	82
Honolulu, Hawaii	60	83	73	Stockholm, Sweden	0	83	68
Johannesburg, South Africa	40	82	65	Sydney, Australia	50	90	72
London, England	30	85	70	Tokyo, Japan	20	93	81
Madrid, Spain	25	95	75	Vancouver, Canada	10	80	68
Manila, P.I.		88	78	Vienna, Austria	5	92	71
Melbourne, Australia	35	95	74	Winnipeg, Canada	-30	90	71

Table 12-2 Recommended outside design conditions for selected localities throughout the world

design temperature of -20°F, the air-conditioning equipment must be capable of maintaining a summer inside design temperature of 78°F at 50% RH and a winter inside temperature of 68°F and 30% RH during the periods of the maximum outside temperatures.

Design Conditions – Outside

Heating. Tables 12-1 and 12-2 give the design outside dry-bulb temperatures used in calculating heating loads. For those cities and localities not listed, use the design temperatures of listed cities that most closely approximate the climate conditions of the desired locality. Use a value of 80% for the design outside humidity condition for a given locality. This value may seem high, but when the relative humidity is compared to the grains of moisture in the air on a low-temperature psychrometric chart, a relative humidity of 80% at most of the winter outside design temperatures represents a minimum number of grains. For example, at a dry-bulb temperature of 5°F and 80% RH, the moisture content of the outside air is less than 4 gr./lb.

Cooling. The design outside dry-bulb and wet-bulb temperatures used to calculate the cooling loads are shown in Table 12-1. For those cities and localities not listed, use the design temperatures of listed cities that most closely approximate the climate conditions in the desired locality.

Design Conditions – Inside

Heating. The inside comfort design temperature for winter heating is 72°–75°F DB.

Cooling. The inside comfort design temperature for summer cooling is 78°F DB and 50% RH in those installations where the primary load is due to people. If the largest part of the load is from a source other than people, the dry-bulb temperature may be increased to 80°F.

LOCATION AND OPERATION	EXPOSURE							
	N	NE	E	SE	S	SW	W	NW
30° Latitude								
10-hour Operation	25	75	90	78	76	88	99	78
24-hour Operation	23	56	70	62	60	74	88	70
40° Latitude								
10-hour Operation	23	67	90	80	77	90	99	75
24-hour operation	22	49	70	63	62	77	88	67
50° Latitude								
10-hour Operation	23	60	90	82	78	94	98	71
24-hour Operation	22	45	70	65	64	81	87	64

Table 12-3 Factors for sensible heat gain through glass for average applications* (Btu/Hr./Sq. ft., Sash Area)

*For rooms or zones with more than one exposure, use the North values for all exposures except the one having the largest solar gain (usually greatest glass area).

(Courtesy Air-Conditioning and Refrigeration Institute)

ITEM	FACTOR					
	COOLING LOAD			HEATING LOAD		
Windows:						
Single glass	---			1.13		
Double glass	---			0.50		
Walls:						
Sunlit	0.30			0.25		
Shaded	0.20			0.25		
Partitions	0.20			0.35		
Ceiling under finished room	0.24			0.24		
Floors:						
Over finished room	0.24			0.24		
Over basement with finished ceiling	0.20			0.20		
On ground	0.00			0.00		
	No Ceiling	Ceiling†	Under Attic†	No Ceiling	Ceiling†	Under Attic†
Roofs:						
Uninsulated, frame, or heavy masonry	1.20	1.07	0.93	0.60	0.35	0.30
Uninsulated, light masonry	1.87	1.60	1.33	0.80	0.40	0.35
Insulated	0.80	0.67	0.67	0.15	0.13	0.13

*These factors are combination factors. Do not confuse with U-values (Btu/sq. ft./°F).
†Factors based on flat roof construction.

Table 12-4 Transmission gain factors*
(Courtesy Air-Conditioning and Refrigeration Institute)

Heat Gain Through Glass (Table 12-3)

Multiplying factors for estimating the sensible heat gain through glass are listed in Table 12-3. The factors are based on data in the Guide published by the American Society of Heating, Refrigerating, and Air-Conditioning Engineers. These factors take into account the effect of haze in the atmosphere, window setback, window sash area, shade effect, and storage in the building.

Transmission (Table 12-4)

For most buildings, the transmission load through walls, roofs, ceilings, floors, and partitions is only a small part of the actual load. Therefore, the transmission load may not require as exact a measurement as the solar load through the glass and roof surfaces.

Heat Gain – Occupants (Table 12-5)

Heat gain from occupants is based on the average number of people expected to be present in the building. The heat load produced by each person depends upon the

Degree of Activity	Typical Application	Total Heat Adult Male Btu/hr.	Adjusted Total Heat Btu/hr.	Adjusted Sensible Heat Btu/hr.	Adjusted Latent Heat Btu/hr.
Seated at rest	Theater–Matinee	390	330	180	150
	Theater–Evening	390	350	195	155
Seated, very light work	Offices, Hotels, Apartments	450	400	195	205
Moderately active office work	Office, Hotels, Apartments	475	450	200	250
Standing, light work; walking slowly	Dept. Store, Retail Store, 10c Store	550	450	200	250
Walking; seated	Drugstore	500	500	200	300
Standing; walking slowly	Bank				
Sedentary (seated)	Restaurant*	490	550	220	330
Light bench work	Factory	800	750	220	530
Moderate dancing	Dance Hall	900	850	245	605
Walking, 3 mph	Factory	1000	1000	300	700
Moderately-heavy work					
Bowling†	Bowling Alley	1500	1450	465	985
Heavy work	Factory				

*The adjusted total heat value for *people dining in restaurant* includes 60 Btu per hour for food per individual (30 Btu/hr. sensible and 30 Btu/hr. latent).

†For *bowling*, one person per alley actually bowling and all others as sitting (400 Btu per hour) or standing (550 Btu per hour).

Note: The above values are based on 80°F room dry-bulb temperature. For 78°F room dry-bulb temperature, the total heat gain remains the same, but the sensible heat values should be increased by approximately 10% and the latent heat values decreased accordingly.

Table 12-5 Heat gain from occupants
(Courtesy Air-Conditioning and Refrigeration Institute)

activity of the person. Table 12-5 lists a wide range of activities from being at rest to heavy work. The total heat values and the sensible heat and latent heat values are shown for each activity.

In Table 12-5, the column "Total Adjusted Heat" lists the heat values per person based on the average number of men, women, and children likely to be present simultaneously in the typical applications shown. To establish these standards, the adult male is considered to produce 100% or 390 Btu/hr. The adult female produces 85% as much heat as the adult male, and the child produces 75% as much heat as the adult male. The combination of these values results in an average of 330 Btu/hr. for each man, woman, or child present in a given situation. If the average number of people present per hour in an apartment is three (one man, one woman, and one child), the total heat load per hour from people is 3 x 330 = 990 Btu.

Heat Gain – Appliances (Table 12-6)

Appliances in common use today are listed in Table 12-6. The Btu values shown are based on average usage in commercial businesses.

Recommended Btu/Hr. for Average Use						
APPLIANCE	ELECTRICAL*			GAS*		
	Sensible	Latent	Total	Sensible	Latent	Total
Coffee Brewer, ½ gallon	900	220	1120	1350	350	1700
Coffee Brewer Unit, 4½ gallon	4800	1200	6000	7200	1800	9000
Coffee Urn, 3 gallon	2200	1500	3700	2500	2500	5000
Coffee Urn, 5 gallon	3400	2300	5700	3900	3900	7800
Coffee Warmer, ½ gallon	230	60	290	400	100	500
Food and Plate Warmer, per sq. ft. of top surface	350	350	700	–	–	–
Food Warmer (only), per sq. ft. of top surface	–	–	–	850	450	1300
Fry Kettle, per sq. ft. of fry area	3500	5000	8500	6000	4000	10000
Griddle, per sq. ft. of fry area	3000	1600	4600	–	–	–
Grille, Meat, per sq. ft. of fry area	4700	2500	7200	10000	2500	12500
Grille, Sandwich, per sq. ft. of fry area	2700	700	3400	–	–	–
Hair Dryer, Blower	2300	400	2700	–	–	–
Hair Dryer, Helmet	1870	330	2200	–	–	–
Hair Dryer System, 5 Helmets	–	–	–	15000	4000	19000
Hair Dryer System, 10 Helmets	–	–	–	21000	6000	27000
Sterilizer for Physicians' Instruments	650	1200	1850	–	–	–
Stoves, Short-Order, per sq. ft. of top:						
Closed-Top	–	–	–	3300	3300	6600
Fry-Top	–	–	–	3600	3600	7200
Open-Top	–	–	–	4200	4200	8400
Toaster, Belt-Driven, 2 slices wide	5100	1300	6400	7700	3300	11000
Toaster, Belt-Driven, 4 slices wide	6100	2600	8700	12000	5000	17000
Toaster, Pop-Up, per slice	37	8	45	–	–	–
Waffle Iron, 20 waffles per hour	1100	750	1850	–	–	–

*When hooded and with adequate exhaust, use 50% of above values.

Table 12-6 Heat gain from appliances
(Courtesy Air-Conditioning and Refrigeration Institute)

Ventilation (Table 12-7)

Ventilation for summer or winter can be established by determining the cubic feet of air per minute (cfm) that is required for each person in the building. Ventilation can also be established by determining the cubic feet of air per minute (cfm) required for each square foot of floor space in the building.

Table 12-7 shows the amount of outside air per person and per square foot of floor space in various types of structures. The values shown in the column labeled "Recommended" include a slight safety factor to insure that the amount of air supplied meets the established ventilation codes. The minimum values given (both columns) are used according to the judgment of the person who is determining the load. This judgment is sometimes based on the desires or specifications of the owner of the building.

Ventilation is usually determined by the cfm per person method because the cfm per sq. ft. of floor method sometimes results in a much larger cfm quantity, depending on the number of people using the space. However, when the number of people is unknown, the cfm per sq. ft. method is often used.

APPLICATION	SMOKING	OUTSIDE AIR		
		Cfm per Person		Cfm per Sq. Ft. of Floor
		Recommended	Minimum*	Minimum*
Apartment	Some	20	15	0.33
Banking Space	Occasional	10	7.5	–
Barber Shops	Considerable	15	10	–
Beauty Parlors	Occasional	10	7.5	–
Cocktail Bars	Heavy	30	25	–
Stores	None	7.5	5	0.05
Drugstores	Considerable	10	7.5	–
Factories	None	10	7.5	0.10
Funeral Parlors	None	10	7.5	–
Hospitals – Private Rooms	None	30	25	0.33
Hospitals – Wards	None	20	15	–
Hotel Rooms	Heavy	30	25	0.33
Meeting Rooms	Very Heavy	50	30	1.25
Offices – Private	None	25	15	0.25
Offices – General	Some	15	10	–
Restaurants	Considerable	15	12	–
Cafeterias	Considerable	12	10	–
Theatres	None	7.5	5	–

The ventilation air should be calculated as follows:

_____ cfm per person x _____ persons = _____ Cfm

or

_____ cfm per sq. ft. x _____ sq. ft. = _____ Cfm

*When minimum values are used, determine the air quantity by cfm per person and by cfm per sq. ft., and then use the larger of the two total air quantities in the load estimate.

Table 12-7 Ventilation
(Courtesy Air-Conditioning and Refrigeration Institute)

PROBLEM 1 CFM PER PERSON METHOD

Given

Restaurant seating a maximum of 100 people

Find

Ventilation air (V) required

Solution

V = cfm per person x number of people

V = 15 cfm x 100 = 1,500 cfm

Note: 15 cfm is shown in Table 12-7 as the recommended quantity for restaurants

PROBLEM 2 CFM PER SQUARE FOOT METHOD

Given

Apartment building with floor area = 3,000 sq. ft.

Find

Ventilation air (V) required

Solution

V = cfm per sq. ft. x number of sq. ft. of floor area

V = 0.33 (Table 12-7) x 3,000 = 990 cfm

Regardless of the method used for any application, the ventilation air quantity must equal or exceed the values established by the ventilation codes and ordinances that apply in the area in which the air-conditioning system is installed. In addition, the quantity of outside air used for ventilation should be sufficient to change half the air per hour in buildings with normal ceiling heights. If the amount of the infiltration air is larger than the ventilation air value, the ventilation air should be increased until it is at least equal to the amount of the infiltration air.

Infiltration (Table 12-8)

Infiltration air is the air that enters a space through window cracks and opened doors. The amount of air entering in this manner can be determined by the crack method or the air change method. The crack method is preferred when greater accuracy is required. In most instances, the air change approach is satisfactory for smaller commercial and residential structures.

Table 12-8 lists the air change factors for summer and winter for a variety of room types and exposures. The equation given in the table will determine the amount of infiltration through windows and walls. Infiltration through doors, as listed at the bottom of the table, must be added to obtain the total infiltration value.

KIND OF ROOM OR BUILDING	AIR CHANGES PER HOUR			
	Summer		Winter	
	Ordinary	Weatherstripping or Storm Sash	Ordinary	Weatherstripping or Storm Sash
No Windows or Outside Doors	0.30	0.15	0.50	0.25
Entrance Halls	1.20 to 1.80	0.60 to 0.90	2.00 to 3.00	1.00 to 1.50
Reception Halls	1.20	0.60	2.00	1.00
Bath Rooms	1.20	0.60	2.00	1.00
Infiltration through Windows:				
Rooms, 1 wall exposed	0.60	0.30	1.00	0.50
Rooms, 2 walls exposed	0.90	0.45	1.50	0.75
Rooms, 3 walls exposed	1.20	0.60	2.00	1.00
Rooms, 4 walls exposed	1.20	0.60	2.00	1.00

The air quantity is computed as follows:

$$\frac{(H) \times (L) \times (W) \times (AC)}{60} = \_\_\_\_ \text{ cfm}$$

Where: H = room height, ft.
L = room length, ft.
W = room width, ft.
AC = air changes per hour

Note: The total room infiltration for an entire building is divided by 1/2 since infiltration takes place on the windward side of a building.

Infiltration Through Doors*

For each person passing through a door leading to the outside or to an unconditioned space, add the following door infiltration to the summer infiltration through windows:

Usage	*Cubic Feet per Minute**
Infrequent	60
Average	50
Heavy	40
36-inch Swinging Door	100

*These figures are based on the assumption that there is no wind pressure and that swinging doors are in use in one wall only. Any swinging doors in other walls should be kept closed to insure air-conditioning in accordance with these recommended standards.

Table 12-8 Infiltration
(Courtesy Air-Conditioning and Refrigeration Institute)

PROBLEM 3 SUMMER INFILTRATION – AIR CHANGE METHOD

Given

Type of building	Apartment (4 units)
Number of rooms	12
Room size (height, length, width)	10 x 15 x 12
Exposure	8 rooms, two walls exposed to outdoors 4 rooms, one wall exposed to outdoors
Entrance hall size	8 x 8 x 7
Door	Average usage
Weatherstripping on all windows	

Find

Infiltration air quantity by the air change method.

Solution

1. Determine the infiltration through the rooms exposed on two walls.

$$\frac{H \times L \times W \times AC}{60} = \frac{10 \times 15 \times 12 \times 0.45 \text{ (Table 12-8)}}{60} = \frac{810}{60}$$

$$= 13.5 \text{ cfm per room}$$

$$13.5 \text{ cfm} \times 8 \text{ rooms} = 108.0 \text{ cfm}$$

2. Determine the infiltration through the rooms exposed on one wall.

$$\frac{H \times L \times W \times AC}{60} = \frac{10 \times 15 \times 12 \times 0.30 \text{ (Table 12-8)}}{60} = \frac{540}{60}$$

$$= 9 \text{ cfm per room}$$

$$9 \text{ cfm} \times 4 \text{ rooms} = 36 \text{ cfm}$$

3. Determine the infiltration through the entrance hall.

$$\frac{H \times L \times W \times AC}{60} = \frac{8 \times 8 \times 7 \times 0.60}{60} = \frac{268.8}{60} = 4.4 \text{ cfm}$$

4. Total infiltration through the walls = 108 + 36 + 4.4 = 148.4 cfm.

   The total infiltration is actually 1/2 x 148.4 cfm since the air infiltrates on the windward side but leaves on the leeward side. Therefore, the air infiltrates through half of the building at a given time.

   Total room infiltration = 148.4 ÷ 1/2 = 74 cfm

5. Determine the infiltration through the door.

   Table 12-8 shows that infiltration through a door with average use = 50 cfm

6. Total building infiltration = total room + door = 74 + 50

   = 124 cfm

As stated earlier, the ventilation air must at least equal the infiltration air to create air pressure in the building and thus prevent infiltration air from entering. The effects of the equalizing pressure allow the ventilation air to offset or balance the effects of the infiltration air on the load.

Exhaust Air

If an exhaust fan is used to remove air from the building, the ventilation air quantity must be large enough to replace the air removed by the exhaust. At the same time, the air pressure must be maintained to offset infiltration as shown in Problem 4.

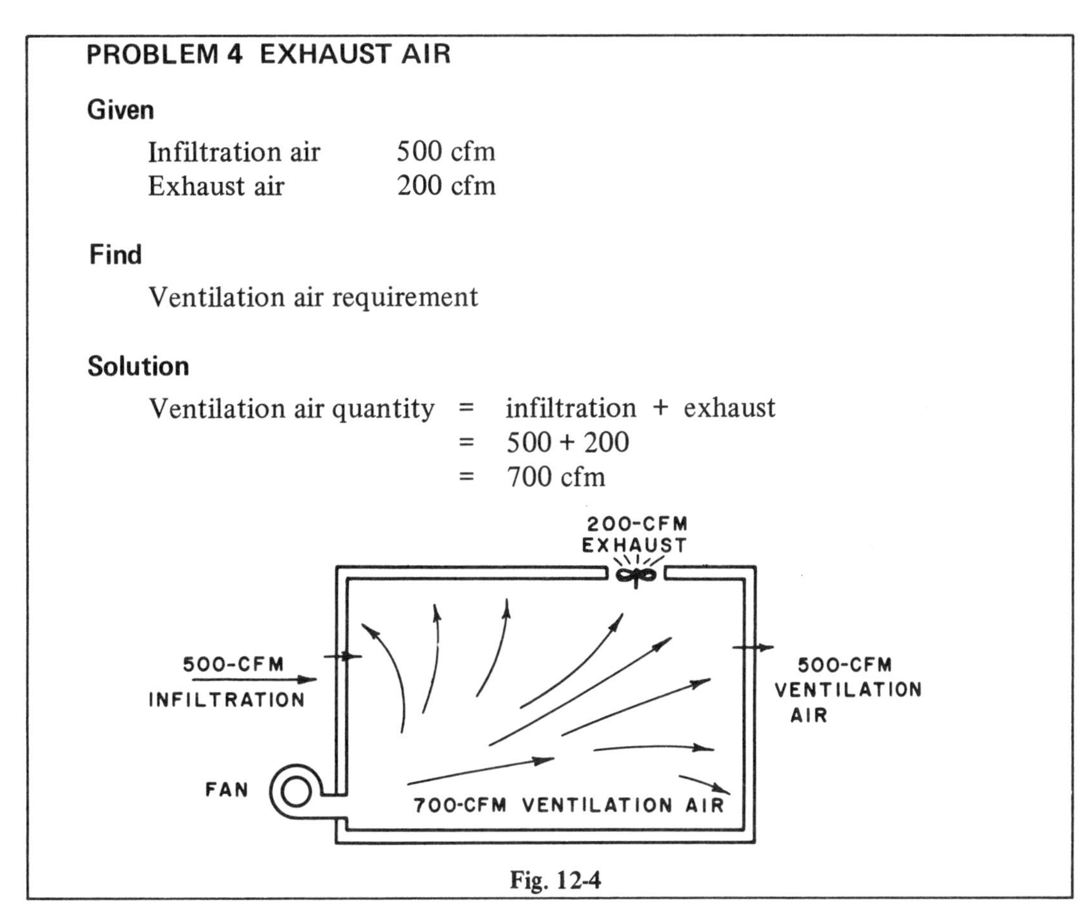

PROBLEM 4 EXHAUST AIR

Given

Infiltration air	500 cfm
Exhaust air	200 cfm

Find

Ventilation air requirement

Solution

Ventilation air quantity = infiltration + exhaust
= 500 + 200
= 700 cfm

Fig. 12-4

Additional Standards

In addition to the desired conditions described previously, there are two additional comfort criteria.

1. Comfort air conditioning is designed to maintain a space temperature that does not vary by more than ±3°F DB.
2. Air velocities should be less than 50 linear ft. per minute between the floor and the 5-ft. level in the room to avoid undesirable draft conditions.

ESTIMATING FORMS

Estimating forms of many different types are used to determine the cooling and heating loads. Although individual manufacturers of air-conditioning equipment develop forms particularly adapted to their needs, all forms follow the same basic pattern. The estimating forms developed by the Air-Conditioning and Refrigeration Institute will be used to help explain the process of estimating cooling and heating loads. There are two types of forms: one for commercial buildings and one for residential buildings. Examples of these forms are included in the back cover envelope. These forms should be removed at this point and used with the following instructions.

COOLING LOAD ESTIMATE

Remove the Cooling Load Estimate Form from the back cover envelope. This basic form was developed by the Air-Conditioning and Refrigeration Institute. Since all of the forms used by equipment manufacturers are basically similar, the ARI form is a convenient aid in learning the procedure for estimating the air conditioning for a commercial building.

Item 1: Design Conditions

a. Enter the outside design temperature conditions from Table 12-1.
b. Determine the grains of moisture from the psychrometric chart and enter this value in the space allotted.
c. Enter the inside design temperature conditions as specified by the customer.
d. Determine the grains of moisture for the inside temperature from the psychrometric chart. Enter this value where indicated.
e. Find the difference between the outside and inside temperatures and record this value.

Item 2: Sensible Heat Gain Through Glass

a. Determine the square feet of window sash area and enter this value in the space indicated.
b. Multiply the square feet of window sash area by the appropriate factor listed in Table 12-3 to obtain the Btu sensible heat load through glass.
c. Enter the Btu totals in the sensible heat column.
d. Enter the same Btu total in the subtotal column on the dotted line provided. (The subtotal is the same since latent heat is not added through glass.)

Item 3: Transmission Gain

a. Determine the total square feet of wall area.
b. Subtract the total window area from the wall area and enter the resulting square feet of net wall area in the space provided.
c. Determine the total area in square feet of the partitions, roofs or ceilings, and floors. Enter these values in the spaces provided.
d. Select appropriate factors from Table 12-4 for partitions, roofs or ceilings, and floors. Enter these values in the spaces provided.
e. Record the design dry-bulb temperature difference after each item listed. Multiply the square feet of area x the factor x the temperature difference.
f. Enter the answer, in each case, in the sensible heat column in the space provided.
g. Add all of the values and enter the answer in the subtotal column. This Btu value represents the transmission gain through the walls, partitions, roofs or ceilings, and floors. (The final Btu subtotal does not include latent heat.)

Item 4: Internal Heat Gains from People and Lights

a. Determine the number of people in the building to be air conditioned and enter this value in the space provided.
b. Determine the degree of activity of the people as listed in Table 12-5.
c. Select the appropriate sensible heat and latent heat Btu values from Table 12-5, according to the type of person and activity. Enter this value in the space provided.
d. Multiply (1) the number of people x sensible heat in Btu, and (2) the number of people x latent heat in Btu. Enter the resulting answers in the appropriate sensible and latent columns, in the spaces provided.

e. Determine the total wattage of lights. Enter this value in the space provided.
f. Multiply the total wattage by 3.4 (as established in a previous unit). Enter the resulting value in the sensible column in the space provided.
g. Add the sensible heat in Btu (from people and lights) and the latent heat in Btu (from people). Enter the sum in the subtotal column in the appropriate space.
The final Btu subtotal (people and lights) includes both sensible heat and latent heat.

Item 5: Internal Heat Gains from Motors and Appliances

a. Determine the total horsepower in use at the design temperature and humidity conditions. Insert the h.p. value in the space indicated.
b. Multiply the horsepower by 3,393. Enter the resulting value in the sensible column.
c. Determine the Btu values for the sensible heat and latent heat for the appropriate appliances (Table 12-6). Enter these values in the sensible and latent columns on the form.
d. Add the total sensible heat to the total latent heat (in Btu) for motors and appliances. Enter the sum in the subtotal column.

Item 6: Ventilation and Infiltration

(Tables 12-7 and 12-8: use the larger quantity only)

a. Determine the ventilation air quantity (cfm) from Table 12-7.
b. Determine the infiltration air quantity (cfm) from Table 12-8.
c. Enter the larger of the two air quantities in two places under the heading "Cubic Feet Per Minute."
d. Multiply the first cfm entry by the design dry-bulb temperature difference (Item 1) and by 1.08. Enter the resulting value (in Btu) in the sensible column.
e. Multiply the second cfm entry by the design specific humidity (Item 1) and by 0.67. Enter the resulting value (in Btu) in the latent column.
f. Add the sensible and latent values (in Btu) and enter the sum in the subtotal column.

Item 7: Total Sensible and Latent Loads (Btu)

a. Add all of the sensible loads listed and enter the total.
b. Add all of the latent loads listed and enter the total.

Item 8: Total Air Cycle Load

a. Add the subtotal loads. Enter the sum in the space at the bottom of the subtotal column. The sum of the latent load + the sensible load should equal the total air cycle load.
b. If the building has one zone, the Btu value at the bottom of the subtotal column represents the air-conditioning load. Refrigeration equipment is selected according to this load. The equipment must be able to handle this load and maintain the required inside temperature and humidity conditions during periods when the following conditions occur simultaneously: (1) outside design conditions, and (2) maximum heat gain inside the building.

Type of Building	Operation	Glass Gain	Transmission, Wall and Roof	People & Lights	Ventilation
Office:					
Very Light Construction	10 Hr.	1.00	0.90	0.75	0.98
	24 Hr.	0.80	0.90	0.60	0.95
Average Construction	10 Hr.	0.80	0.90	0.75	0.95
	24 Hr.	0.80	0.90	0.50	0.91
Very Heavy Construction	10 Hr.	0.83	0.90	0.75	0.92
	24 Hr.	0.83	0.90	0.45	0.88
Apartments and Hotels	24 Hr.	0.75	0.90	0.45	0.85

Table 12-9 Refrigeration load factors for multizone buildings
(Courtesy Air-Conditioning and Refrigeration Institute)

Item 9: Total Multizone Refrigeration Load

a. To determine the size of the refrigeration load if the building has more than one zone, use the following procedure:
   (1) Calculate the load for the entire building in the same manner as that given for a single zone. (Consider the whole building as one zone.)
   (2) Multiply the subtotal heat loads by the appropriate factor from Table 12-9. Enter this value in the refrigeration load column.
   (3) Add the refrigeration load in Btu and enter the total on the estimate form at the bottom of the refrigeration load column across from Item 9.

b. The refrigeration equipment is selected according to this load. The air system is then designed to supply individually controlled air to each zone.
   Note: If the building has more than one zone and each zone must have a separate refrigeration source (self-contained or packaged equipment), then each zone is estimated separately. Separate equipment is selected according to the load required for each zone.

HEATING LOAD ESTIMATE

The heat gain from lights, people, motors, and solar radiation is not included in heating load estimates. These sources are variable and do not necessarily occur during the maximum load periods. A systematic method of determining the heating load is presented in the Heating Load Estimate Form, figure 12-5. The item numbers in the following procedure correspond to the item numbers on this form.

Item 1: Design Conditions

a. Determine the design dry-bulb temperature, wet-bulb temperature, and specific humidity (grains of moisture) from Table 12-1. Enter these values in the spaces provided on the form.

b. Enter the specified or required inside conditions in the spaces provided.

c. Determine the difference between the outside and the inside conditions and enter this value in the appropriate spaces.

Item 2: Transmission Loss

a. Determine the area (in sq. ft.) of the windows, walls, roof, and floor. Enter the results in the separate spaces provided. Remember that the wall area does not include the window area.

1. DESIGN CONDITIONS	Dry-Bulb Temp., °F	Wet-Bulb Temp., °F	Specific Humidity, Gr./lb.
Outside (Table 1)	______	______	______
Inside	______	______	______
Difference	______	______	______

ITEMS						HEATING LOAD Btu/Hr.
2. TRANSMISSION LOSS (Table 3)	Sq. Ft.		Factor		Dry-Bulb Temperature Difference	
Windows	______	X	______	X	______	= ______
	______	X	______	X	______	= ______
Walls	______	X	______	X	______	= ______
	______	X	______	X	______	= ______
	______	X	______	X	______	= ______
	______	X	______	X	______	= ______
Roof	______	X	______	X	______	= ______
Floor	______	X	______	X	______	= ______
Other	______	X	______	X	______	= ______
	______	X	______	X	______	= ______
3. VENTILATION OR INFILTRATION (Tables 6 and 7; use the larger quantity only.)						
	Cubic Feet Per Minute		Dry-Bulb Temperature Difference			
Sensible Load	______	X	______	X	1.08	= ______
	Cubic Feet Per Minute		Specific Humidity Difference			
Humidification Load	______	X	______	X	0.67	= ______
4. TOTAL HEATING LOAD, Btu/Hr.						

Fig. 12-5 A heating load estimate form
(Courtesy Air-Conditioning and Refrigeration Institute)

b. Determine the appropriate factor for the windows, walls, roof, and floor from Table 12-4. Enter the value in the space provided.
c. Record the dry-bulb temperature difference (Item 1) after each item.
d. Multiply the following: square feet of area x factor (from Table 12-4) x dry-bulb temperature difference.
e. Record the product (in Btu) in the heating load column for each item.

Item 3: Ventilation or Infiltration

a. Determine the ventilation air quantity (in cfm) from Table 12-7.
b. Determine the infiltration air quantity (in cfm) from Table 12-8.
c. Enter the larger of the two quantities in the two spaces provided under the heading "Cubic Feet Per Minute."
d. Multiply the first entry by the dry-bulb temperature difference and the factor 1.08. The resulting answer is the ventilation sensible heat load. This value is to be entered in the heating load column.
e. Multiply the second entry by the grains of moisture difference and the factor 0.67. The resulting answer is the humidification load and should be entered in the heating load column in the space provided.

Item 4: Total Heating Load (Btu/Hr.)

a. Add the transmission loads to the ventilation loads. Enter the total value in Btu at the bottom of the heating load column. The result of this addition is the total heating load. The selection of the heating equipment is based on this load. The equipment must be able to handle the load and maintain the required temperature and humidity conditions during those periods when the following conditions occur simultaneously:
   1. outside design conditions, and
   2. maximum heating load inside the building.

RESIDENTIAL COOLING LOAD ESTIMATE FORM

Cooling Load

The Residential Cooling Load Form, such as the sample provided in the back cover envelope, helps the estimator to establish a load in a systematic manner. The form is suitable for all normal residential applications. Although the form is similar to the commercial form, it is more compact and incorporates features that permit a faster but adequate estimate. The item numbers in the following procedure correspond to the item numbers on the form.

The first step is to determine the outside design conditions from Table 12-1 and enter these values at the top of the form. The specified or required inside design conditions are also entered on the form. The difference between the inside and outside dry-bulb temperatures is then computed and recorded in the space provided at the top of the form.

Item 1A: Windows, Heat Gain from Sun

a. Determine the window area (in sq. ft.) for each exposure. Enter these values in the space provided on the form.
b. Select the appropriate factor listed on the form for each exposure according to the type of shading or shading device used.
c. Multiply the window area for each exposure by the appropriate factor. Enter this value in the column under the heading Load for Each Exposure.
   (Note: For glass block areas, reduce the factor by 50%; for double-pane glass areas, reduce the factor by 15% before multiplying by the appropriate factor. For glass areas partially shaded by the roof overhang, use the following example as a guide in determining the area directly exposed to the sun before multiplying by the factor.)
d. Select the largest of the exposure loads and enter this value in the appropriate column (Btu/Hr.) in the space provided.

PROBLEM 5 CALCULATING SHADED AND UNSHADED AREAS OF WINDOWS UNDER OVERHANGING ROOFS

Given

South wall, window 5 ft. x 4 ft., Latitude 35°

Find

Shaded window areas

Solution

1. From the table, the factor is 2.24.
2. The overhang is 2 ft. as shown in figure 12-7.

   2 x 2.24 = 4.48 = the distance below the bottom edge of overhang covered by shadow.
3. Then: 4.48 ft. - 1 ft. = 3.48 ft. of window (from top) shaded.
4. 3.48 ft. x 4 ft. (width of window) = 13.9 sq. ft. of window shaded.
5. 4 ft. x 5 ft. = 20 sq. ft. (total area of window).
6. 20 - 13.9 = 6.1 sq. ft. of window unshaded.

OVERHANG SHADOW FACTOR				
Direction Glass Faces	Latitude			
	30°	35°	40°	45°
E or W	0.60	0.60	0.59	0.57
SE or SW	1.55	1.43	1.24	1.15
S	2.90	2.24	1.80	1.52
NE or NW	(Shadow insignificant)			

Fig. 12-6

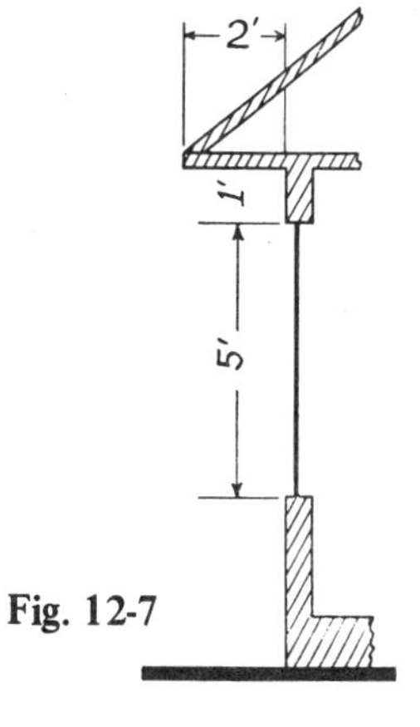

Fig. 12-7

Item 1B: Windows, Heat Gain Caused by Temperature Difference

a. Multiply the total single-pane glass window area by the factor listed under the appropriate Design Dry-Bulb Temperature Difference heading on the form.
b. Enter this value in the column under the heading Btu/Hr. in the space provided.
c. Multiply the total double-pane glass or glass block area by the factor listed under the appropriate Design Dry-Bulb Temperature Difference on the form.
d. Enter this value in the column under the heading Btu/Hr. in the space provided.

Item 2: Walls

a. Determine the wall area in square feet according to the type of construction listed on the form. Enter this value in the area column. Do not include the window area.
b. Multiply the wall area entries by the factor listed under the appropriate Design Dry-Bulb Temperature Difference and determined according to the type of wall construction.
c. Enter these values in the Btu/Hr. column in the space provided.

Item 3: Partitions

a. Determine the area of the wall partitions (in sq. ft.) between the conditioned and the unconditioned spaces (garages and unconditioned rooms).
b. Multiply the partition area by the appropriate temperature difference factor and enter this value in the Btu/Hr. column.

Item 4: Roofs

a. Determine the area in square feet of the ceiling under the roof.
b. Enter these values in the area column.
c. Multiply the ceiling areas by the appropriate temperature difference factors and enter the resulting values in the Btu/Hr. column.

Item 5: Ceiling Under Unconditioned Rooms

a. Determine the ceiling area (in sq. ft.) under the unconditioned rooms and enter this value in the area column.
b. Multiply the ceiling area by the appropriate temperature difference factor. Enter this value in the Btu/Hr. column.

Item 6: Floors

(Over an unconditioned room or an open crawl space. Omit Item 6 if the floor is installed over the basement, the enclosed crawl space, or directly on the ground.)

a. Determine the floor area (in sq. ft.) and enter this value in the area column.
b. Multiply the floor area by the appropriate temperature difference factor and enter the result in the Btu/Hr. column.

Item 7: Outside Air

(Based on the floor area regardless of the floor construction)

a. Determine the floor area (in sq. ft.) and enter this value in the area column.
b. Multiply the floor area by the appropriate temperature difference factor. Enter the result in the Btu/Hr. column.

Item 8: People

(Use a minimum value of five)

a. Determine the average number of people who are to occupy the residence. Multiply this value by 200. Enter the result in the Btu/Hr. column.

Item 9: Btu Subtotal (Sensible Heat)

a. Add all of the values in the Btu/Hr. column. Enter the sum in the space provided.
b. The sensible heat load per room is determined by proportioning the total sensible heat according to the area (in sq. ft.) of the floor for each room.

Item 10: Latent Heat Allowance

a. Multiply the Btu/Hr. subtotal by 30%. Enter the result in the Btu/Hr. column in the appropriate space.

Item 11: Total Cooling Load (Btu/Hr.)

a. Add the Btu/hr. subtotal and the Btu/hr. latent heat allowance. Enter the sum at the bottom of the Btu/hr. column.
b. Equipment is selected according to the total Btu value listed in Item 11. The equipment must be able to handle the load and maintain the inside design conditions when the outside design conditions prevail.

RESIDENTIAL HEATING LOAD ESTIMATE

In normal residential applications, the load from solar radiation, people, and appliances is not included in the estimate since it is variable and cannot be predicted for a particular time.

The item numbers in the following procedure correspond to the item numbers in Table 12-10. Use Table 12-1 to determine the outside design dry-bulb temperature.

ITEM	FACTORS						
	Design Outside Dry-Bulb Temperature						
	-30F	-20F	-10F	0F	10F	20F	30F
1. WINDOWS							
Single-glass	113	102	90	79	68	57	45
Double-glass or glass block	50	45	40	35	30	25	20
2. WALLS							
No insulation (brick veneer, frame, stucco, masonry)	25	23	20	18	15	13	10
1 in. insulation or 25/32 in. insulating sheathing	20	18	16	14	12	10	8
2 in. or more insulation	10	9	8	7	6	5	4
3. ROOFS							
Pitched or flat with vented air space, and:							
No insulation	31	28	25	22	19	16	12
2 in. insulation	10	9	8	7	6	5	4
4 in. insulation	7	6	6	5	4	4	3
Flat with no air space, and:							
No insulation	50	45	40	35	30	25	20
25/32 in. insulation	25	23	20	18	15	13	10
1 1/2 in. insulation	15	14	12	11	9	8	6
3 in. insulation	10	9	8	7	6	5	4
4. FLOORS							
Over:							
Basement	0	0	0	0	0	0	0
Enclosed crawl space	23	20	18	16	14	11	9
Open crawl space	34	31	27	24	20	17	14
On ground							
No edge insulation	81	73	65	57	49	41	32
1 in. edge insulation	68	61	54	48	41	34	27
2 in. edge insulation	55	50	44	39	33	28	22
5. OUTSIDE AIR	108	97	87	76	65	54	43

Table 12-10 Residential heating load estimate factors
(Courtesy Air-Conditioning and Refrigeration Institute)

Item 1: Windows

a. Determine the total window area (in sq. ft.) in the house.
b. Multiply the window area by the appropriate outside design dry-bulb temperature factor (from Table 12-10, page 137). Record the result on a piece of paper. Obtain the appropriate outside design dry-bulb temperature from Table 12-1.

Item 2: Walls

Note: Do not include the window area.

a. Determine the wall area in square feet (omit windows). Multiply this value by the appropriate outside design dry-bulb temperature factor (from Table 12-10).
b. Record the value determined in step a. The appropriate factor is determined according to the outside design temperature (Table 12-1) and the type of wall construction.

Item 3: Roofs

a. Determine the roof area (in sq. ft.). Multiply the roof area by the appropriate outside design temperature factor and record the product.
b. The appropriate factor is determined by the outside design temperature and the type of roof construction.

Item 4: Floors

a. For floors installed over basement or crawl spaces:
   (1) Determine the floor area (in. sq. ft.). Multiply this value by the appropriate outside design temperature factor (from Table 12-10).
   (2) Record the result.
b. For floors installed on a slab on ground:
   (1) Determine the floor perimeter (in linear feet). Multiply this value by the appropriate outside design temperature factor.
   (2) Record the result.

Item 5: Outside Air

Infiltration air usually satisfies the ventilation requirements in residences. The infiltration air quantity must be equal to local code requirements. In addition, the filtration air should also provide at least 1 to 1 1/2 air changes per hour.

a. Determine infiltration air (cfm) using Table 12-8. Multiply this value by the outside design temperature factor.
b. Record the result.

Total Heating Load (Btu/Hr.)

a. The total heating load is the sum of all of the values recorded for the heating load estimate. The heating equipment is selected to handle the total heating load. Thus, the equipment must be able to maintain the required inside design conditions during those periods when the maximum inside load and the outside design conditions occur simultaneously.

The load estimate forms discussed in this unit cover the basic steps in determining the air-conditioning loads. Manufacturers of air-conditioning equipment usually provide survey and estimate forms which follow procedures similar to the ones presented in this unit. In later units, some of these forms will be used to complete an actual estimate.

SUMMARY

- An air-conditioning survey is a review of a particular structure to determine the best possible system for that structure and the best method of installing the system.
- A survey usually includes:
  1. building orientation
  2. building size
  3. building shape
  4. materials of construction
  5. glass surface area
  6. infiltration quantity
  7. average number of people occupying the structure
  8. number and size of motors, lights, and appliances
  9. ventilation quantity
  10. equipment location
  11. water, gas, and electrical service sources
  12. local and state code requirements
- The load estimate is based on design conditions inside the building to be air conditioned and the conditions of the outside atmosphere surrounding the building.
- The load estimate determines the size of the heating and cooling load; therefore, this estimate is used to establish the equipment size.
- The outside design conditions are the extremes of temperature occurring in a specific locality.
- Winter design heating load includes heat lost by conduction through building surfaces, and heat and moisture required to establish and maintain comfort conditions.
- Summer design cooling load includes heat gained by conduction through building surfaces and heat gained from sun radiation.
- Use 80% relative humidity for design outside humidity condition in winter.
- Inside design temperature for winter heating is 72°F to 75°F DB.
- Inside design temperature for summer cooling is 78°F DB and 50% RH. If the largest load is from sources other than people, the design temperature may be increased to 80°F.
- Transmission load through walls, roofs, ceilings, and floors is usually only a small part of the total load.

- Heat gain from people is based on the average number of people present in the building at a given time.
- Heat gain from appliances is based on the power rating and the rate of usage.
- Ventilation air requirements can be established on a cubic feet of air per person basis, or per square foot of floor area basis.
- Ventilation air quantity must equal or exceed the local ventilation codes and ordinances, and provide at least one-half air change per hour in the building.
- Infiltration air quantity can be determined by the crack method or by the air change method.
- Ventilation air must at least equal infiltration air, to offset (by equalizing pressure) the effects of infiltration on the load.
- If an exhaust system is used, ventilation air quantity must be large enough to replace the exhaust air and simultaneously offset infiltration.
- Inside space temperature should not vary more than ±3°F.
- Air velocities should be less than 50 linear feet per minute up to the 5-ft. wall level.
- A commercial building cooling load estimate should include:
    1. design outside conditions
    2. sensible heat gain through glass
    3. transmission heat gain
    4. internal heat gain from people and lights
    5. internal heat gain from motors and appliances
    6. ventilation and infiltration heat gain
    7. total sensible and latent loads (Btu/hr.)
    8. total air cycle load (Btu/hr.)
    9. total multizone refrigeration load
- A commercial building heating load estimate should include:
    1. outside and inside design conditions and temperature difference
    2. transmission loss
    3. ventilation or filtration load
    4. total heating load (Btu/hr.)
- A residential cooling load estimate should include:
    1. heat gain through windows
    2. heat gain through walls
    3. heat gain through partitions
    4. roof heat gain
    5. ceiling heat gain
    6. heat gain through floors
    7. heat gain from outside air
    8. heat gain from people
    9. total sensible heat gain
    10. total latent heat gain (30% of sensible)
    11. total cooling load (Btu/hr.)

- A residential heating load estimate should include the following losses in Btu/hr.
  1. window heat loss
  2. wall heat loss
  3. loss through roof
  4. loss through floors
  5. outside air load
  6. total heating load (Btu/hr.)

REVIEW

Solve the following problems, using the ARI tables provided in this unit.

1. What outdoor design conditions are used for:
   a. Sacramento, California
   b. Daytona Beach, Florida
   c. Mexico City, Mexico
2. Three windows, each 3 ft. x 6 ft. are located on an outside wall which is 30 ft. long x 10 ft. high. The wall faces south and the building is located in 32° latitude. What is the heat gain through the glass in a 24-hr. period?
3. What is the transmission gain through the wall of Problem 2 if it is shaded?
4. A theater seats 800 people. What is the total heat gain if the theater is filled for an evening performance?
5. A bowling center has 200 persons bowling while another 100 people are seated watching. Find the total heat gains from each group.
6. A restaurant has the following gas-burning equipment: two coffee brewers, 1/2 gal. each; and one steam table with a 6 ft. x 2 ft. 6 in. surface. What are the sensible and latent heat gains from this equipment?
7. The restaurant in Problem 6 also has a 4,500-watt, belt-driven, two-slice toaster, one 18-in. x 24-in. food and plate warmer, five 300-watt fluorescent fixtures, and one 1/2-h.p. motor. What is the heat gain from this equipment?
8. A club meeting room in which there is a heavy amount of smoking seats 125 people. How much outdoor air should be introduced for ventilation?
9. Name the items usually included in the following:
   a. An air-conditioning survey.
   b. A commercial building cooling load estimate.
   c. A commercial building heating load estimate.
   d. A residential cooling load estimate.
   e. A residential heating load estimate.

SECTION 4
AIR DISTRIBUTION

UNIT 13
AIR DISTRIBUTION—DUCTS

OBJECTIVES

After completing the study of this unit, the student will be able to

- list the primary factors to be considered in determining the fan and fan motor sizes for an air-conditioning system and the amount of air pressure required.
- describe each of the following items and state its effect on the air-conditioning system –

propeller fan	tip speed
centrifugal fan	fan efficiency
fan velocity	fan noise

- list the basic duct shapes and indicate preferred usages for each shape.
- list the types of duct systems.
- describe the duct systems listed previously.

Air distribution systems direct the air from the air-conditioning equipment to the space to be conditioned and return the air to the equipment. The best system is usually the simplest combination of fans, ductwork, and outlets.

A simple air distribution system includes a fan assembly with the fan outlet opening connected to a straight duct run. A supply outlet or terminal is installed at the end of the duct. The duct run has no turns and does not change in size. The return duct in this system is a straight run of duct from the return outlet to the fan assembly housing.

The simple system becomes more complex as the following additions are made: elbows, bends, size reductions, and other restrictions such as dampers, louvers, turning vanes, and intricately designed outlets. Each of these components adds resistance to the system. There are two major factors in determining the size of the fan and fan motor and the amount of air pressure that is required for a system, figure 13-1. These factors are the total resistance of the components to the flow of air and the friction resistance of the air passing over the surface of the straight duct run.

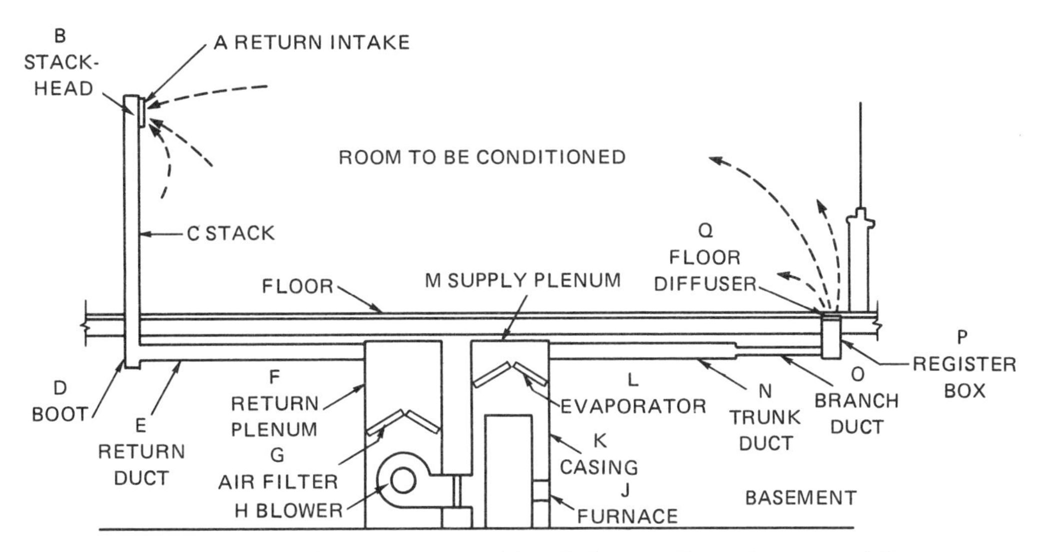

Fig. 13-1 Air passage in a duct system; identified parts offer resistance to airflow.

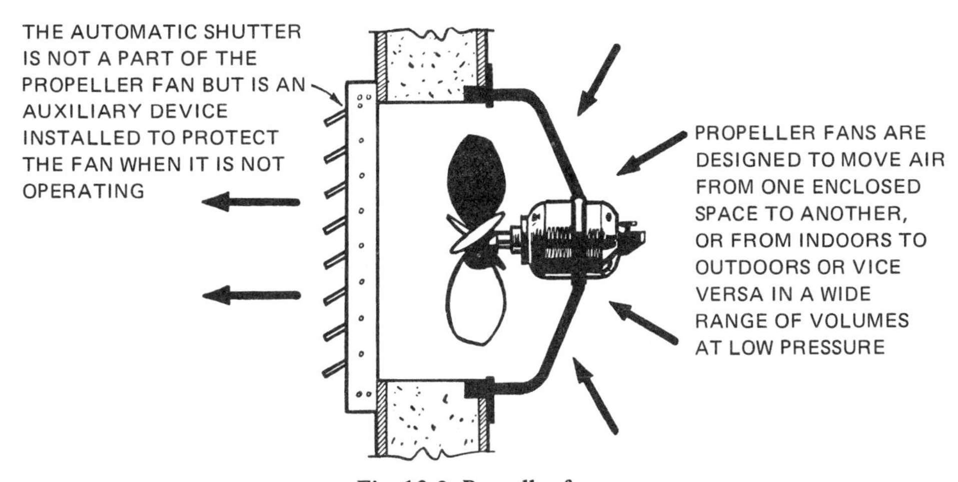

Fig. 13-2 Propeller fan

Most air distribution systems contain elbows, dampers, and various fittings to change the size or shape of the duct run. Thus, these systems are somewhat more complex as compared to the very simple system covered in this unit.

FANS

There are two types of fans commonly used in air-conditioning systems: the propeller fan and the centrifugal fan.

The propeller fan, figure 13-2, is used in installations where the supply duct is short or where a supply duct is not used at all, such as in window air-conditioning units.

The centrifugal fan, figure 13-3, page 144, is commonly used in commercial and residential system installations. This fan consists of a scroll, a shaft, and a wheel. The scroll is the housing for the shaft and wheel. The shaft serves as an axle for the wheel. The wheel is cylindrical and has many small rectangular blades arranged in parallel around the circumference of the cylinder.

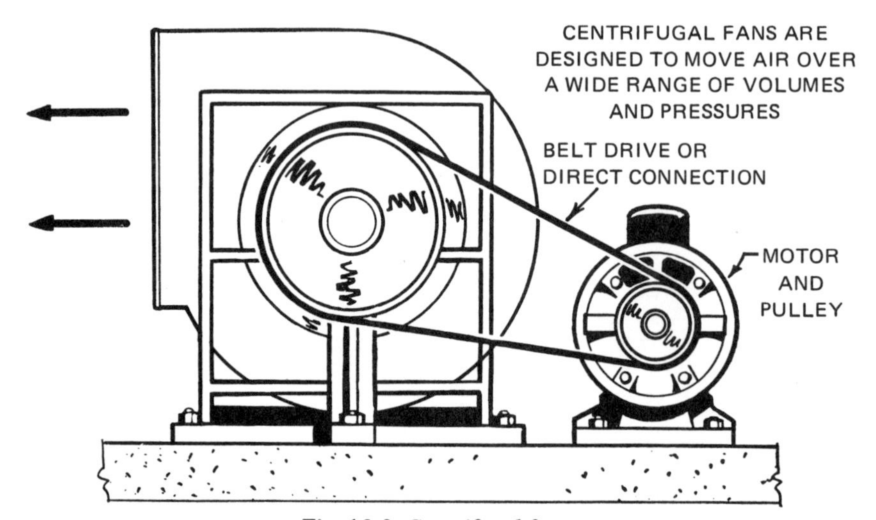

Fig. 13-3 Centrifugal fan

In an installation, the entire centrifugal fan assembly is enclosed in a housing or plenum. The housing must be large enough to permit free airflow into the fan scroll so that maximum efficiency and satisfactory air delivery are achieved by the fan. If the housing is too small, the fan does not perform to system specifications.

Centrifugal fans are available with forward- or backward-curved blades. For foward-curved blades, figure 13-4, the blade curve is in the direction of rotation of the fan wheel. A backward-curved blade, figure 13-5, is curved in a direction opposite to the direction of rotation of the wheel.

A fan wheel with forward-curved blades can deliver a required air quantity at low fan speeds. The operation of this type of fan wheel requires that the resistance of the duct system be determined accurately. If the resistance in the ducts is calculated to be a higher value than actually exists, then the fan selected has a higher capacity than is actually required. Thus, the fan runs at the same speed but delivers more air than needed. As a result, the fan requires a greater horsepower value than is actually necessary.

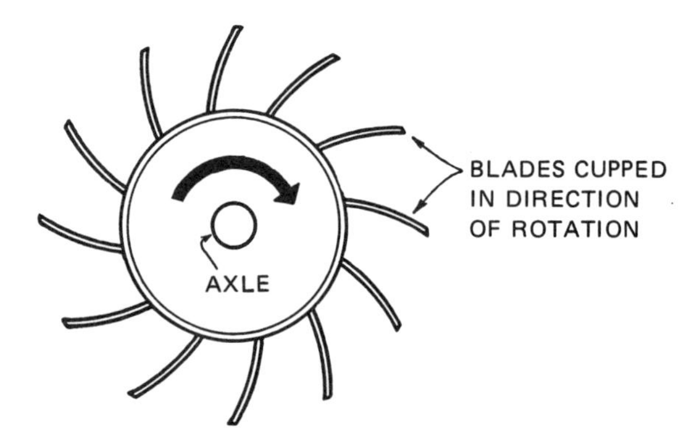

Fig. 13-4 Forward-curved blade

The operation of a fan wheel with backward-curved blades does not depend as closely on the duct resistance. This type of fan runs at a set speed for a motor with a given horsepower. In other words, the speed and horsepower relationship never change. In addition, there is less chance of overloading with a fan having backward-curved blades.

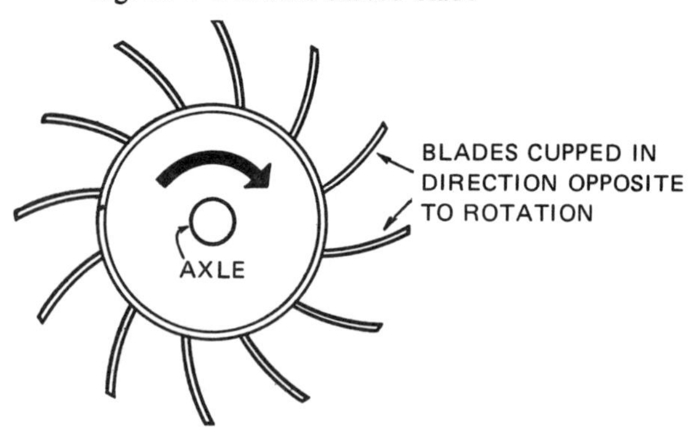

Fig. 13-5 Backward-curved blade

Fan Velocity

Fan velocity is the speed of the air as it leaves the fan. The air speed, in feet per minute (fpm), is measured at the outlet opening in the fan scroll. The air velocity and the speed of the fan wheel are important factors in determining the efficiency of the fan. The speed at which the fan wheel rotates is measured at the tip of the fan blades. Thus, the fan wheel speed is sometimes called the *tip speed.* For a given fan blade, the tip speed is measured in revolutions per minute (rpm).

In addition to the outlet velocity of the air and the fan wheel tip speed, the efficiency of the fan is affected by the construction of the fan scroll and the size of the fan assembly housing. Fan scrolls are constructed so that the areas of the inlet and outlet openings allow a suitable amount of air to pass through the fan without overloading the fan motor or causing undue strain on the fan wheel.

Sound Level (Noise)

The scroll construction, fan outlet air velocity, and tip speed affect the level of the sound (noise) generated by the fan. A high tip speed and high air velocity usually cause a higher noise level. When fans are used in residential installations where the sound level may be an important factor, the air velocity at the fan outlet must be kept as low as possible to minimize the noise. A suitable velocity range is between 1,000 and 1,500 rpm for those installations in which no sound absorbing materials or devices are used. If sound absorbing techniques are used in the system, higher air velocities may be permitted.

The location of the air-conditioning equipment can cause a significant variation in the intensity and pitch of the sound of the fan and the fan motor. The return air register usually has the most noticeable sound level. Since the equipment is placed ideally at a central location, the return air register generally is closest to the fan housing. Thus, there may be only a few feet of duct between the register and the fan, resulting in a higher noise level. Some of the noise can be absorbed by a longer run of duct. If the equipment is located far enough from the conditioned space, the duct surfaces absorb a significant part of the noise. In some instances, the noise level is reduced to the point that it is difficult to determine when the fan is running. Remote equipment locations, of course, are not always possible nor are they always practical. If a remote location is used, more duct is required; thus, the duct system is more expensive.

Figure 13-6, page 146, illustrates two methods of reducing the noise level. In duct A, sound insulation is installed on the inside surface of the duct. A more efficient method is to relocate the return intake by placing it at the ceiling, as shown for duct B. This location lengthens the duct and the addition of elbows and bends further reduces the noise level. The speed of the air through the room supply outlet can also cause objectionable noise.

DUCT SHAPES

Ducts are available in circular, rectangular, or square cross sections, figure 13-7, page 146. From an economic point of view, circular ducts are preferred because the circular shape can carry more air in less space. The circular shape means that there are reductions in the amount of duct material, duct surface, duct surface friction, and

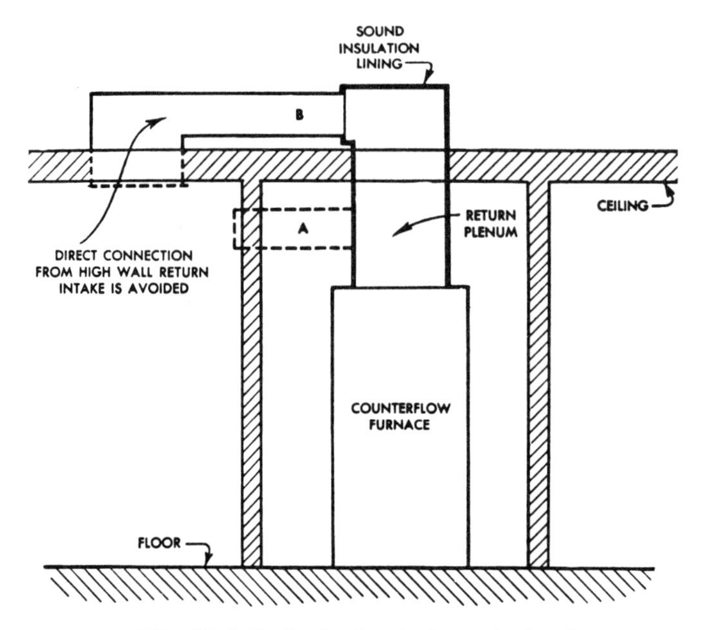

Fig. 13-6 Methods of reducing noise level

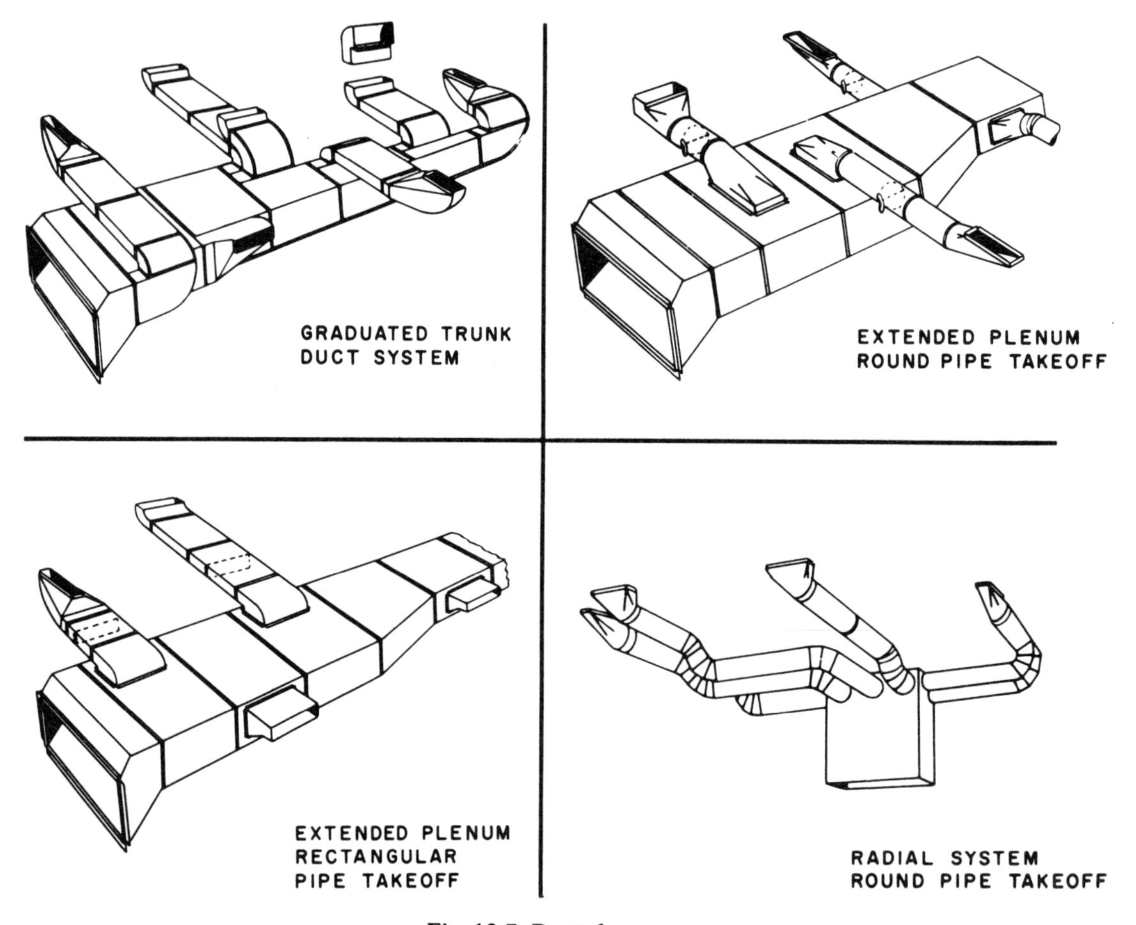

Fig. 13-7 Duct shapes
(Courtesy Reynolds Metals Company)

insulation as compared to other duct shapes. As far as appearance is concerned, the rectangular duct is preferred because it presents a flat surface that is easier to work with on the finish surface of the room or space. The rectangular duct is more practical in installations where space limitations in the ceiling or between the floor joists or wall studs require that a minimum duct height be used.

It is common practice to use rectangular duct for the plenum section of a duct system and round duct for branch runs. A combination of round and rectangular duct is used for takeoff fittings between a rectangular plenum and the round duct branch runs. A rectangular plenum section and rectangular branch runs are also commonly used. In this case, the takeoff fitting is also rectangular. If space permits, a round plenum and a round branch run can be used.

In some installations, the space between the floor joists (crawl space) is used as part of the duct system, figure 13-8. In such a system, the joist space must be sealed to prevent air leakage. Other techniques use the space above the ceiling or under the floor as plenum chambers. The only requirements for these approaches is that the space used be sealed (airtight), insulated, and vapor sealed. These techniques must also comply with local codes. When the space between the floor joists is properly sealed, it is suitable for use as part of the return air system. However, this space is rarely used as part of the supply air system.

Fig. 13-8 Crawl space installation

DUCT MATERIALS

Ducts are available in a wide variety of materials, including sheet metal, aluminum, fiberglass, tile, and cement. Although there is some overlapping in their usage, several of the materials are preferred for specific applications. For example, sheet metal is heavier than aluminum, but it may cost less. Aluminum and fiberglass are less subject to corrosion. Tile may be more suitable for slab construction. Plastic ducts are commonly used because of their flexibility and resistance to heat.

DUCT FITTINGS

Figures 13-9 through 13-11, pages 148-150, illustrate the variety of duct fittings available, including takeoffs, bends, elbows, turning vanes, reducers, transformations, collars, flexible connections, dampers, offsets, end caps, stack fittings, boots, register

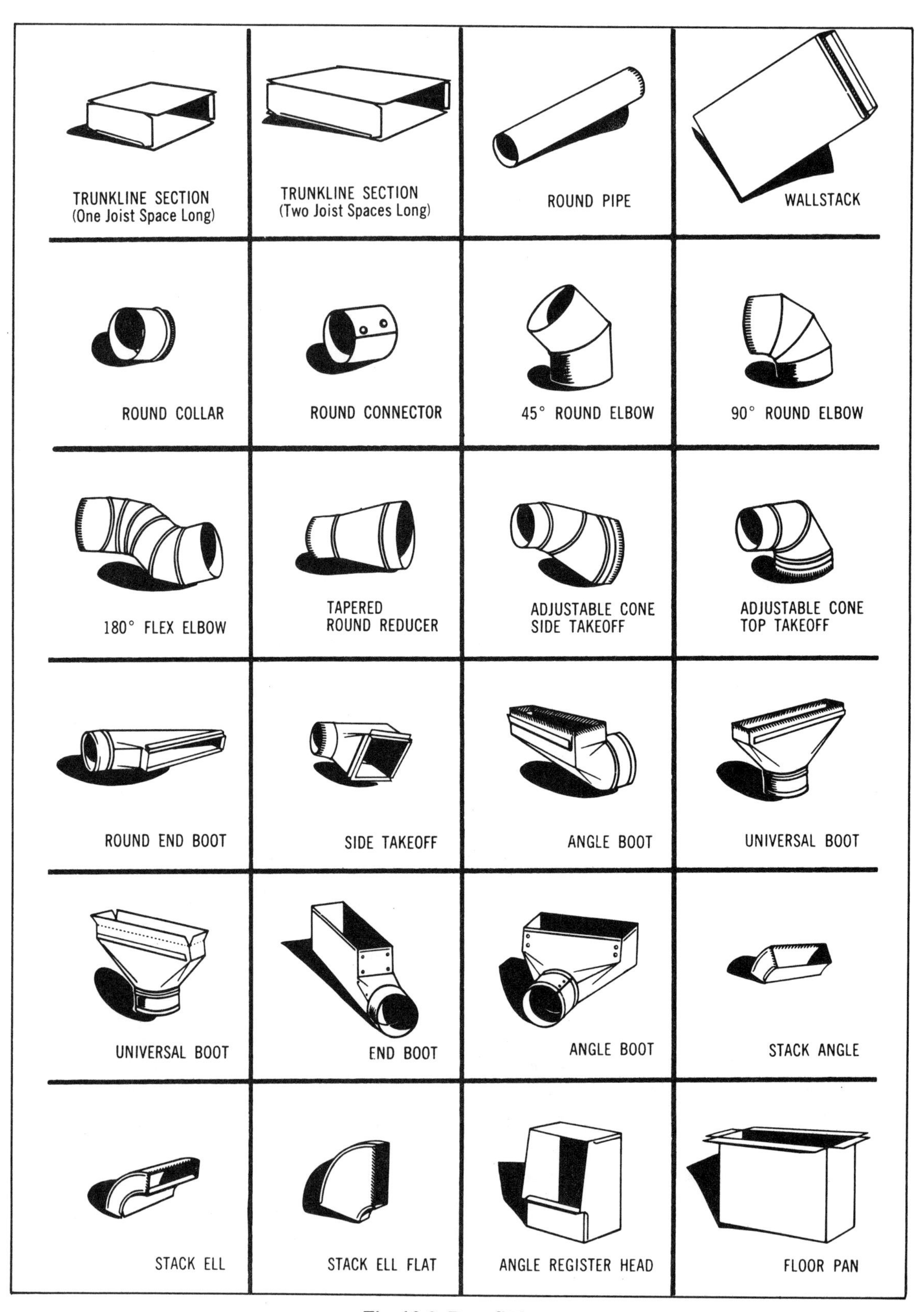

Fig. 13-9 Duct fittings
(Courtesy Reynolds Metals Company)

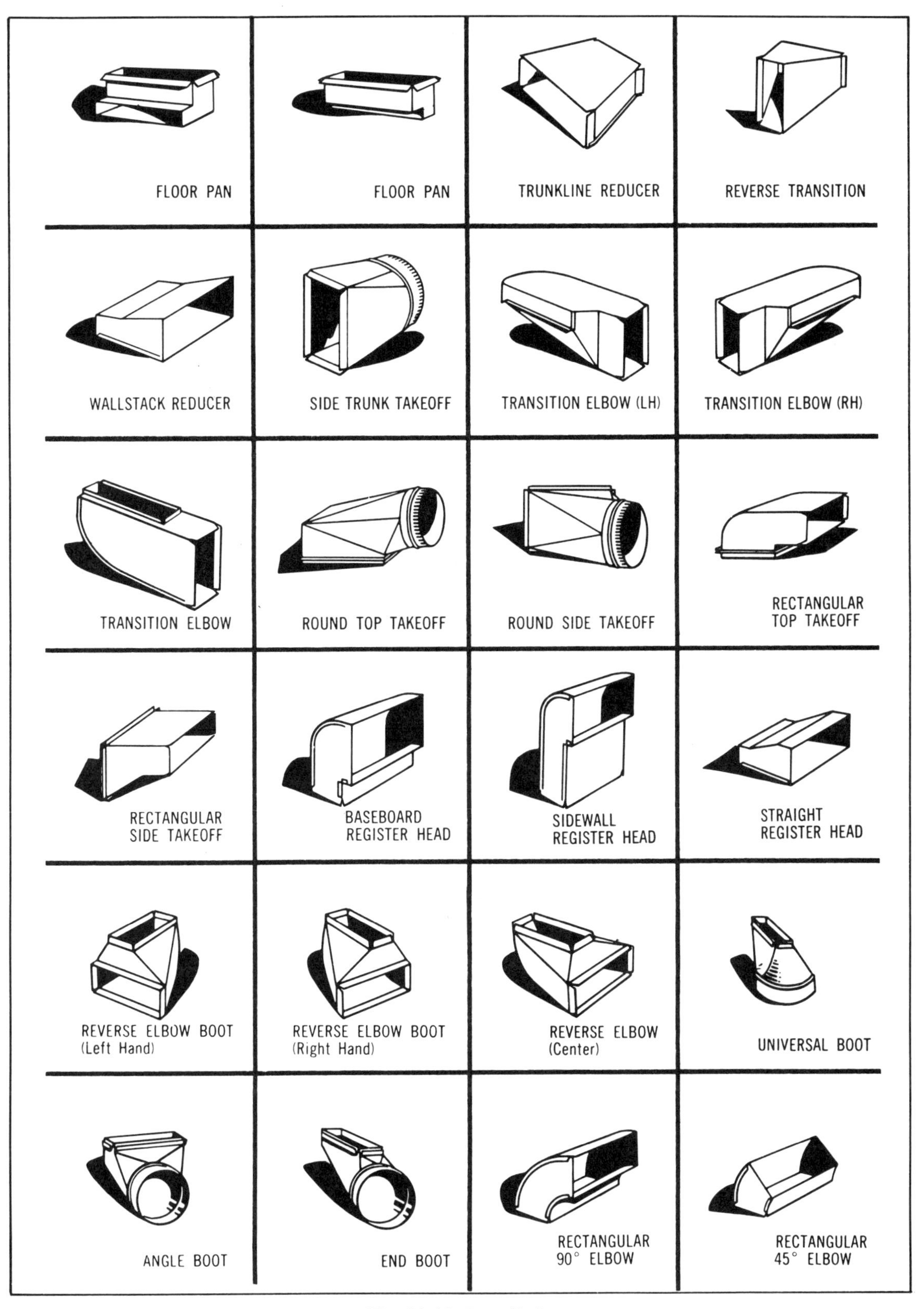

Fig. 13-10 Duct fittings
(Courtesy Reynolds Metals Company)

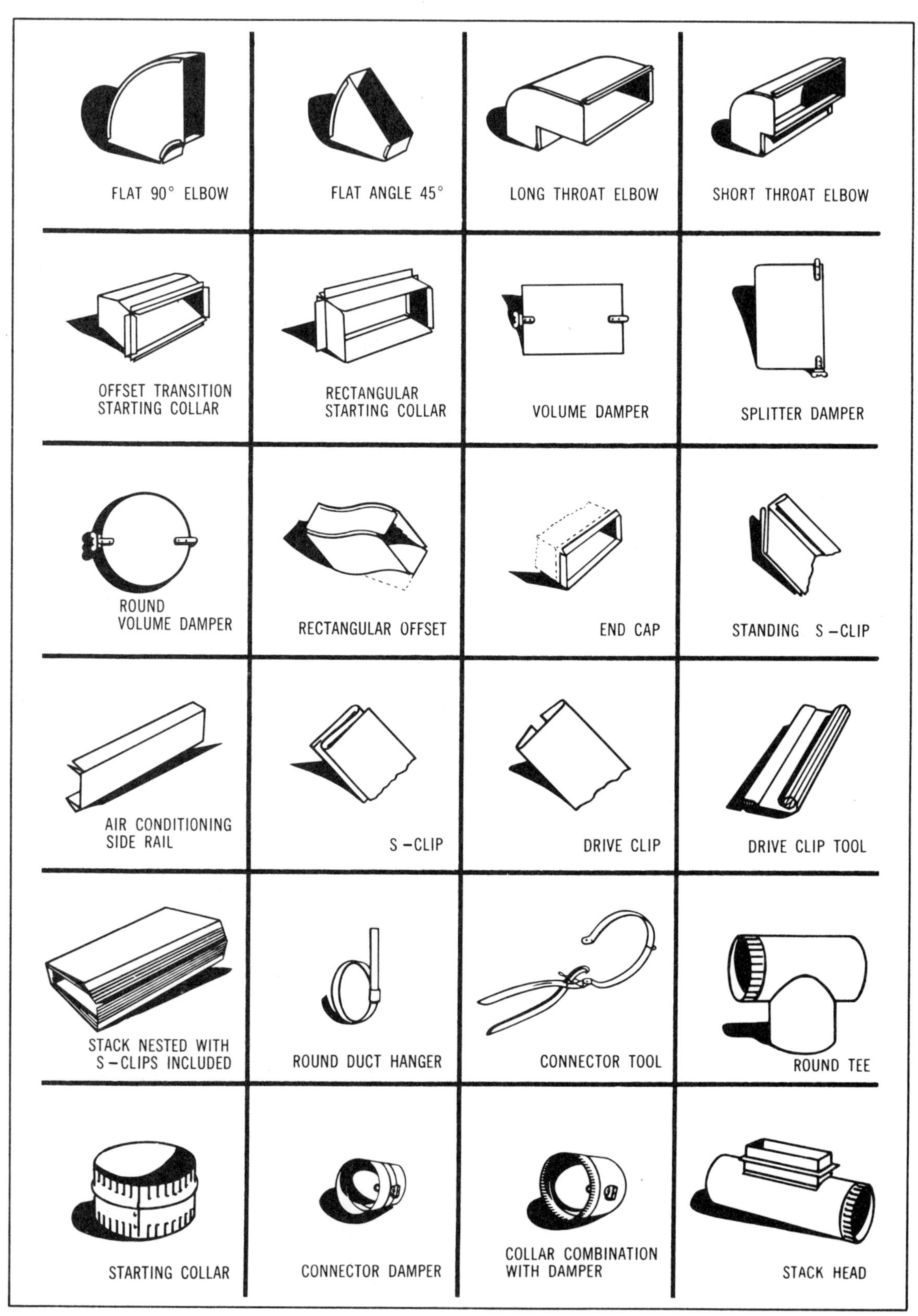

Fig. 13-11 Duct fittings
(Courtesy Reynolds Metals Company)

heads, floor pans, and combinations of these fittings. When making up a duct system, the use of fittings that change the shape, size, or direction of the ductwork should be kept to a minimum. This does not mean that essential fittings must be omitted. It does mean that the system should be planned to keep the plenum, branches, and stacks as simple as possible and still provide satisfactory air quantities and air distribution. In other words, the designer should avoid unnecessary and complicated combinations of fittings and poor planning in locating the plenum, branch runs, and stacks. Good design prevents increases in materials, installation time, and system friction.

DUCT SYSTEMS

Duct systems are divided into two general categories: (1) those systems designed for areas in which the heating season is the major consideration, and (2) those systems designed for areas in which the cooling season is the major consideration. Perimeter systems that supply air near or at the floor level and along the outside walls of the structure are considered most suitable for localities in which heating is the major factor. Overhead systems that supply air from the inside or outside wall and at the ceiling or high wall level are suitable for localities in which cooling is the major factor.

Loop Perimeter System

A loop perimeter system is shown in figure 13-12. The supply duct of this system is installed in a continuous, closed loop around the perimeter of the building. The supply air passes from a central furnace or air-conditioning equipment plenum to the perimeter loop through several feeder ducts. The feeder ducts are evenly spaced along the perimeter loop. In some applications, the ducts are designed to enter the loop at the point where the greatest discomfort is experienced in the room. In a heating application, the registers are placed under the windows where cold downdraft

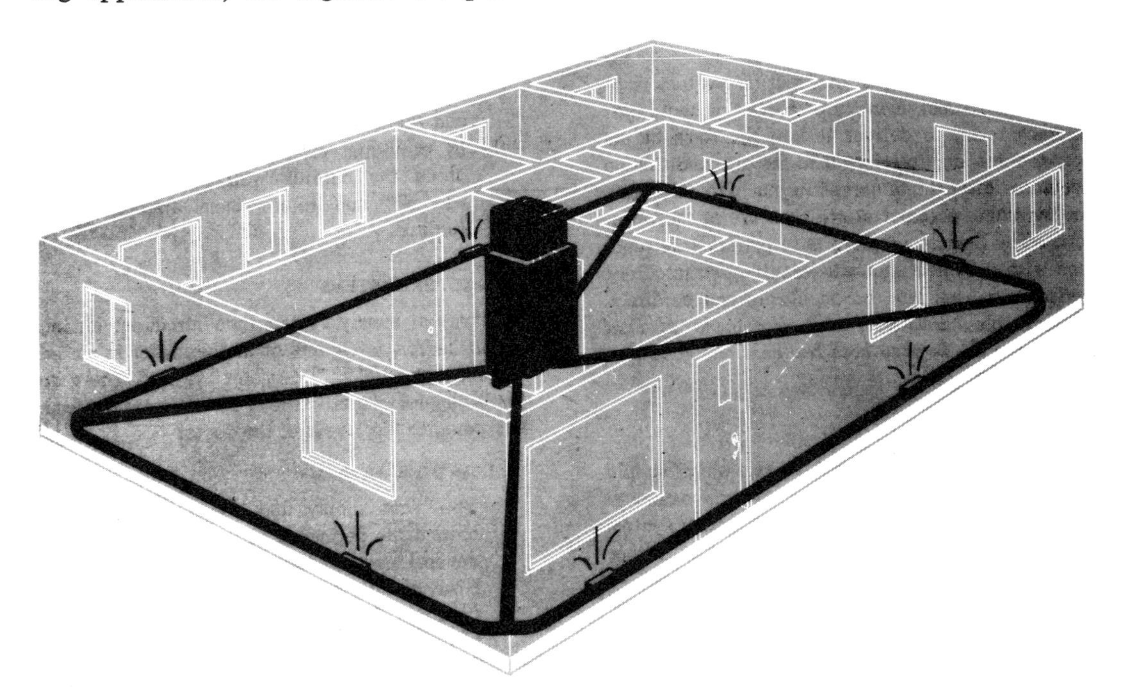

Fig. 13-12 Loop perimeter system

is critical. In a cooling application, the registers are installed in the areas of greatest room heat concentration. A perimeter loop system is ideal for slab construction since the heat from the feeder ducts running under the floor can help to offset the cold floor condition that may be present.

Radial Perimeter System

In a radial perimeter system, figures 13-13 and 13-14, the supply air is delivered from a central plenum through separate ducts running from the plenum to each outlet. The supply ducts extend from the plenum in all directions. Thus, the radial perimeter system is suitable only for crawl space or slab construction. If it is used in a structure with a basement, additional basement height is required since some of the supply ducts must run across the joists as well as parallel to the joists. Since the supply ducts must be located below the joist level for this reason, there will be less headroom in the basement unless a height allowance is made.

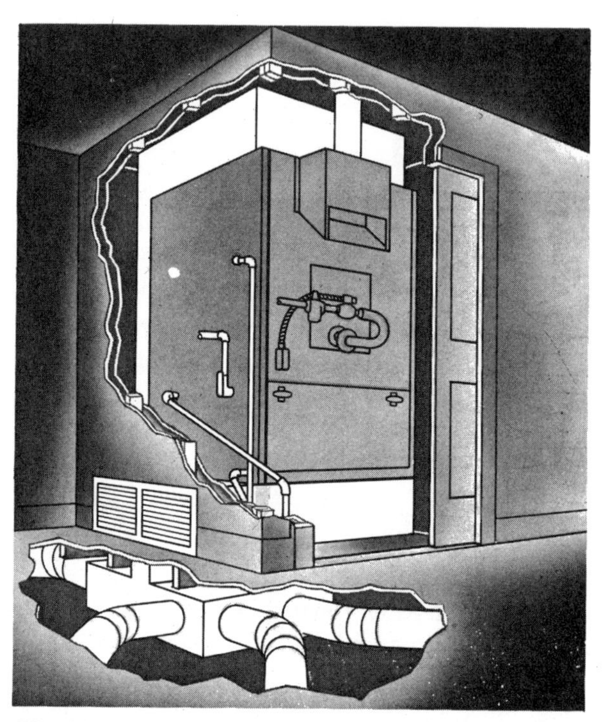

Fig. 13-13 Radial perimeter system – details at unit

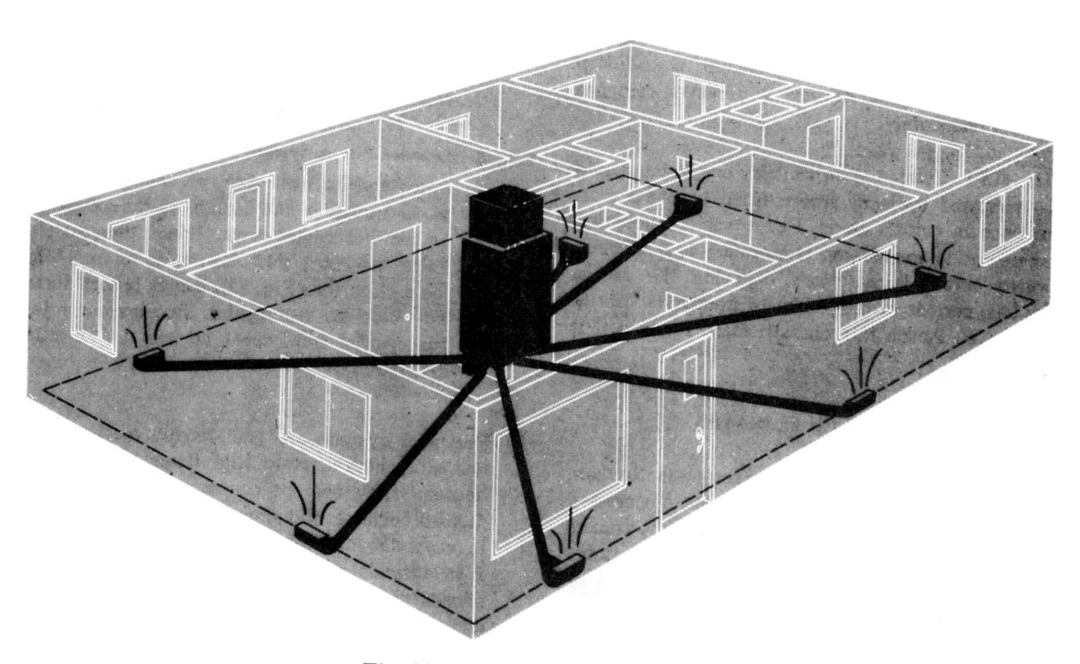

Fig. 13-14 Radial perimeter system

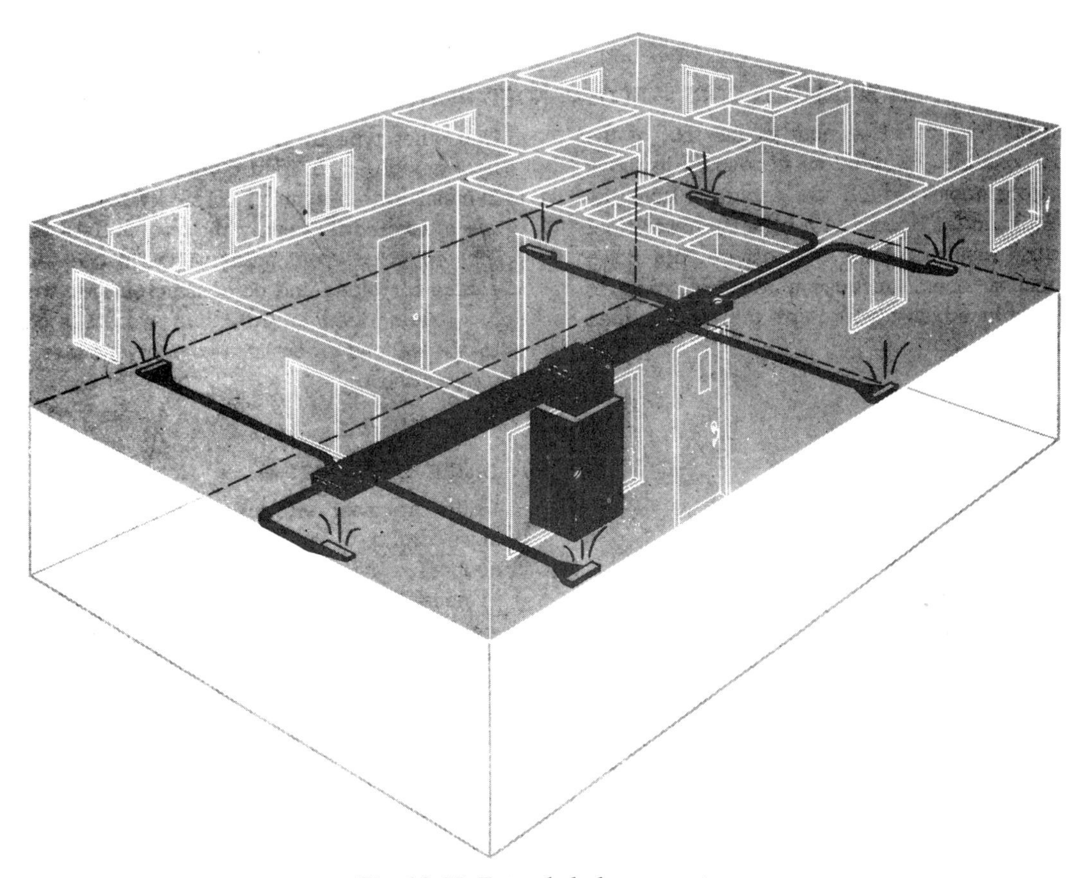

Fig. 13-15 Extended plenum system

Extended Plenum System

An extended plenum system, figure 13-15, has a rectangular plenum extended from one or both sides of the heating or cooling source. The plenum usually is located along the center beam of the structure. In some instances, the plenum extends the full length of the beam. As in the radial perimeter system, each outlet has a separate supply duct. In the extended plenum system, the supply ducts are taken off at various points along the plenum. Therefore, all supply ducts can be installed parallel to and between the joists. This type of system permits more headroom if the installation is made in the basement. The extended plenum system can also be installed in a crawl space or in the attic.

Perimeter systems generally are not used on second and succeeding floors. It is difficult and impractical to install duct risers to the second floor in the outside wall. The standard space between the wall studs is not large enough to handle the duct and the required insulation. If sufficient insulation is not installed, the heat loss is prohibitive. In addition, it is not practical to run the supply duct between the floor joists of the second floor since ducts are difficult to install in this area. In most installations, the second floor heating is handled by stacks and outlets in the inside walls.

A variation of the perimeter system for a second floor heating application is shown in figure 13-16, page 154. In this case, the outside walls and windows are blanketed by supply air from outlets located in adjacent partition walls. These outlets discharge air along and across the outside walls at a low level.

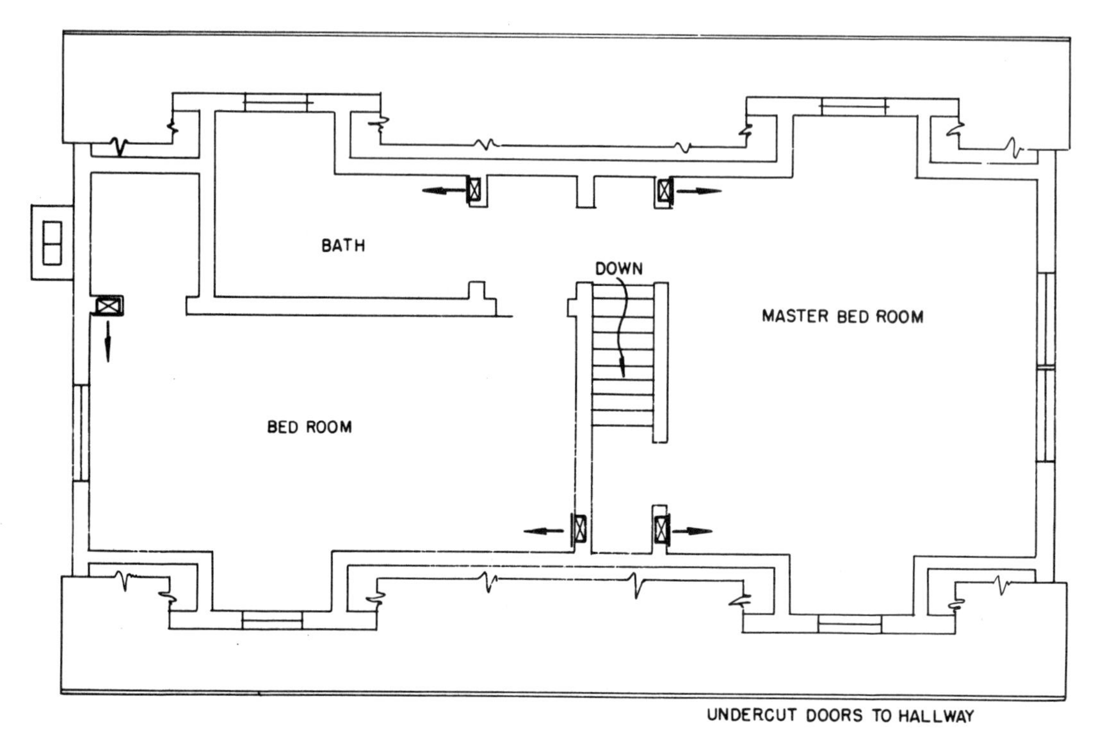

Fig. 13-16 Recommended method of air distribution for the second floor

OVERHEAD SYSTEMS

Overhead Trunk

An overhead trunk system is shown in figure 13-17. The main plenum or trunk in this system is extended in a false ceiling which usually runs along a center hall. Short duct runs or extended collars direct the air from the plenum to high wall outlets on the inside walls of the rooms. The main plenum can also be located in the attic. In this case, the supply air is directed from the plenum to the inside wall outlets through wall stacks.

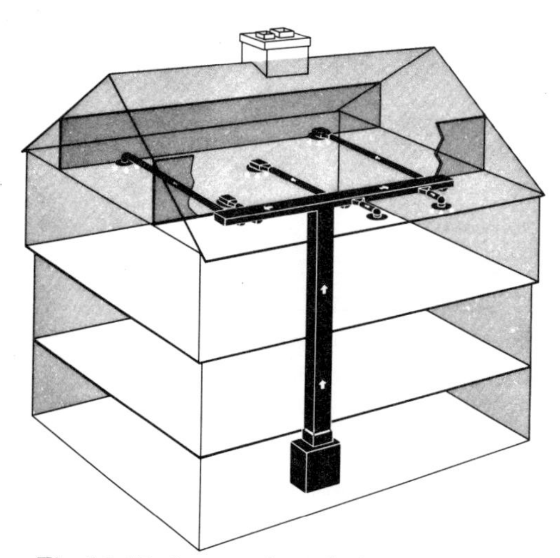

Fig. 13-17 Overhead trunk duct system

The overhead extended trunk system is also used with ceiling outlets. The supply air flows from the extended plenum to the outlets through supply ducts in the attic.

Overhead Radial Duct System

The overhead radial system, figures 13-18 and 13-19, does not have an extended plenum. The supply air flows directly from the central plenum through ducts located in a false hall ceiling to the inside wall outlets. This system can also be used with ceiling outlets where the duct runs are located in the attic. The attic location, however, is more expensive and requires more complicated installation techniques.

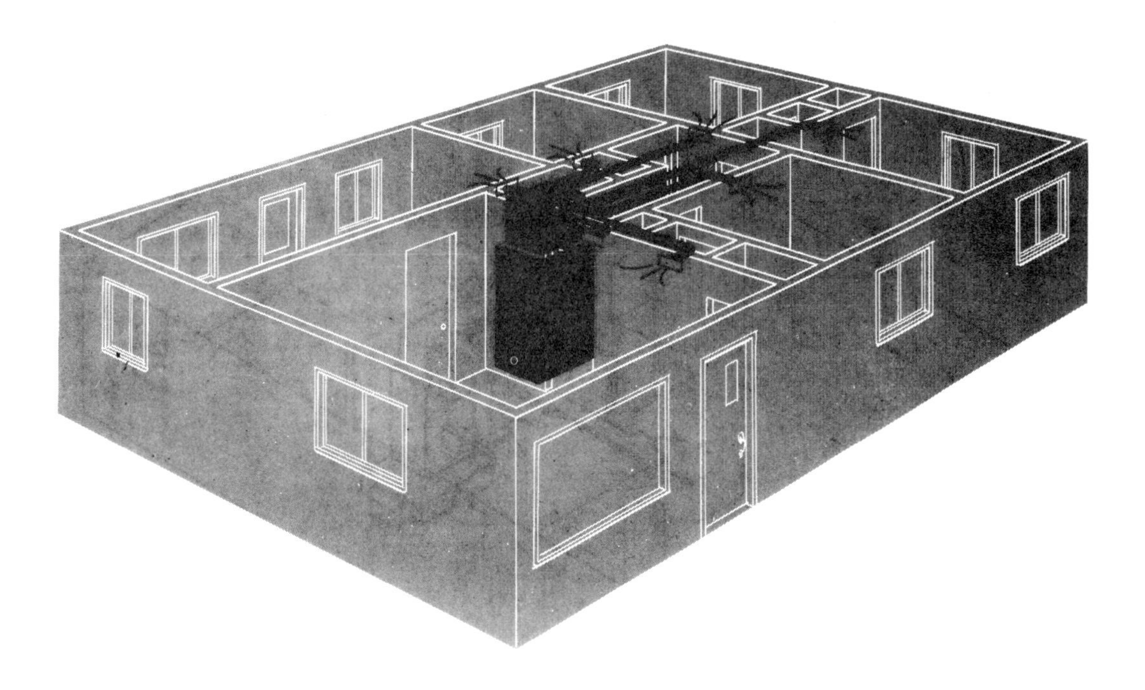

Fig. 13-18 Overhead radial duct system

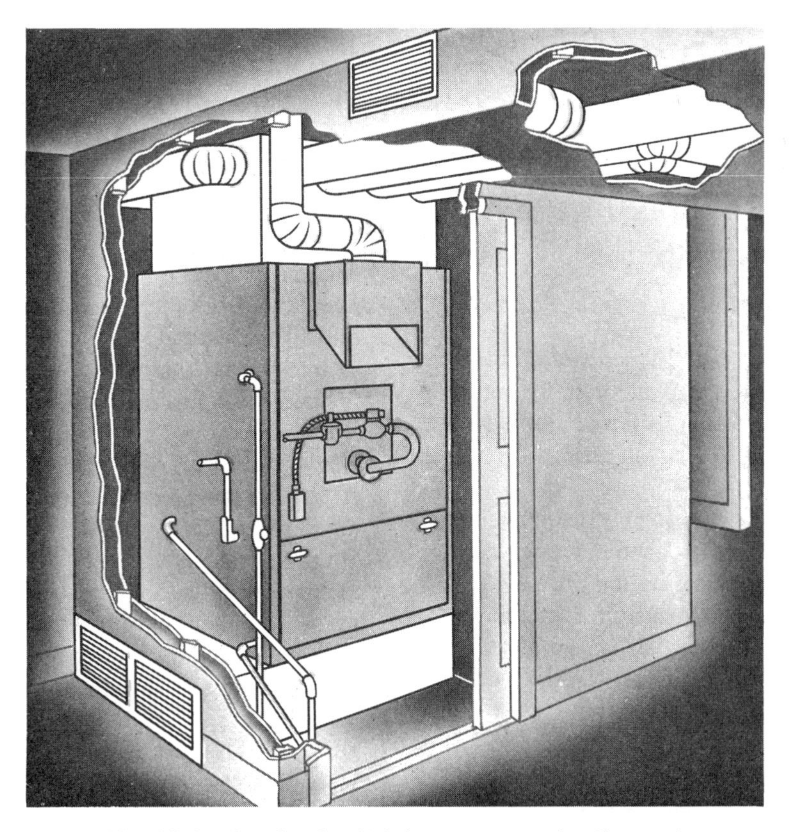

Fig. 13-19 Overhead radial duct system – details at unit

SUMMARY

- Air distribution systems direct air from the air-conditioning equipment to the space to be conditioned and return the air to the equipment.
- Propeller fans and centrifugal fans are used in air conditioning.
- A simple air distribution system includes a fan, a straight run of duct to supply outlets in the space to be conditioned and a straight run of duct from the return outlets to the fan.
- Fans with forward-curved blades run at low speeds; fans with backward-curved blades have nonoverloading characteristics.
- The sound or noise level in a room is affected by the fan outlet air velocity, fan wheel tip speed, fan location, and the supply air velocity through the room outlet.
- Round ducts carry more air in less space than other duct shapes. As a result, there is a decrease in material, duct surface, friction, and insulation.
- The basic types of duct systems are as follows:
  1. Graduated trunk
  2. Extended plenum – round takeoff
  3. Extended plenun – rectangular takeoff
  4. Radial – round takeoff
- Duct materials are sheet metal, aluminum, fiberglass, tile, cement, and plastic.
- Duct systems are designed to supply air around the perimeter of a structure or from the inside walls.
- Perimeter systems generally are not used for second floor installations in residences.
- The basic duct system designs are as follows:
  1. Perimeter loop
  2. Radial perimeter
  3. Extended perimeter
  4. Trunk
  5. Overhead trunk
  6. Overhead radial

REVIEW

Insert in the space provided the word or phrase which correctly completes each statement.

1. The propeller-type fan is used for ____________ (high) (low) operation.
2. The centrifugal-type fan is used for __________________ (high) (low)-resistance operation.
3. A fan with forward-curved blades is designed to operate at a __________ (high) (low) speed.

4. A nonoverloading fan is one with _______________ (backward) (forward)-curved blades.

5. Loop or radial perimeter duct systems are _____________________ (satisfactory) (unsatisfactory) for slab construction.

6. Outside wall stack and outlets ________ (are) (are not) recommended for second floor air distribution systems.

7. The attic cooling installation shown in figure 13-20 represents the _____________ (overhead radial) (overhead trunk) type of duct design.

8. Add arrows to figure 13-20 to trace the path of the warm room air from the point of return until it is distributed through the supply outlets.

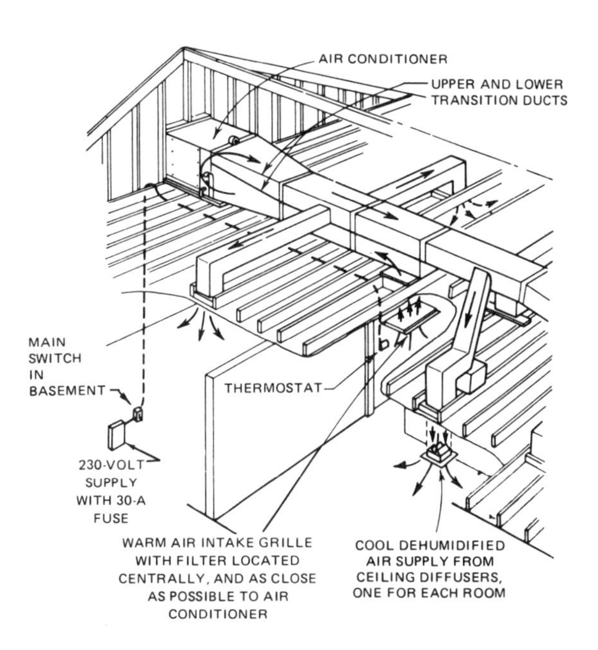

Fig. 13-20

Briefly answer each of the following questions.

9. Why is a radial perimeter duct system more suitable for crawl space or slab construction than for full basement construction?

10. State three reasons why round duct is more efficient than other duct cross sections.

11. For each of the duct system designs listed, describe its identifying features and suggest the type of installation for which it is best used.

    a. Overhead trunk
    b. Radial perimeter
    c. Overhead radial
    d. Perimeter loop
    e. Extended plenum

UNIT 14
AIR DISTRIBUTION—OUTLETS

OBJECTIVES

After completing the study of this unit, the student will be able to

- list the functions for which outlets are designed.
- describe the following types of commonly used outlets –

supply outlet	register
return outlet	fixed louver register
ceiling diffuser	adjustable louver register

- state the factors that determine the locations of the outlets for heating and for cooling applications.
- state the functions and advantages of the following types of outlets –

low wall outlet	baseboard outlet
high wall outlet	ceiling outlet
floor outlet	

- list the factors to be considered in the selection of outlets.

Outlets are another major part of the air distributtion system. They are extremely important from the point of view of appearance, function, and performance.

Many types of outlets are available but all are designed to provide the properly controlled method of distributing air to the room or removing air from the room. Outlets are designed to:

- attractively conceal the duct opening
- throw the conditioned air no less than three-quarters of the distance to the opposite wall
- deflect or diffuse the air
- adjust the airflow rate
- prevent dirt streaking and noise.

TYPES OF OUTLETS

The following list covers the basic types of commonly used outlets.

- The *supply outlet opening* is a wall, ceiling, or floor opening through which warm, cool, humidified, or dehumidified air is delivered to a room.
- A *return outlet opening* is a wall, ceiling, or floor opening through which room air is either exhausted to the outdoors or returned to the fan housing.

- The *ceiling diffuser* is a square or circular facing device that covers the supply air opening in a ceiling. Most ceiling diffusers cause air to enter the room in multiple layers, figure 14-1.
- The *grille* is a covering for any opening through which air passes.
- A *register* is a grille fitted with a damper to control the quantity of air passing through it.
- The *free area* is the area of the openings in the outlet through which air can pass, figure 14-2.

Fig. 14-1 Air pattern from a diffuser

- The *core area* is the total area of the grille opening.
- The *fixed louver register* is a register with stationary bars or louvers that are factory set to direct the air in a specific unchangeable pattern, figure 14-3, page 160.
- The *adjustable louver register* is a register with movable bars or louvers. The louvers are horizontal or vertical or both. Thus, an adjustable louver register can be varied to direct the air upward and to one side or the other, or downward and to one side or the other, figure 14-4, page 160. Some adjustable louvers are completely flexible and can be made to direct the air upward, downward, and to both sides at the same time.

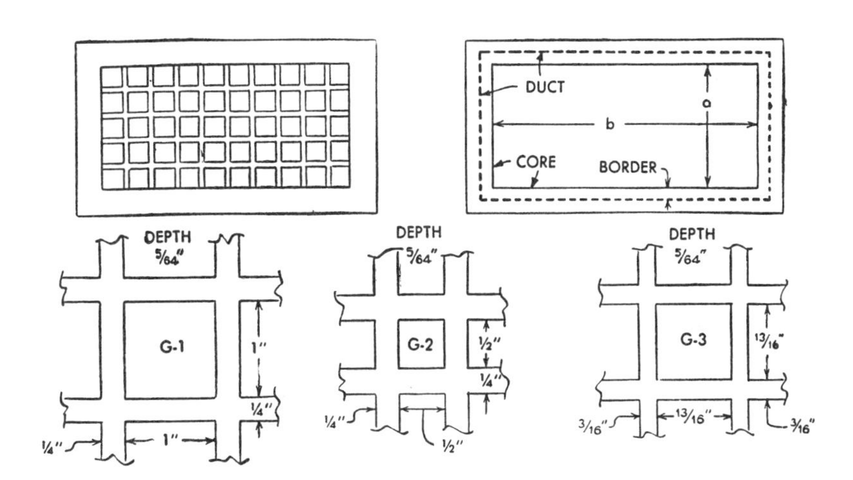

Fig. 14-2 Perforated grilles

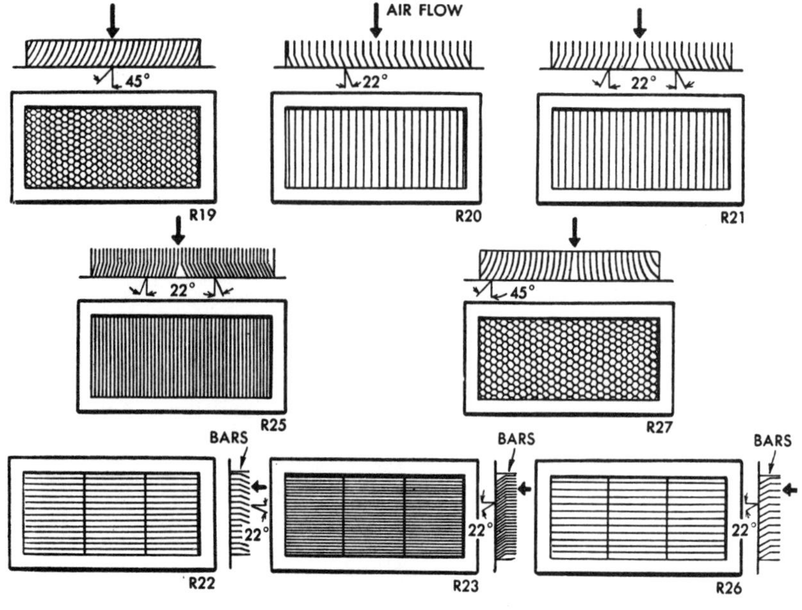

Fig. 14-3 Fixed deflecting-type registers

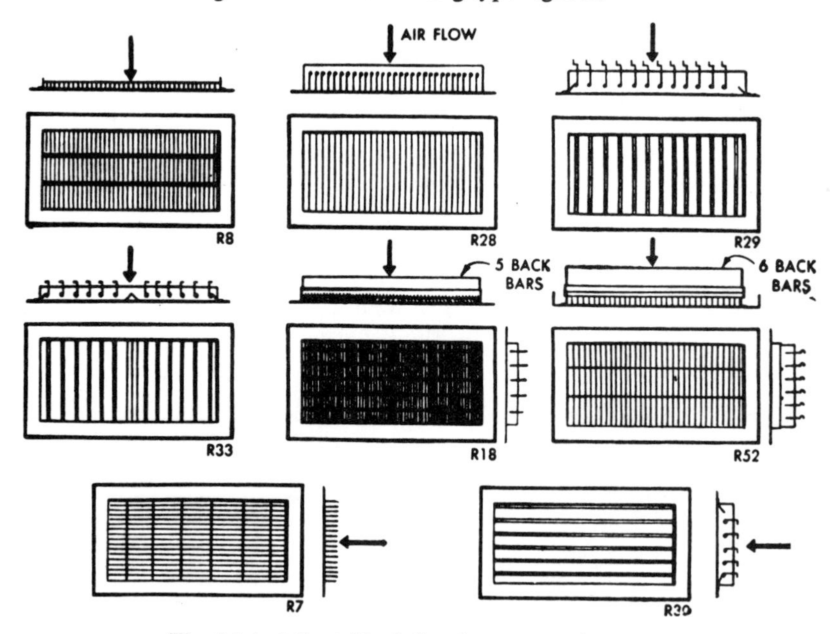

Fig. 14-4 Adjustable deflecting-type registers

LOCATION OF OUTLETS

Supply and return air openings in the conditioned space must be located to insure that sufficient air is supplied to establish and maintain comfort conditions in the room, figure 14-5. These conditions must be as uniform as possible to prevent discomfort from hot or cold drafts.

The locations of the outlets are determined by the system application: whether for heating or cooling or both. If the system is to be used for both heating and cooling, the application that is used most of the time during the year should be the most important factor.

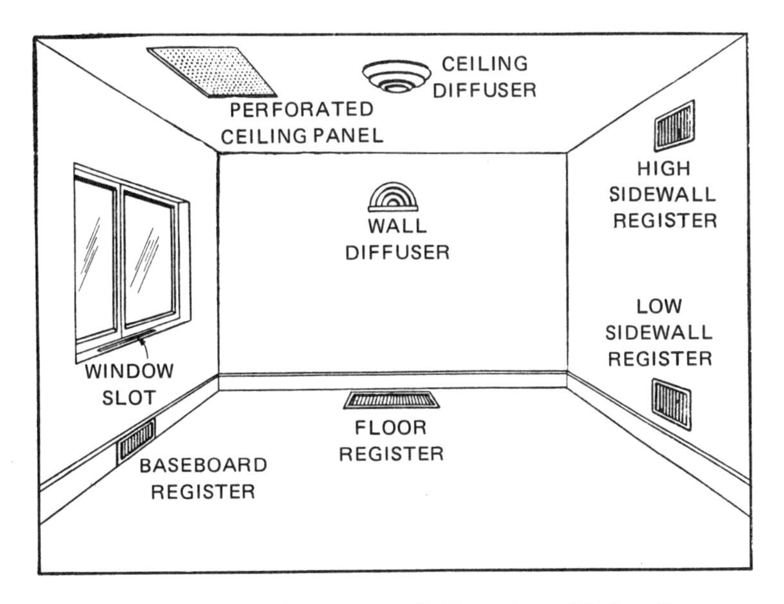

Fig. 14-5 Types of registers available and possible locations

Heating

The chief cause of discomfort in heating applications is cold downdrafts from cold surfaces, especially windows. If the cold downdraft is not checked, the cold air falling down the window spills out across the floor. To overcome the downdraft, supply air outlets for heating are located on the outside wall under windows or on the floor under windows, figure 14-6. The warm air from the outlet is controlled by the register so that a blanket of air sweeps up the window. As a result, the cold downdraft is checked. This method is effective because the cold air from the window surface is warmed before it is circulated in the room. Another method of reducing the cold drafts is not as effective as the warm air blanket. In the second method, return air outlets are located under the window. This means that the cold downdraft flows into the return outlet and directly into the heating chamber. Since the cold air is not warmed in the room, some of it spills across the floor before it flows into the return air outlet.

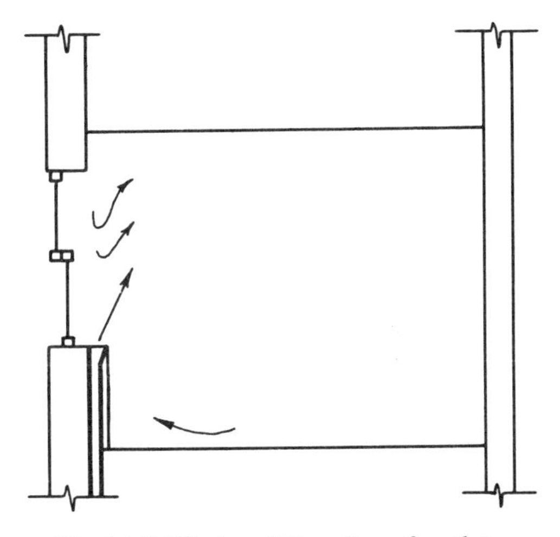

Fig. 14-6 Windowsill location of outlet

Cooling

The main causes of discomfort in cooling applications are warm or stagnant air spots caused by poor air distribution and concentrated heat sources from people, lights, or appliances. Cold downdrafts from windows do not exist in cooling applications. Therefore, there is a wider choice of supply outlet locations and simpler duct systems.

Low Wall Outlet

A low wall outlet generally is located in the outside wall under a window. In many systems, such an outlet is fitted with a register to spread the airstream and direct it upward. Thus, the installation of a low sidewall outlet and the proper register is ideal for both heating and cooling. Downdrafts are prevented in winter because the wall and windows are blanketed with warm air. This type of installation also provides good circulation in summer since it directs the cool air upward and blankets the ceiling. Since cool air is heavier than warm air, it begins to drop toward the floor as it spreads over the ceiling, resulting in good circulation.

If a low wall outlet is fitted with a register that directs the supply air in a fixed horizontal pattern, the combination of outlet location and register is satisfactory for some heating applications. However, this combination is not satisfactory for cooling.

High Wall Outlet

High wall outlets generally are placed on an inside wall. In most cases, it is not possible to locate the outlet on an outside wall. This is due to the fact that the stack duct leading to the outlet requires insulation against the colder outside temperatures. However, the space between the studs is just large enough for the stack and does not provide space for insulation. Since the heat loss from an uninsulated stack duct may be prohibitive, most high wall outlets are located on inside walls.

A high wall outlet is satisfactory for cooling applications because the registers designed for this location usually provide a horizontal air distribution pattern. The high wall outlet location for heating is satisfactory if the return outlet is located under the window on the outside wall, figure 14-7. High wall outlets should not be used for heating applications in slab floor construction. This restriction is true especially in colder climates, where a heat source is usually required in the slab.

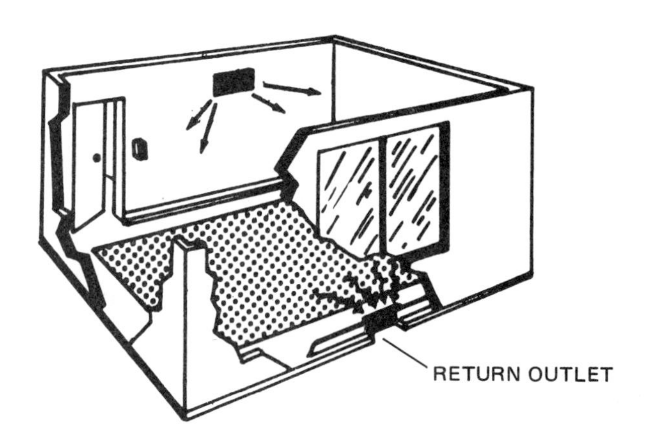

Fig. 14-7 High sidewall outlet location

If beams extend from the ceiling, the high sidewall outlet should be installed low enough to permit a free horizontal flow of air across the room. If an adjustable louver register is used, the airflow can be directed downward to miss the beam. However, the resulting airflow pattern is lowered and a hot draft condition is often created at the opposite side of the room from the outlet.

Stratification is another condition that must be considered when using high wall locations for outlets. Since hot air rises, the heat tends to stratify in the upper part of the room. The lower portion of the room remains uncomfortably cool. As a solution to this problem, the installation of a low return outlet on the outside wall not only decreases the downdraft from the window, but also tends to pull down the hot air, thus decreasing the stratification.

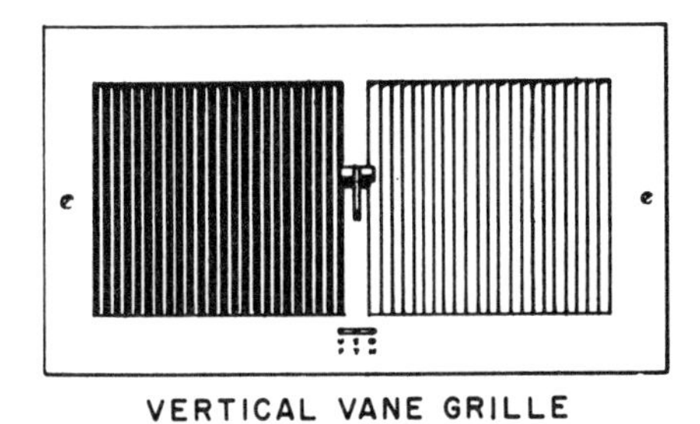

VERTICAL VANE GRILLE

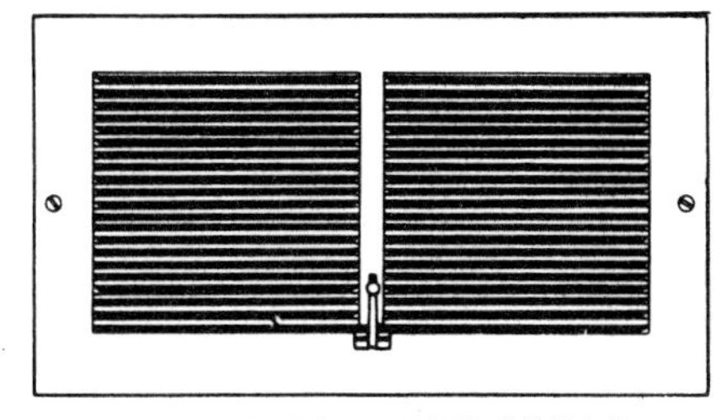

HORIZONTAL VANE GRILLE

Fig. 14-8

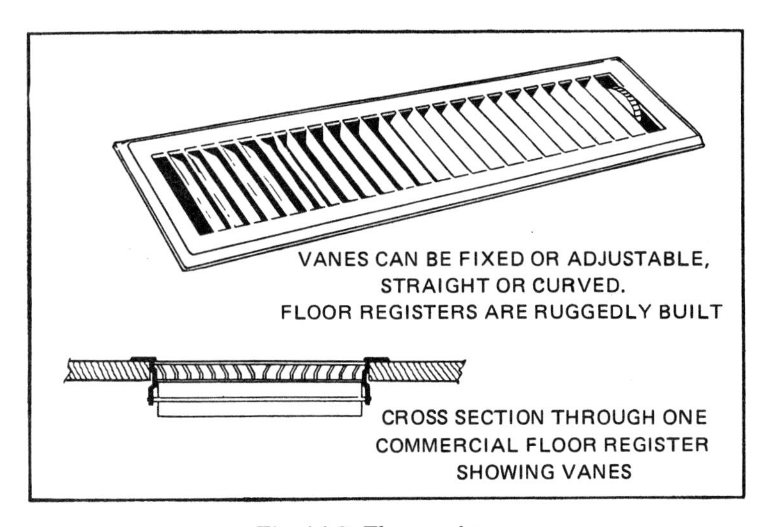

Fig. 14-9 Floor registers

Floor Outlets

A floor outlet, figure 14-9, is located along the outside wall, usually under a window. A floor outlet is good for both heating and cooling applications if the register provided fans the air so that it spreads over the wall surface. In cooling applications, air supplied through a floor outlet must have sufficient velocity to carry it to or near the ceiling level. If the air speed is not great enough, cold air will stratify at the floor level, resulting in a discomfort condition. Floor outlets are more commonly used in residential than in commercial installations. However, floor outlets can be used in commercial applications if special care is taken to locate the outlets to prevent the discharge of the supply air into pedestrian traffic lanes.

Baseboard Outlet

A baseboard outlet, figure 14-10, page 164, is located in the baseboard along the outside wall, usually under a window. This outlet is satisfactory for both heating and cooling applications if the register used fans the air over the wall and window surface, figure 14-11, page 164. The air velocity in a cooling application must be great enough to carry the air to or near the ceiling to prevent cold air stratification at the floor level.

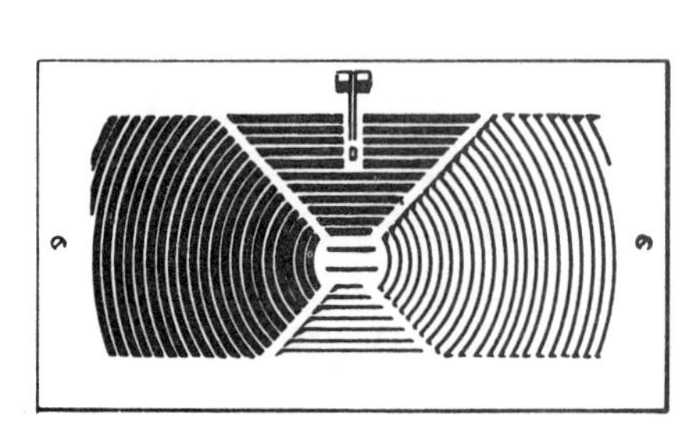

Fig. 14-10 Baseboard Outlets

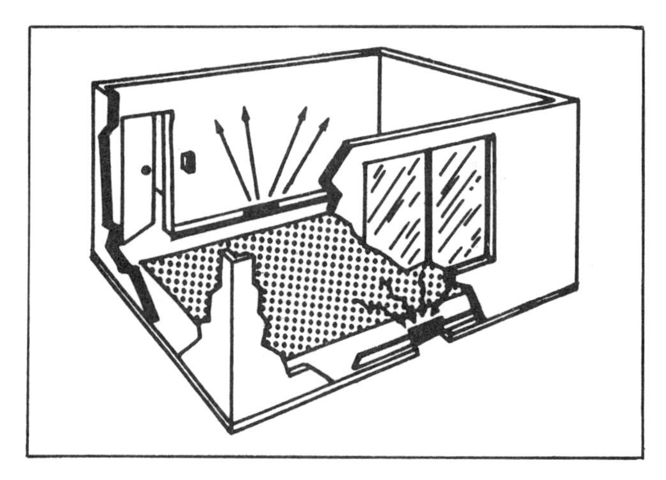

Fig. 14-11 Baseboard supply and return

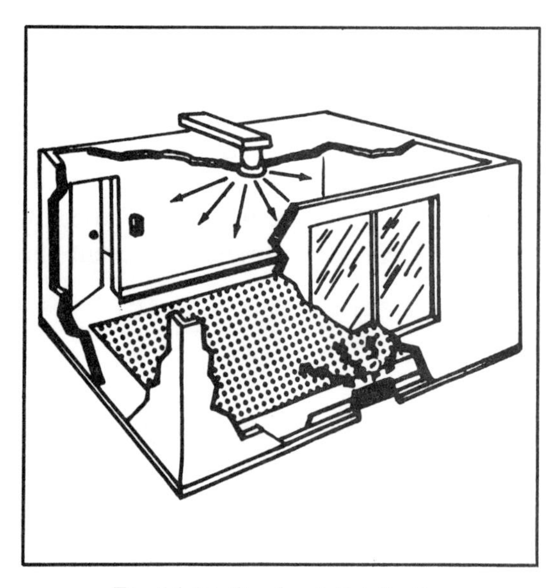

Fig. 14-12 Overhead distribution

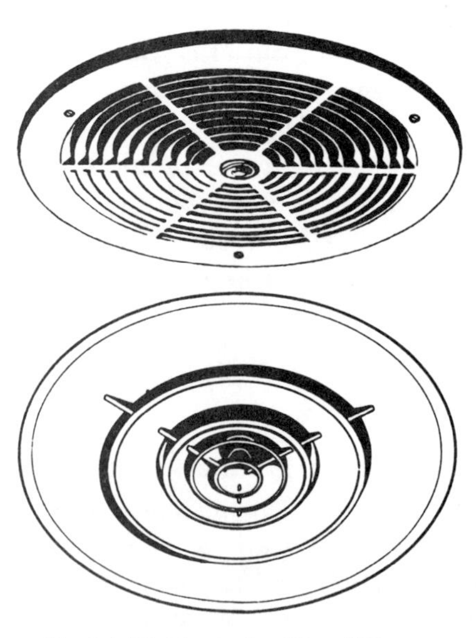

Fig. 14-13 Typical ceiling diffusers

Ceiling Outlets

Ceiling outlets are usually placed in the exact center of the room or space to be conditioned, figure 14-12. These outlets are fitted with ceiling diffusers, figure 14-13. If more than one outlet is needed, the ceiling area is divided equally. Each outlet is then located in the center of one of the divided areas. Ceiling outlets and diffusers are used principally in commercial buildings where they are suitable for most heating and cooling applications. This type of outlet is not recommended for installation in slab construction in colder climates.

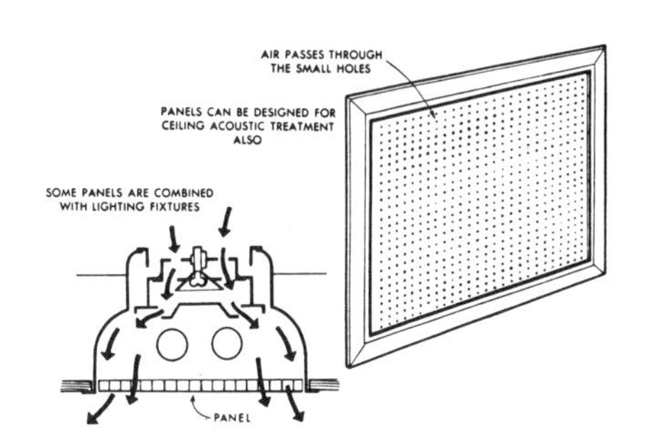

Fig. 14-14 Perforated panels

Perforated panels, figure 14-14, are used to provide a large airflow to areas without drafts. These panels can be placed at any point, but are usually installed as ceiling or high wall outlets.

Figure 14-15 is a comparison of the five basic outlet locations. The diagram for each outlet also indicates the register design that is usually provided with the outlet.

	FLOOR	BASEBOARD	LOW SIDEWALL	HIGH SIDEWALL	CEILING
COOLING PERFORMANCE	Excellent	Excellent if used with perimeter systems	Excellent if designed to discharge upward	Good	Good
HEATING PERFORMANCE	Excellent	Excellent if used with perimeter systems	Excellent if used with perimeter systems	Fair—should not be used to heat slab houses in Northern climates	Good—should not be used to heat slab houses in Northern climates
INTERFERENCE WITH DECOR	Easily concealed because it fits flush with the floor and can be painted to match	Not quite so easy to conceal because it projects from the baseboard	Hard to conceal because it is usually in a flat wall	Impossible to conceal because it is above furniture and in a flat wall	Impossible to conceal but special decorative types are available
INTERFERENCE WITH FURNITURE PLACEMENT	No interference—located at outside wall under a window	No interference—located at outside wall under a window	Can interfere because air discharge is not vertical	No interference	No interference
INTERFERENCE WITH FULL-LENGTH DRAPES	No interference—located 6 or 7 inches from the wall	When drapes are closed, they will cover the outlet	When located under a window, drapes will close over it	No interference	No interference
INTERFERENCE WITH WALL-TO-WALL CARPETING	Carpeting must be cut	Carpeting must be notched	No interference	No interference	No interference
OUTLET COST	Low	Medium	Low to medium, depending on the type selected	Low	Low to high—wide variety of types are available
INSTALLATION COST	Low because the sill need not be cut	Low when fed from below—sill need not be cut	Medium—requires wall stack and cutting of plates	Low on furred ceiling system; high when using under-floor system	High because attic ducts require insulation

Fig. 14-15 Comparison of outlets for residences
(Courtesy Carrier Corporation)

DEFINITION OF TERMS

- *Primary Air* – air delivered to the room from the supply duct.
- *Secondary Air* – room air that is drawn into the primary airstream.
- *Outlet Velocity* – the speed of the air measured at the face of the register.
- *Temperature Difference* – the difference in temperature between the primary air and room air.
- *Temperature Variation* – the difference in temperature between the various points or areas in a room.
- *Throw or Blow* – the distance that the supply air travels into the room after it leaves the supply air register and before its speed slows down to 50 fpm. Figures 14-16 and 14-17 show long and short throws with no deflection.
- *Drop (Cool Air)* – the difference in height between the lower edge of the airstream at the register and the lower edge of the airstream at the end of its throw, figure 14-18.
- *Rise (Warm Air)* – the difference in height between the upper edge of the airstream at the register and the upper edge of the airstream at the end of its throw.
- *Spread* – the amount the airstream expands (vertically and horizontally) after it leaves the register.

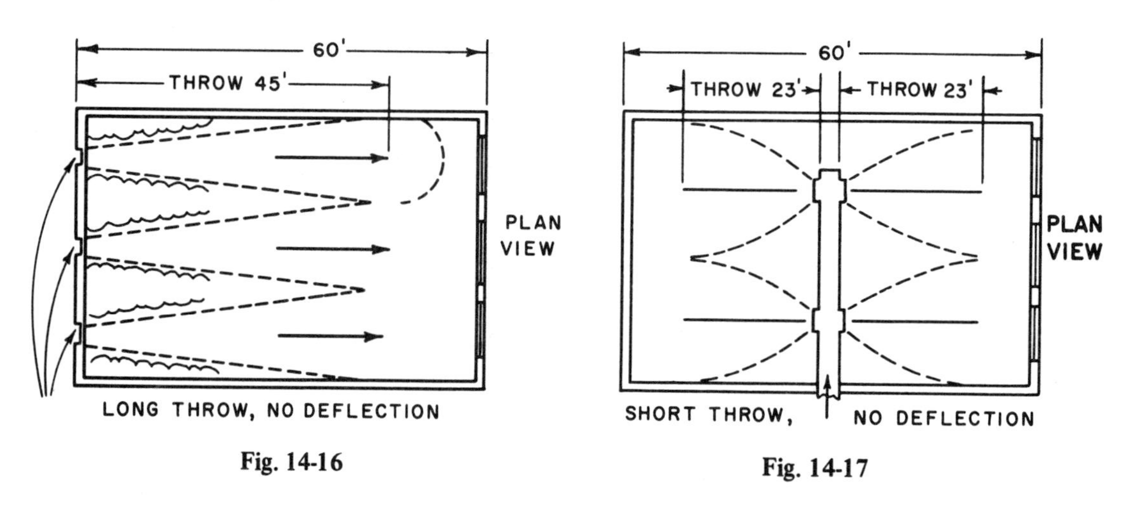

Fig. 14-16

Fig. 14-17

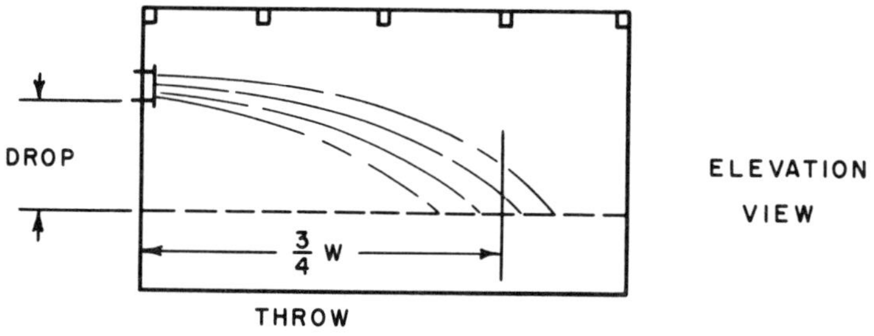

DROP

ELEVATION VIEW

$\frac{3}{4}$ W

THROW

Fig. 14-18 Throw and drop (cool air)

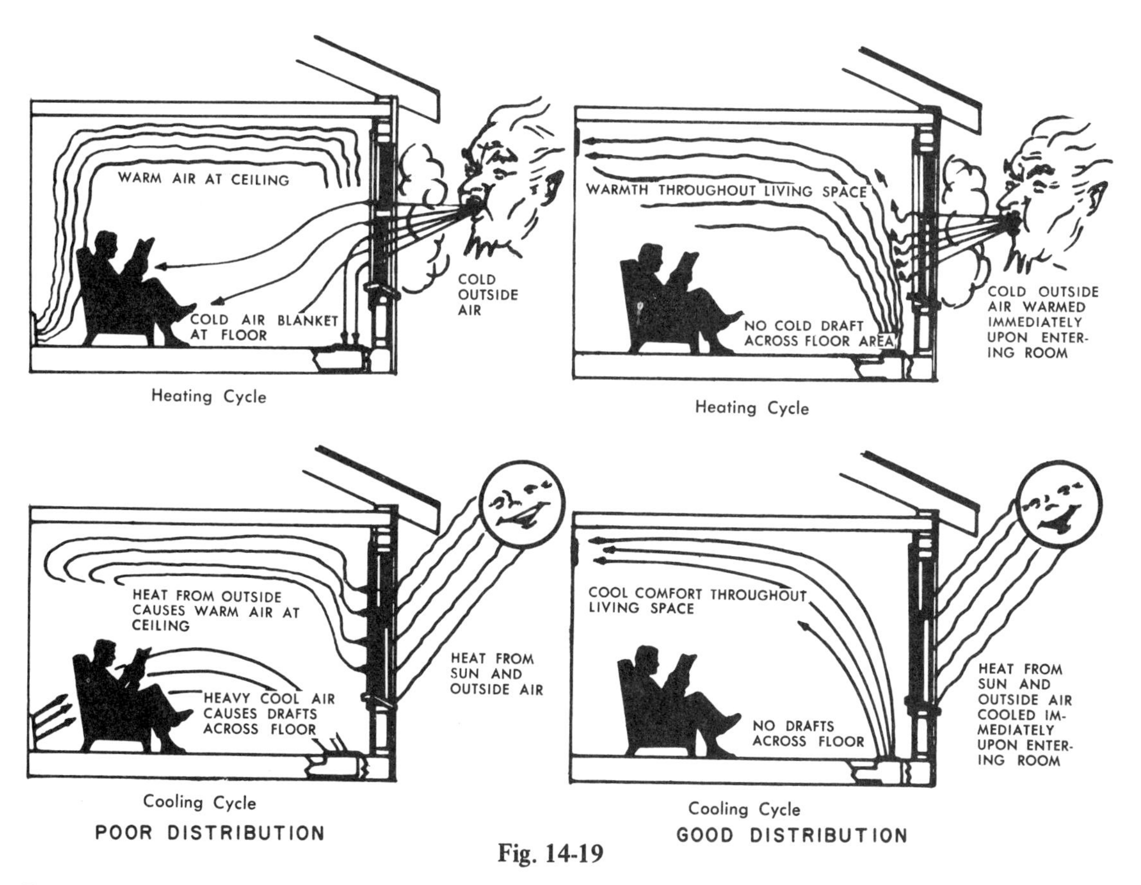

Fig. 14-19

Return Outlets

Floor or baseboard return outlet locations are used in residential installations. Figure 14-19 shows the differences between poor and good air distribution for these outlets. High wall or ceiling return outlet installation is also acceptable if the heat source (for a heating application) is at the floor level around the perimeter of the building. If the heat source is a ceiling or high wall outlet and the return is also in the ceiling or high wall, then there is a possibility of short-circuiting the heated air from the supply outlet directly to the return outlet. This situation results in poor air circulation and reduces the heating capacity of the air. If a perimeter heating or cooling system is used, only one centrally located return outlet is generally required.

Many residential systems for two-story structures have a return outlet in the stairway. Cold air flows down the stairs and passes into the return outlet. Cold floor drafts are reduced by this method in much the same manner as they are when the return outlet is located under a window or on an outside wall.

Commercial installations have satisfactory results from ceiling or high wall return or exhaust outlets. Floor outlets are not recommended for commercial use because they trap dirt.

For small multiroom buildings, the return outlet for each room may be in the form of a louver in the bottom part of the door or it may be an undercut door.

SELECTION OF OUTLETS

A number of important factors are to be considered in the selection of outlets.

- Review the building plan to obtain a general impression of the room size and shape.

- Determine the air quantity required for each room using the load estimate form or by proportioning the total ventilation or infiltration load according to the floor area of each room. Proportioning is required only if the load estimate does not show the load for each room.
- Select and locate the type of supply outlet register (high wall, baseboard, floor, or ceiling) that best suits the load and draft conditions in the room and the air quantity required for the room.
  1. The diffusion or spread pattern and the throw of the register must be considered in relation to the air velocity and the distance available in the room or space.
  2. The throw pattern of high wall outlet registers should extend three-fourths of the distance across the room.
  3. The diffusion pattern of low wall, baseboard, or floor outlet registers should be sufficient to blanket the window or wall.
  4. The diffusion pattern of a ceiling diffuser should be sufficient to blanket the area in which it is installed, without seriously overlapping the air pattern of neighboring diffusers (when more than one diffuser is required in the space).

  When selecting the outlet register and establishing the proper location, consult manufacturers' catalogs. Variations in design and air patterns can have a considerable effect on location.
- The supply outlet registers are to be sized from the manufacturers' catalogs according to the air quantity, discharge velocity, length of throw, and noise level. The sizing of registers will be covered in more detail in Unit 15.

Fig. 14-20 Air patterns from baseboard supply outlets

(Courtesy Lima Register Company)

- Drafts from obstacles are an important factor affecting outlet placement.
- Stratification can be caused by inadequate air velocity.
- Drafts can result from too great a temperature difference between the supply air and the room air or from the placement of the outlets too low on the wall.

SUMMARY

- Outlet locations are determined by the shape, size, usage, and load concentration of the room. Another consideration is whether the system is used in a cooling or heating application or both.
- Cold downdraft areas and ceiling stratification must be given special consideration in heating applications.
- Stratification of cold air at the floor and stagnant air areas must be given special consideration in cooling applications.
- Supply air outlet registers may be located at low wall, high wall, baseboard, floor, or ceiling positions.
- Return air outlet grilles may be located at the floor or ceiling levels. Return air outlet location is not as critical as the location of supply air outlets.
- When selecting the supply and return outlet registers and grilles:
  1. Study the building plans for the size and shape of rooms.
  2. Determine the air quantity for each room.
  3. Select and locate the outlets according to the requirements and conditions in the room or space. Use manufacturers' recommendations.
- The supply outlet registers and return grilles are sized according to the manufacturers' catalogs, taking into account the following conditions: air quantity, discharge velocity, length of throw, and noise level.

REVIEW

1. In a cooling operation, what happens when supply air is delivered with insufficient velocity from a low wall, floor, or baseboard outlet?
2. In a heating operation, how can a cold downdraft be avoided?
3. Describe two ways to prevent stratification.
4. Describe the three major factors which determine the number, location, and type of supply outlets.
5. Describe two ways in which doors may serve as return outlets.
6. What type of duct system is most satisfactory for slab construction? Where should the supply outlets be located for this duct system? Explain.

7. a. Using the grid provided, make a simple sketch to describe throw and drop.
   b. What is the throw if a room is 12 ft. x 12 ft.?

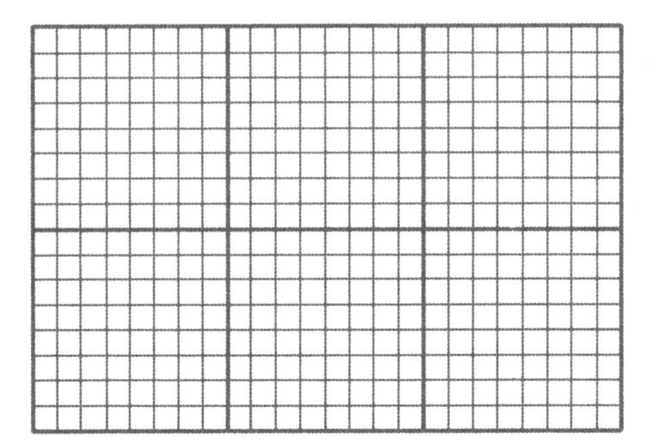

UNIT 15
DUCT SIZING

OBJECTIVES

After completing the study of this unit, the student will be able to

- state the reason(s) why it is necessary to coordinate the fan size, duct size, supply outlet register size and location, and the return outlet grille size and location.
- describe how a manometer and a Pitot tube are used to measure the total pressure and the static pressure in a duct.
- list the factors that result in the decrease in the pressure of the airflow due to friction loss.
- list and explain the steps recommended for the design of a duct system to meet specified conditions.
- describe the two methods used to size the supply and return duct system.
  1. static regain method
  2. equal friction method
- define the phrase "equivalent length of straight duct" and state how it is used in determining the total friction of a duct system.

A basic approach to designing a duct system is presented in this unit. In addition, several basic concepts are presented which are essential to a clear understanding of why it is necessary to coordinate the fan size, duct size, supply outlet register size and location, and the return outlet grille size and location.

PRESSURES IN DUCTS

Air exerts pressure – as shown when a balloon or an automobile tire is inflated. Air pressure is present in ductwork. Since the pressure does not depend on air movement, it is called static (stationary) pressure. Static pressure pushes against the walls of the duct. If the pressure is great enough, the duct walls will burst.

Since the pressures in a duct system are seldom very large, standard pressure measuring instruments cannot be used. In their place, sensitive gauges are installed which measure the effect of air pressure on a column of water. One type of gauge is called a U-tube manometer and is shown in figure 15-1, page 172.

One leg of the instrument is connected to the air duct and the other leg is open to the atmosphere. The greater pressure inside the duct forces the water to rise in the open leg. The amount of rise (head) is measured in inches. Thus, pressures in ducts are measured in terms of the *inches of water column* (wc) rather than in pounds per square inch (psi): 1 psi = 27.7-in. wc.

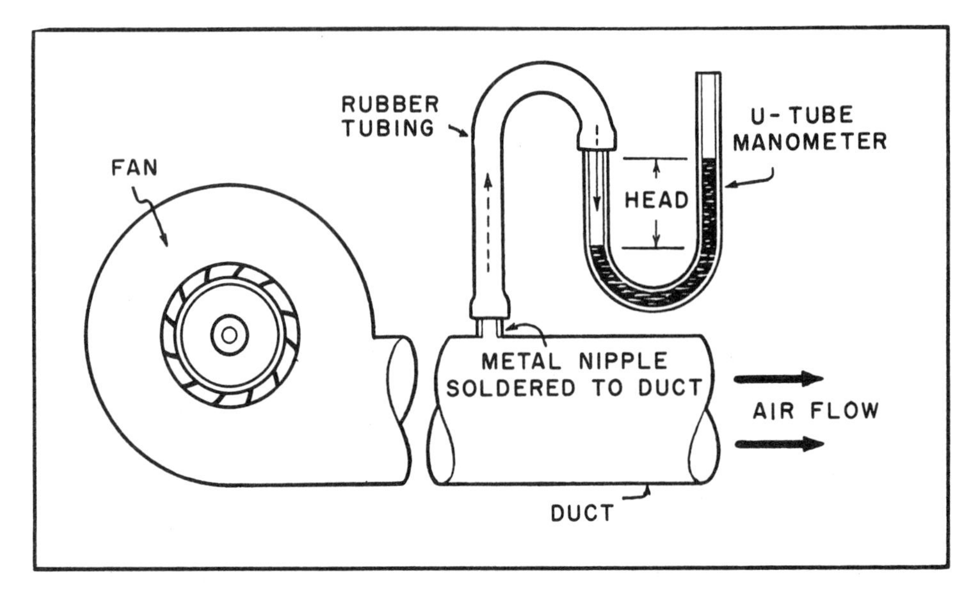

Fig. 15-1 The U-tube manometer measures pressure in the air duct

TOTAL PRESSURE

When air is moving it creates a greater pressure than when it is standing still. If the manometer leg is placed so that the flow of air through the duct is directly against the instrument, the pressure is greater than the static pressure alone. This additional pressure is called the velocity pressure. The total pressure consists of both the static pressure and the velocity pressure.

An instrument called the Pitot tube is used to measure the total pressure and the static pressure. The velocity pressure is obtained by subtracting the static pressure from the total pressure.

If the static pressure alone is to be measured, the total pressure tube of the Pitot tube is disconnected and the static pressure tube is connected to the cup of the manometer. If a measurement of the total pressure alone is desired, the static pressure tube is disconnected from the manometer. Thus, the total pressure is indicated on the gauge.

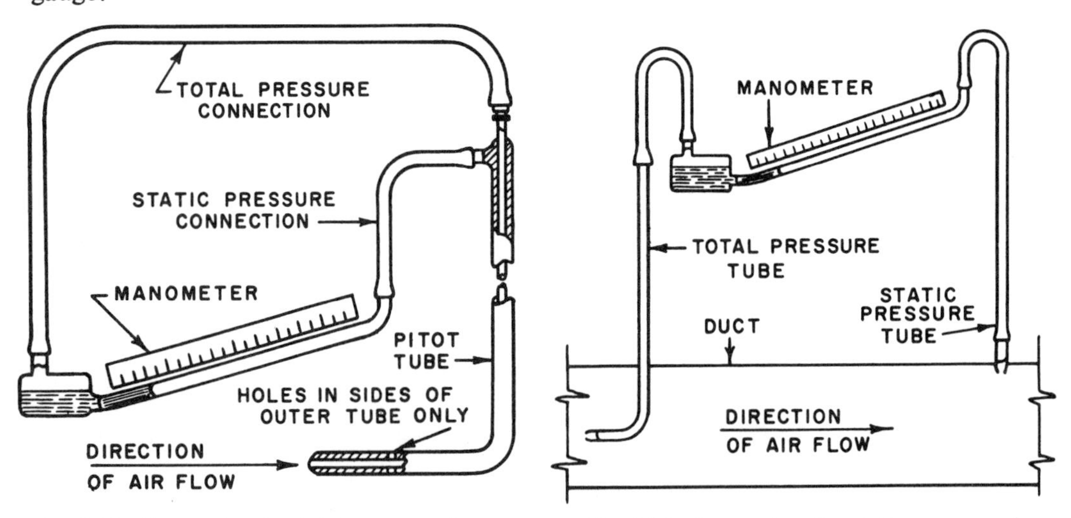

Fig. 15-2 Connection for total pressure alone

Fig. 15-3 Connection for static pressure alone

FRICTION LOSS

As the air flows through the duct, some pressure is lost due to the resulting friction of the air against the sides of the duct. This friction loss increases as the rate of airflow increases.

Pressure losses are always present in any duct system. The following conditions increase the pressure losses:

- High air velocities
- Small diameter ducts
- Large airflow
- Long lengths of ducts
- Changes in the direction of airflow
- Sudden contractions or expansions in the airstream

BUILDING A DUCT SYSTEM

A good air distribution system insures that rapid and satisfactory balancing takes place while providing the best possible comfort conditions in the room. The following steps are recommended for the design of a duct system to meet specified conditions.

Step 1. Select the type of system: loop perimeter, radial perimeter, extended plenum perimeter, overhead trunk, or overhead radial.

Step 2. Select the supply and return outlet locations according to the needs of the rooms or zones.

Step 3. Determine the air quantities (in cfm) to each supply outlet.

Step 4. Size the supply air outlets and return outlets.

Step 5. Sketch the final supply and return outlet locations on the building plans. Record the airflow (in cfm) to be handled and the size of the register at each outlet.

Step 6. Size the supply and return duct system.

In general, the design of a duct system for a particular structure follows the steps given. In some instances, however, the sequence may vary. The following sections investigate in detail each of the steps in the design of a duct system.

Selecting the Duct System

The building plans must be reviewed to help in selecting the type of system most suitable for the particular structure. The following items must be considered: type of foundation (slab, crawl space, or full basement), number of stories, area of the house, total heat losses, location of partitions, attic space available, and the placement of individual rooms.

When these considerations are combined with the design factors given in Units 12 and 13, the resulting information can be tabulated as in figure 15-4.

The need for zones should be established at this point. In general, residential heating loads above 100,000 Btu/hr. and residential cooling loads above 50,000 Btu/hr. require zoning. Commercial loads above 160,000 Btu/hr. (heating) and 75,000 Btu/hr. (cooling) also require zoning.

Selecting Outlet Locations

Unit 13 covered the factors affecting the location of outlets for the supply and return ducts. These locations should be roughed in on the building plans.

RECOMMENDED AIR DISTRIBUTION SYSTEMS FOR RESIDENCES
(Winter Design Temperature Below + 15F)

TYPE OF HOUSE				SUPPLY SYSTEM	SUPPLY OUTLETS	RETURN SYSTEM	REMARKS
SLAB	ONE STORY	A		Loop perimeter system.	Perimeter floor diffusers or perimeter baseboard outlets in all rooms except possibly kitchens and baths. High sidewall outlets recommended in rooms where floors are likely to be washed down often.	Central return system. All doors must be undercut.	Avoid running feeder ducts close together for long distances. Perimeter loop systems are limited to houses with heat losses of 100,000 Btu/hr or less. Radial systems to houses of less than 1200 square feet.
		B		Radial perimeter system.			
	TWO STORY	C	First Floor	Same as one story slab house.	Same as one story slab house.	Central return on first floor. Undercut doors.	Same as one story slab house.
		D	Second Floor	Risers in partition walls to supply second floor rooms.	Sidewall outlets in partition walls.	No returns necessary on second floor except in large houses, in isolated rooms, or in rooms with slightly lower floor levels. Doors to second floor rooms should be undercut.	Place central return opening on first floor as near as possible to foot of stairs.
				Trunk duct furred in along first floor ceiling. Branch ducts between joists to perimeter outlets for second floor rooms.	Perimeter floor or perimeter baseboard outlets except in bathrooms; high sidewall outlets recommended in bathrooms.		
				Vertical trunk duct through closet into attic. Branch ducts in attic to supply second floor rooms from overhead.	Ceiling diffusers.		
CRAWL SPACE	ONE STORY	E		Radial perimeter system.	Perimeter floor diffusers, perimeter baseboard outlets, or perimeter type low side-wall outlets in all rooms except kitchens and baths. High sidewall outlets in rooms where floors are likely to be washed down often.	Central return system. All doors must be undercut.	Radial system most economical for central unit location in houses under 1200 sq. ft.; extended plenum system for unit location at one end of house or for larger houses.
		F		Extended plenum system.			
	TWO STORY	G	First Floor	Same as one story crawl space house.	Same as one story crawl space house.	Same as two story slab house.	Same as one story crawl space house and two story slab house.
			Second Floor	Space as second floor of two story slab house.	Same as second floor of two story slab house.		
BASEMENT	ONE STORY	H		Extended plenum system.	Same as one story crawl space house.	Central return system. All doors must be undercut.	Same as one story crawl space house.
		I		Radial perimeter system.			
	TWO STORY	J	First Floor	Same as one story basement house.	Same as one story crawl space house.	Same as two story slab house.	Same as one story crawl space house and two story slab house.
		K	Second Floor	Risers in partitions to supply second floor rooms.	Sidewall outlets in partition walls.		
				Vertical trunk duct into attic. Branch ducts in attic to supply second floor rooms from overhead.	Ceiling diffusers.		

Fig. 15-4 Recommended air distribution systems for residences (Continued)
(Courtesy Carrier Corporation)

(Winter Design Temperature Above +15F)							
TYPE OF HOUSE				SUPPLY SYSTEM	SUPPLY OUTLETS	RETURN SYSTEM	REMARKS
SLAB	ONE STORY	L		Overhead radial or trunk and branch system in furred space below ceiling.	High sidewall outlets.	Central return system. All doors must be undercut.	
		M		Overhead radial or trunk and branch system in attic.	Ceiling diffusers.		
		N		Duct system imbedded in slab, same as Northern one story slab house.	Same as Northern one story slab house.		Same as Northern one story slab house.
	TWO STORY	O	First Floor	Same as one story slab system.	Same as those used with the various systems for one story slab house.	Central return system. Undercut doors no returns necessary on second floor except in large houses, isolated rooms, or in rooms with slightly lower floor levels. Doors to second floor rooms should be undercut.	Place central return on first floor as near as possible to foot of stairs.
		P	Second Floor	Risers in partition walls to supply second floor rooms.	Sidewall outlets in partition walls.		
				Vertical trunk duct into attic. Branch ducts in attic to supply second floor rooms from overhead.	Ceiling or high sidewall outlets.		
				Trunk duct furred in along first floor ceiling. Risers in partition walls to supply second floor rooms.	Sidewall outlets in partition walls.		
CRAWL SPACE	ONE STORY	Q		Overhead radial or trunk and branch system in furred space below ceiling.	High sidewall outlets.	1. Central return system. Undercut doors. 2. Use crawl space as a return air plenum, where codes permit.	
		R		Overhead radial or trunk and branch system in attic.	Ceiling diffusers.		
		S		Underfloor duct system in crawl space. Same as Northern one story crawl space house.	Same as Northern one story crawl space house.		Same as Northern one story crawl space house.
	TWO STORY	T	First Floor	Same as one story crawl space house.	Same as those used with the various systems for one story crawl space house.	1. Same as two story, slab house. 2. Crawl space as a return air plenum, where codes permit.	
		U	Second Floor	Same as second floor of two story slab house.	Same as second floor of two story slab house.		
BASEMENT	ONE STORY	V		Perimeter system same as Northern one story basement house.	Same as Northern one story, crawl space house.	1. Central return system. Undercut doors. 2. Basement as a return air plenum, where codes permit.	Same as Northern one story crawl space house.
		W		Extended plenum system with plenum and round branch ducts to each riser or individual round ducts run parallel as much as possible to each riser. Risers in partitions.	High sidewall outlets.		
	TWO STORY	X	First Floor	Same as one story basement house.	Same as one story basement house.	1. Same as two story slab house. 2. Basement as a return air plenum, where codes permit.	Same as Northern one story crawl space house.
		Y	Second Floor	Same as second floor of two story slab house.	Same as second floor of two story slab house.		

Fig. 15-4 Recommended air distribution systems for residences

DETERMINING THE AIR QUANTITIES

Cooling or heating load estimate forms usually provide values for determining the air quantities (in cfm) to be delivered to each room. If this information is not given, however, one of the following three methods can be used.

Method 1.

$$\text{cfm} = \frac{\text{Btu/hr./room (sensible load)}}{1.08 \text{ x temperature difference between the supply air and the room air}}$$

For example, assume that the sensible load for a room is determined to be 7,000 Btu/hr. and the temperature difference between the supply air to the room and the room air temperature is 50°F. The cfm value is determined as follows:

$$\text{cfm} = \frac{7{,}000}{1.08 \text{ x } 50} = 129 \text{ cfm}$$

Method 1 is simplified by the use of the chart shown in figure 15-5, page 176. This chart enables the designer to determine quickly the air quantity for many combinations of sensible loads and temperature differences.

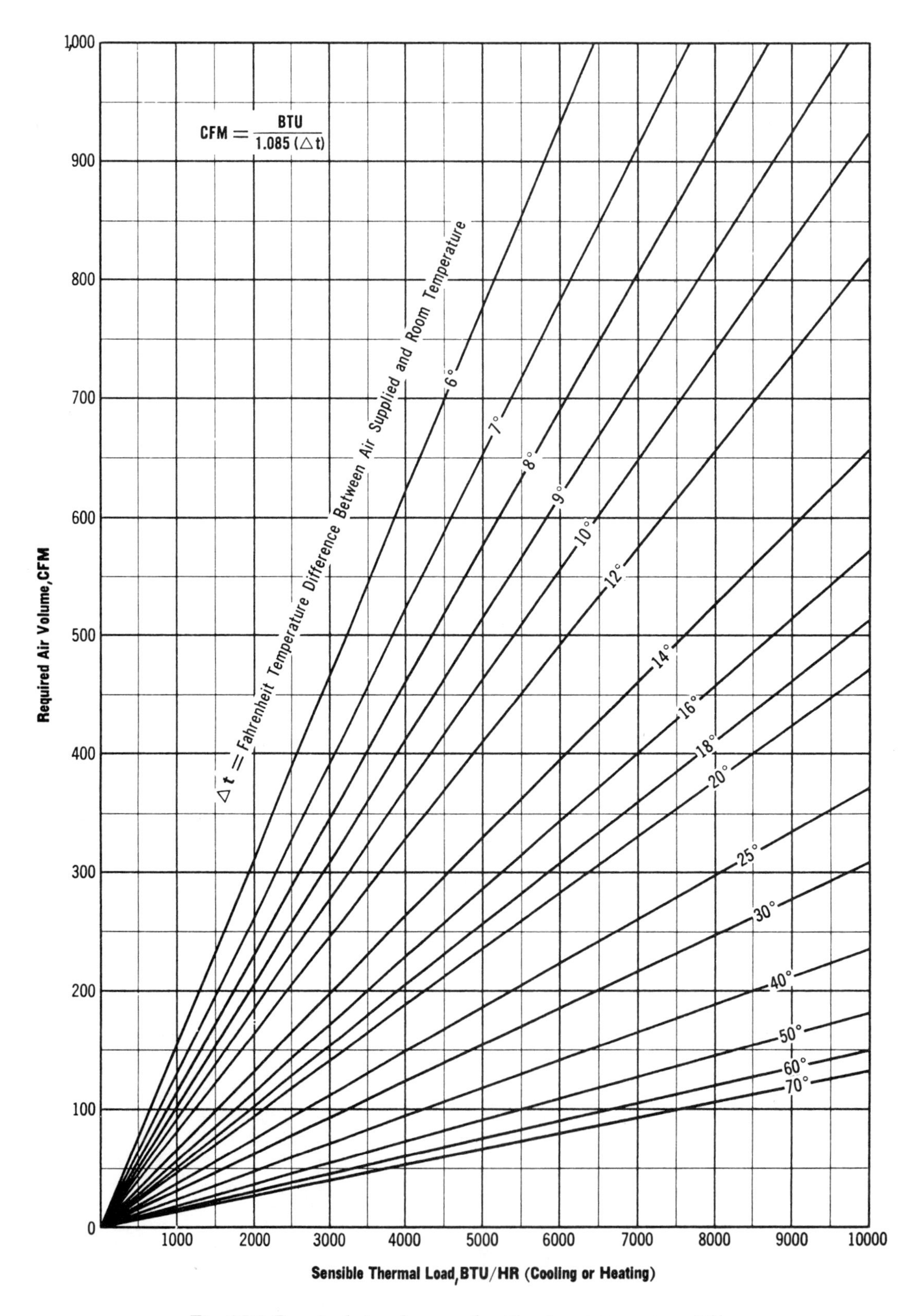

Fig. 15-5 Required air volume vs. heat load vs. temperature difference
(Courtesy Reynolds Metals Company)

Refer to figure 15-5 and note where the temperature difference line intersects with the sensible heat load line. The air volume (in cfm) that must be supplied to the room is read at this intersection point. The cfm value is then recorded on the layout next to the discharge outlet. If more than one outlet is provided for a room, the cfm value determined for the room is divided between the outlets.

To demonstrate the way in which the chart is used, locate the given conditions in figure 15-5 and determine the required air volume. For a sensible heat load of 7,000 Btu/hr. and a temperature difference of 50°F, the air volume is 125 cfm.

For a sensible heat load of 64,000 Btu/hr. and a temperature difference of 19°F, the air volume is 3,100 cfm. (This reading is possible because both scales of the chart can be multiplied by 10 or 100, as necessary.)

Method 2. One estimate form gives the following equations to be used to determine the air quantity.

$$\text{Room cfm} = \frac{\text{Btu/room (sensible load)}}{20}$$

$$\text{Total cfm} = \frac{\text{Total Btu}}{30}$$

The total Btu includes the heat gain in the ducts, the fan heat, and any additional factors.

Method 3. This method uses the previously established procedures for determining the quantities of ventilation or infiltration air. If more than one outlet is required to satisfy the comfort conditions of a room, the required air quantity for that room is divided between the number of outlets used.

Sizing the Supply Air Outlet Registers and Return Grilles

The best method of sizing the registers and grilles is to follow the manufacturers' recommendations. Tables published by various manufacturers show the register size according to the cfm requirements and usually include the throw and spread for each register size. Typical values are shown in Tables 15-1 through 15-4.

	Single Deflection Multi-Shutter Volume Damper		Double Deflection Opposed Blade Volume Damper	
CFM	Size	Throw-Feet	Size	Throw-Feet
60	10 x 4	10	8 x 4	5-10
80	10 x 6	12	8 x 4	7-12
100	14 x 6	14	12 x 4	7-12
120	14 x 6	16	12 x 4	8-14
140	16 x 6	16	10 x 6	8-16
160	16 x 6	18	10 x 6	10-17
180	20 x 6	14	14 x 6	10-17
200	20 x 6	16	14 x 6	10-19
220	20 x 6	18	14 x 6	11-20
240	20 x 6	20	20 x 6	11-19
260	24 x 6	20	20 x 6	11-20
280	30 x 6	18	20 x 6	12-22
300	30 x 6	20	20 x 6	13-23

NOTES:
1. Sizes are based on the recommended velocities for residential applications.
2. The sizes shown for the single deflection, multi-shutter volume damper are minimum sizes.
3. Double-deflection, opposed blade volume damper sizes are on registers set to the maximum (55°) deflections. For a straight flow, select the next smaller size and the throw will be doubled.
4. The throw should equal 3/4 of the distance to the wall opposite the register.
5. For further information, reference should be made to the manufacturer's catalogs.

Table 15-1 High side wall registers for residential use
(Courtesy Carrier Corporation)

	Low Side Wall		Floor		Continuous Baseboard	
CFM	Size	Spread	Size	Spread	Length	Spread
60	10 x 6	10 ft.	2 1/4 x 10	7 ft.	4 ft.	12 ft.
80	10 x 6	11 ft.	2 1/4 x 12	11 ft.	4 ft.	14 ft.
100	10 x 6	13 ft.	2 1/4 x 14	11 ft.	6 ft.	16.5 ft.
120	12 x 6	13 ft.	4 x 10	11 ft.	6 ft.	18 ft.
140	14 x 6	13 ft.	4 x 10	13 ft.	8 ft.	19.5 ft.
160	14 x 6	14 ft.	4 x 12	13 ft.	10 ft.	22 ft.

NOTE: The spread figure represents the horizontal length that will be effectively blanketed with air.

Table 15-2 Perimeter outlets
(Courtesy Carrier Corporation)

CFM	Size
100	6″
120	8″
140	8″
160	8″
180	8″
200	10″
220	10″
240	10″

NOTES:
1. Diffusers should be located in the center of the ceiling of each room.
2. In cases where the room is rectangular and the long side is more than one and one-half times the length of the shorter side, two diffusers should be used.

Table 15-3 Ceiling outlets
(Courtesy Carrier Corporation)

CFM	Sq. In. Free Area	Side Wall Return Grilles	Floor Grilles		
60- 140	40	10 x 6	4 x 14		
140- 170	48	12 x 6	4 x 18	6 x 10	
170- 190	55	10 x 8	4 x 18	6 x 12	
190- 235	67	12 x 8		6 x 14	
235- 260	74	18 x 6		6 x 16	8 x 14
260- 370	106	12 x 12			8 x 20
370- 560	162	18 x 12			8 x 30
560- 760	218	24 x 12	10 x 30	12 x 24	
760- 870	252	18 x 18		12 x 30	
870- 960	276	30 x 12		12 x 30	
960-1,170	340	24 x 18			14 x 30
1,170-1,470	423	30 x 18	18 x 30		
1,470-1,580	455	24 x 24		20 x 30	
1,580-1,770	510	36 x 18			22 x 30
1,770-1,990	572	30 x 24	24 x 30		
1,990-2,400	690	36 x 24	24 x 36		
2,400-3,020	870	36 x 30		30 x 36	

Table 15-4 Return air grilles for residential use
(Courtesy Carrier Corporation)

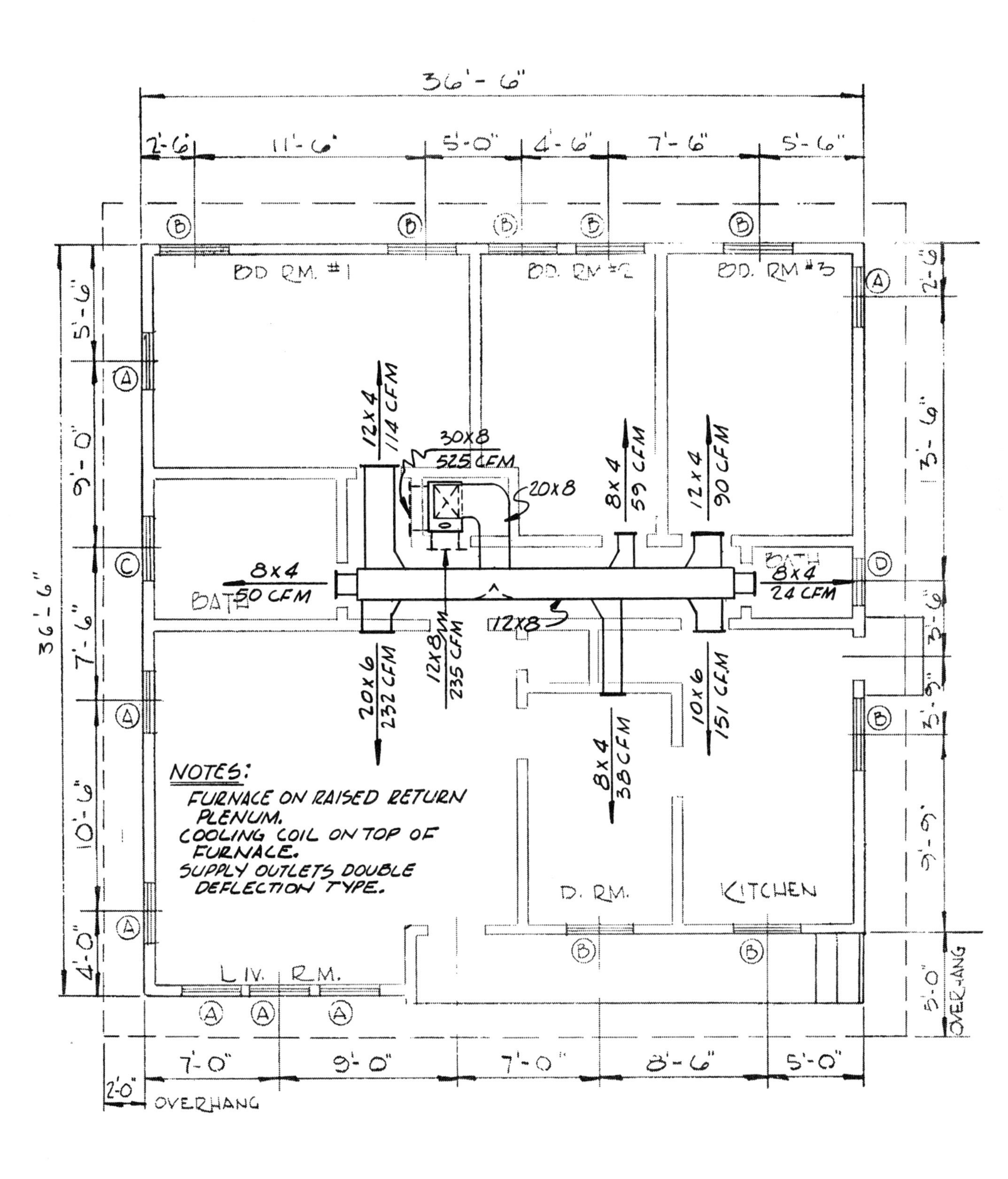

Fig. 15-6 Outlet locations sketched on building plan
(Courtesy Carrier Corporation)

Locating Outlets on the Building Plan

Figure 15-6, page 179, shows the locations sketched on the building plan for the final supply and return outlets. In addition, the cfm to be handled and size of the register at each outlet are also recorded.

Sizing the Supply and Return Duct System

For residential applications, the information required to size the supply and return ducts can be obtained from tables supplied by the manufacturer. Typical examples of manufacturer supplied data are shown in Tables 15-5 and 15-6. Ducts for commercial applications are sized by the static regain method or by the equal friction method. These methods can also be used for residential applications if suitable sizing tables are not available.

	Supply Ducts			Return Ducts		
CFM	Round	Rectangular	Riser	Round	Rectangular	Riser
50	5"	8 x 6	10 x 3 1/4	6"	8 x 6	10 x 3 1/4
75	6"	8 x 6	10 x 3 1/4	7"	8 x 6	12 x 3 1/4
100	6"	8 x 6	10 x 3 1/4	8"	8 x 6	14 x 3 1/4
125	7"	8 x 6	12 x 3 1/4	8"	8 x 8	
150	7"	8 x 6	14 x 3 1/4	9"	8 x 8	
175	8"	8 x 6		9"	10 x 8	
200	8"	8 x 8		10"	8 x 10 10 x 8	
250	9"	8 x 8		12"	10 x 10 12 x 8	
300	10"	8 x 10 10 x 8		12"	10 x 10 14 x 8	
350	10"	8 x 10 10 x 8		12"	12 x 10 16 x 8	
400	10"	10 x 10 12 x 8		12"	14 x 10 16 x 8	
500	12"	10 x 10 14 x 8		14"	16 x 10 20 x 8	
600	12"	12 x 10 14 x 8		16"	18 x 10 24 x 8	
700	12"	14 x 10 16 x 8		16"	20 x 10 26 x 8	
800	14"	16 x 10 20 x 8		16"	22 x 10 30 x 8	
900	14"	16 x 10 22 x 8		18"	24 x 10 32 x 8	
1,000	16"	18 x 10 24 x 8		18"	26 x 10 22 x 12	
1,200	16"	22 x 10 28 x 8		20"	30 x 10 24 x 12	
1,400	18"	26 x 10 22 x 12		20"	36 x 10 28 x 12	
1,600	18"	28 x 10 24 x 12		22"	40 x 10 32 x 12	
1,800	20"	32 x 10 24 x 12		22"	28 x 14 36 x 12	
2,000	20"	34 x 10 26 x 12		24"	32 x 14 38 x 12	
2,500	22"	40 x 10 32 x 12		26"	38 x 14 46 x 12	
3,000	24"	32 x 14 38 x 12		26"	44 x 14 38 x 16	
3,500	24"	36 x 14 44 x 12		28"	50 x 14 42 x 16	
4,000	26"	40 x 14 48 x 12		30"	56 x 14 48 x 16	

Table 15-5 Minimum supply and return duct sizes for residential application

CFM	Supply or Return Sizes				High Side Wall Registers		Ceiling Diffuser Inches	Return Grille
	Round	Rectangular			Size	Throw Feet		
50	5″	8 x 6			8 x 4	5- 8	6″	10 x 6
75	5″	8 x 6			8 x 4	6-12	6″	10 x 6
100	6″	8 x 6			8 x 4	9-16	6″	10 x 6
125	6″	8 x 6			10 x 6	8-14	6″	10 x 6
150	7″	8 x 6			10 x 6	9-16	6″	10 x 6
175	7″	8 x 6			10 x 6	11-20	8″	10 x 6
200	7″	8 x 6			10 x 6	12-22	8″	10 x 6
250	8″	6 x 10	8 x 8	6 x 12	14 x 6	13-23	8″	10 x 8
300	9″	6 x 10	8 x 8	6 x 12	14 x 6	15-28	10″	12 x 8
350	9″	8 x 10	8 x 8	6 x 12	14 x 6	18-32	10″	18 x 6
400	9″	8 x 10	10 x 8	6 x 12	20 x 6	17-30	10″	12 x 12
500	10″	10 x 10	12 x 8	8 x 12	20 x 6	21-38	12″	12 x 12
600	12″	10 x 10	12 x 8	8 x 12	20 x 8	22-40	12″	18 x 12
700	12″	12 x 10	14 x 8	10 x 12	20 x 8	26-45	15″	18 x 12
800	12″	12 x 10	16 x 8	10 x 12	30 x 8	24-43	15″	24 x 12
900	14″	14 x 10	18 x 8	12 x 12	30 x 8	27-47	15″	24 x 12
1,000	14″	14 x 10	18 x 8	12 x 12	30 x 8	30-52	15″	24 x 12
1,200	14″	16 x 10	22 x 8	14 x 12				18 x 18
1,400	16″	18 x 10	24 x 8	16 x 12				24 x 18
1,600	16″	20 x 10	28 x 8	18 x 12				24 x 18
1,800	16″	24 x 10	30 x 8	20 x 12				30 x 18
2,000	18″	26 x 10	32 x 8	20 x 12				30 x 18
2,500	18″	30 x 10	20 x 14	24 x 12				30 x 24
3,000	20″	36 x 10	24 x 14	28 x 12				36 x 24
3,500	22″	40 x 10	28 x 14	32 x 12				36 x 30
4,000	24″	26 x 16	32 x 14	38 x 12				36 x 30

NOTES:
1. Shorter throws indicated for the high sidewall registers refer to 55° deflection settings, while the longer throws indicated are for straight deflection.
2. The figures for high sidewall registers refer to those of the double deflectional type.

Table 15-6 Minimum air distribution sizes for commercial application
(Courtesy Carrier Corporation)

The Static Regain Method. The static regain method requires a reduction in the velocity at the beginning of each major duct section. This reduction tends to equalize the static pressure at the outlets. As a result, there is a regain of enough static pressure to offset the friction loss in the ducts. The static regain method is somewhat complicated, but many designers feel that the best and most economical air system results from the use of this method. It is probably most suitable for systems having several long duct runs with many outlets. The static regain method is thoroughly explained in the ASHRAE Guide.

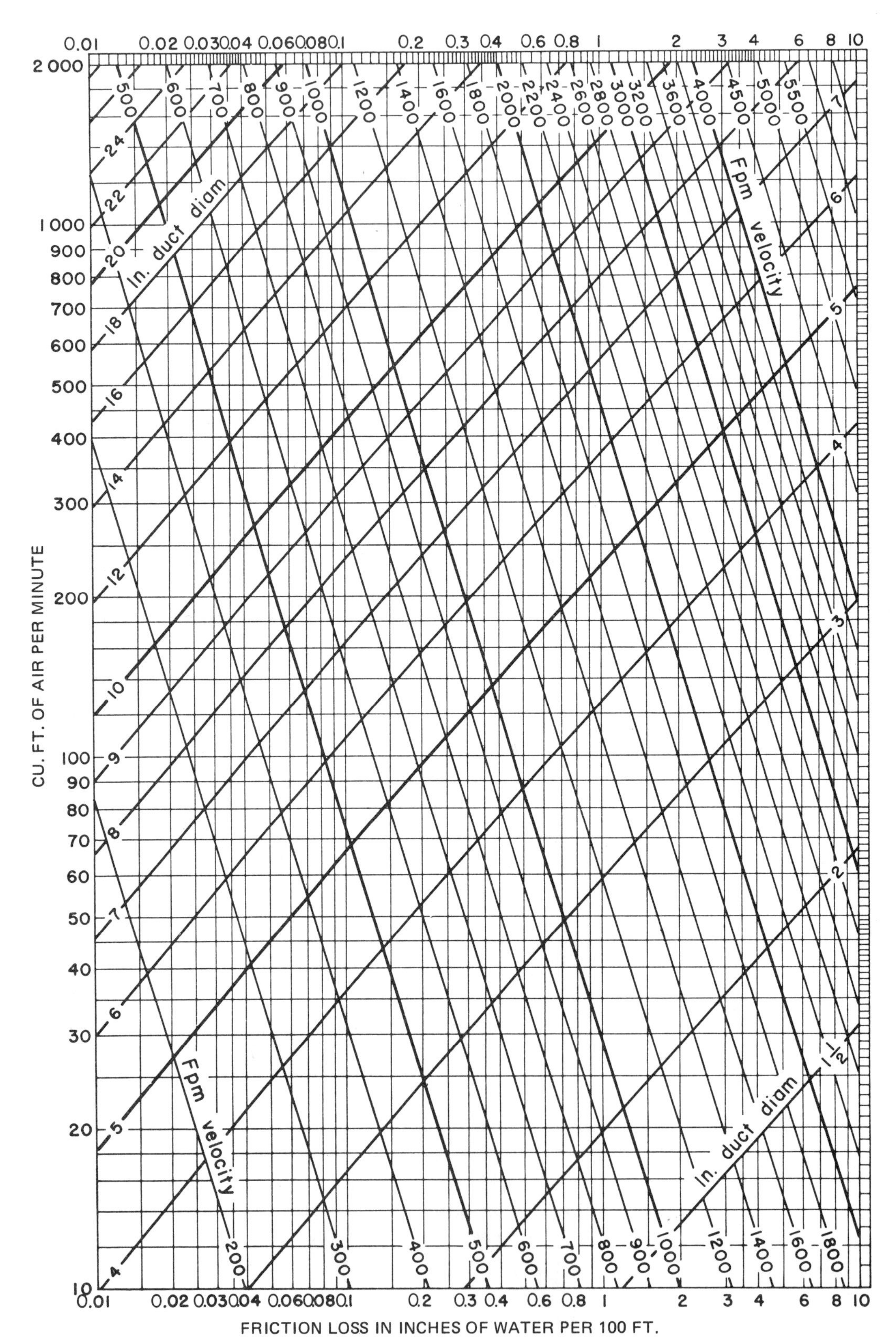

Fig. 15-7 Friction loss chart

(From ASHRAE Guide, 1965, by permission)

Equal Friction Method. The equal friction method of sizing ducts is used by many air-conditioning contractors because it is quick and fairly simple. This method establishes an equal friction drop for each 100 feet of duct throughout the duct system. The equal friction method follows a definite sequence and requires the use of a standard friction loss chart, figure 15-7.

There are four sets of values on the chart. Each set of values is represented by a group of horizontal, vertical, or diagonal lines.

The horizontal lines represent the amount of air moved in cubic feet per minute (cfm). The scale for these lines is located along the left side of the chart. In figure 15-7, the cfm scale begins at a value of 10 and ends at a value of 100,000 cfm in figure 15-8. Although this information is given here in the form of two charts to make it easier to use, the two charts are sometimes combined into a single chart.

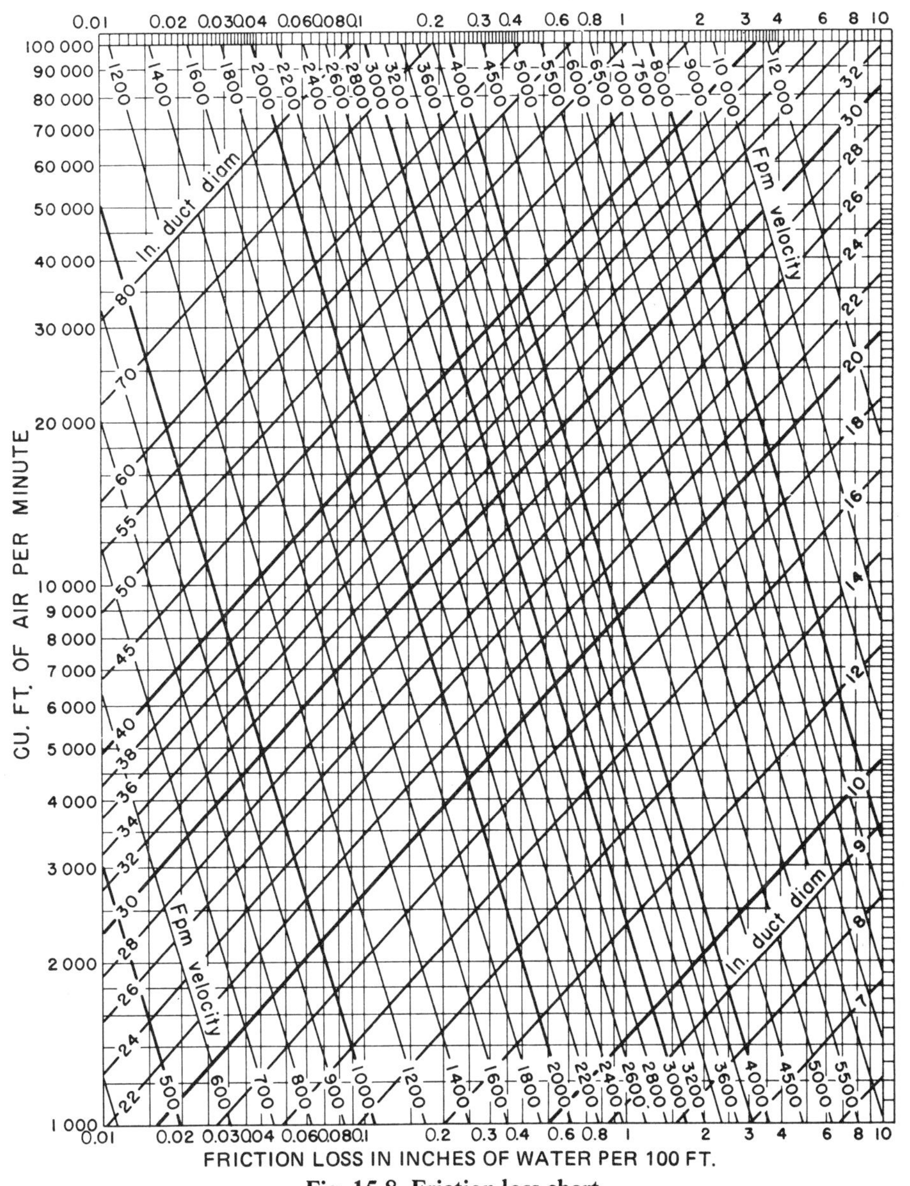

Fig. 15-8 Friction loss chart

(From ASHRAE Guide, 1965, by permission)

The vertical lines represent the friction loss in inches of water per 100 feet. The friction loss scale runs from left to right across the bottom of the chart.

The diagonal lines extending from the upper left of the chart to the lower right represent the air velocity in feet per minute. The air velocity is the actual speed of the air as it flows through the duct system. For example, to find the air velocity at the conditions of 70 cfm of air flowing and 0.1 in. of water friction loss, locate the intersection of these values with the air velocity lines and note a value of 500 fpm.

The diagonal lines extending from the lower left to the upper right of the chart represent round duct sizes. Problem 1 is a typical application of the friction loss chart using two given conditions to find the remaining design values.

PROBLEM 1 RESIDENTIAL APPLICATION

Given

Air required at bedroom outlet	70 cfm
Air velocity permitted	500 fpm

Find

a. Supply branch duct size to the bedroom outlet
b. Friction loss

Solution

1. Find 70 cfm on the cfm scale at the left of the chart.
2. Move horizontally to the right to the 500 fpm air velocity line.

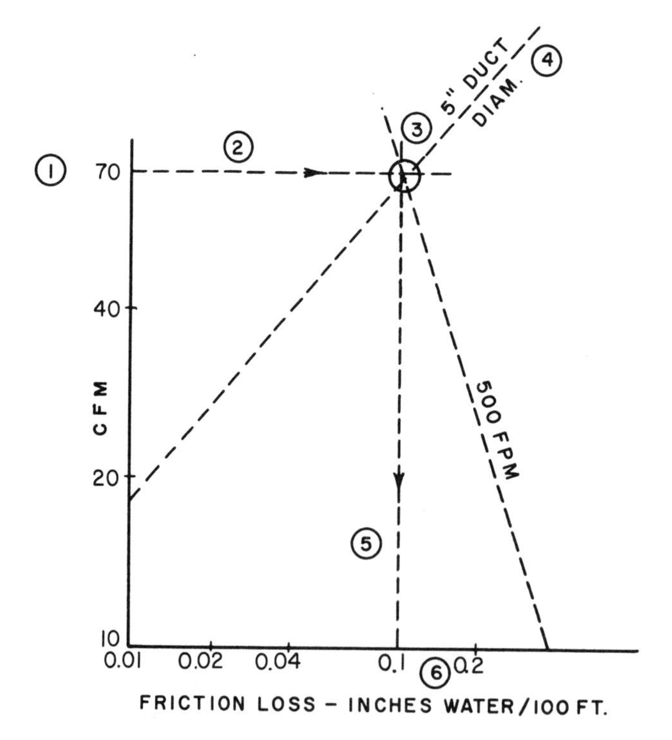

Fig. 15-9

3. At the intersection of 70 cfm and 500 fpm, locate the duct diameter line.
4. A 5-in. duct diameter is indicated.
5. Move downward from the point of intersection to the friction loss scale.
6. A 0.1-in. water friction loss per 100 feet is indicated.

If the duct to be used is rectangular, then the duct size can be determined from the chart shown in figure 15-10. For the conditions in Problem 1, find the 5-in. diam. round duct line in figure 15-10. A number of combinations of rectangular duct sizes can be read along this line, such as: 3 1/2 in. x 6 in., 5 in. x 4 in., 4 in. x 5 in., or 6 in. x 3 1/2 in.

The 500-fpm air velocity given in Problem 1 is obtained from the values listed in figure 15-11, page 186, for the maximum air velocities through various parts of the

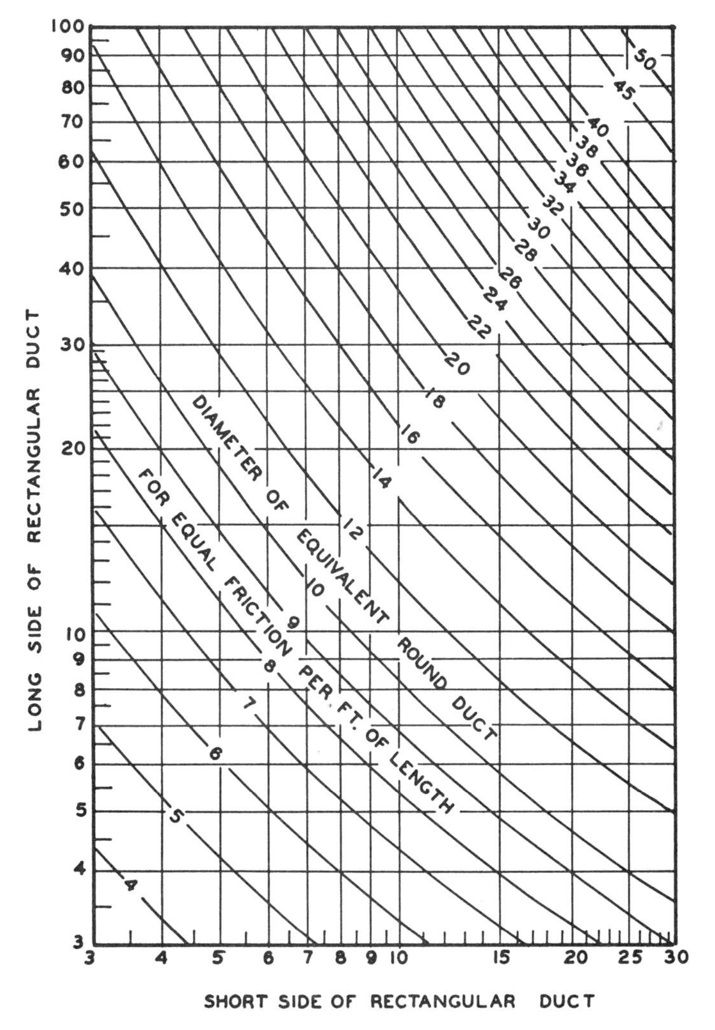

Fig. 15-10

Structure	Supply Outlet	Return Openings	Main Supply	Branch Supply	Main Return	Branch Return
Residential	500-750	500	1,000	600	800	600
Apartments, Hotel Bedrooms, Hospital Bedrooms	500-750	500	1,200	800	1,000	800
Private Offices, Churches, Libraries, Schools	500-1,000	600	1,500	1,200	1,200	1,000
General Offices, Deluxe Restaurants, Deluxe Stores, Banks	1,200-1,500	700	1,700	1,600	1,500	1,200
Average Stores, Cafeterias	1,500	800	2,000	1,600	1,500	1,200

Fig. 15-11 Maximum air velocities through gross area, ft./min.

duct system. The maximum velocities are established as an upper limit. Below this limit, it is possible to maintain acceptable sound levels (noise control) for the various velocities, according to the types of applications listed in the table. The 500-fpm value used for the branch supply duct in Problem 1 is satisfactory since it is below the maximum velocity of 600 fpm shown in the table.

PROBLEM 2 COMMERCIAL APPLICATION (General Office)

Given

Air required in main supply duct	4,000 cfm
Air velocity permitted	1,600 fpm (This value is less than the maximum velocity recommended for a general office)
Length of main supply duct	100 ft.

Find

Main supply duct size
Friction loss

Solution

1. Find 4,000 cfm on the cfm scale at the left of the friction loss chart.
2. Move horizontally to the right to the 1,600 fpm line.
3. At the intersection of 4,000 cfm and 1,600 fpm locate the nearest duct diameter.
4. The indicated supply duct diameter is 22 in.
5. Move downward from the point of intersection to the friction loss scale.
6. The water friction loss reading is 0.15 in. per 100 feet of duct.

Equivalent Length of Straight Duct

Duct fittings such as elbows, boots, and transformations must be considered when determining the total friction of a duct system. These fittings are estimated in terms of the equivalent number of feet of straight duct added to the system. For

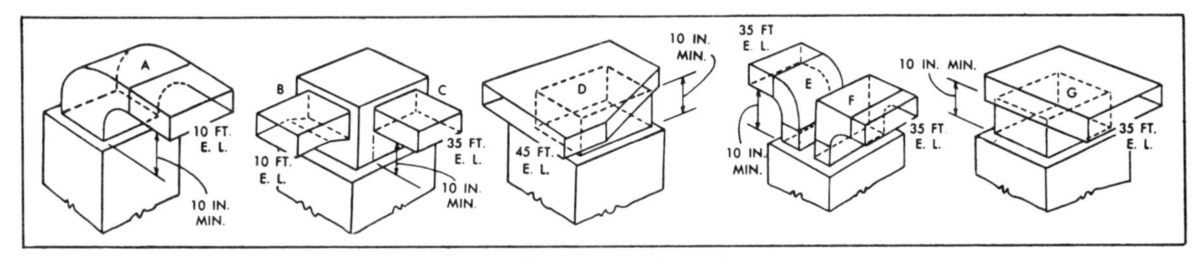

(A) WARM AIR AND RETURN AIR BONNET OR PLENUM
EQUIVALENT LENGTHS BASED ON 8" DEPTH OF DUCT

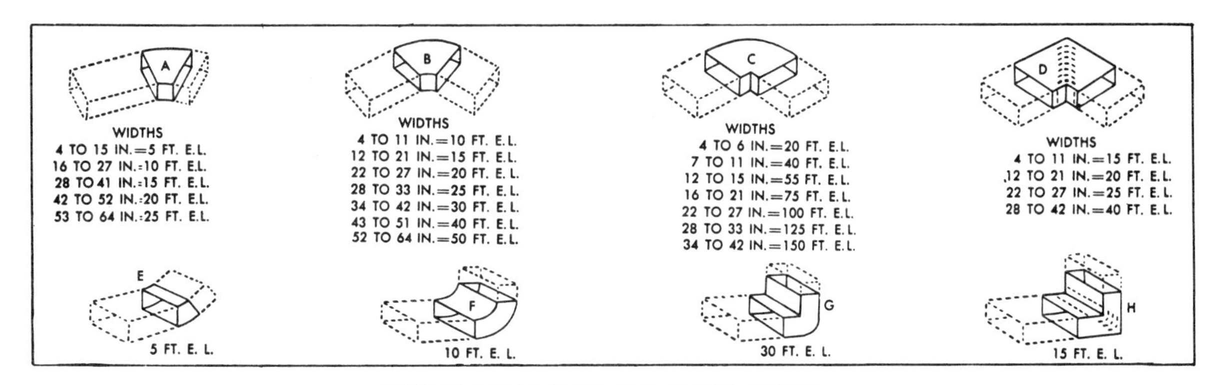

(B) ANGLES AND ELBOWS FOR TRUNK DUCTS
EQUIVALENT LENGTHS BASED ON 8" DEPTH OF DUCT

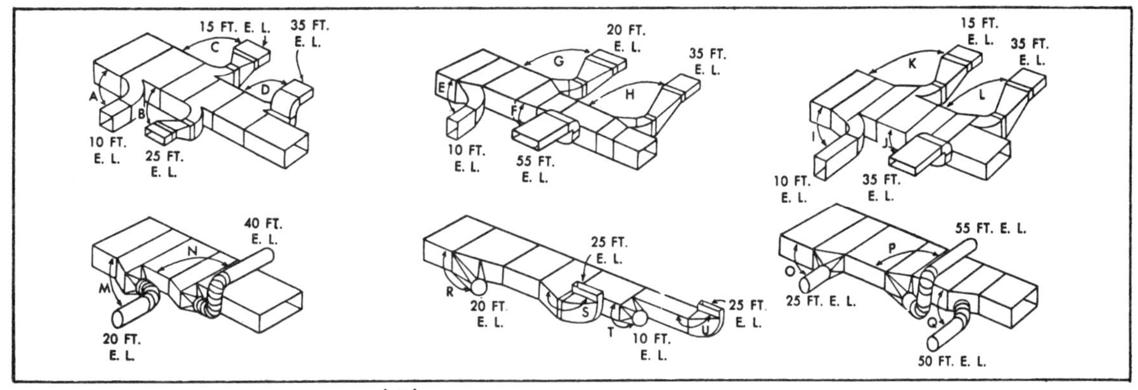

(C) TRUNK DUCT TAKEOFFS

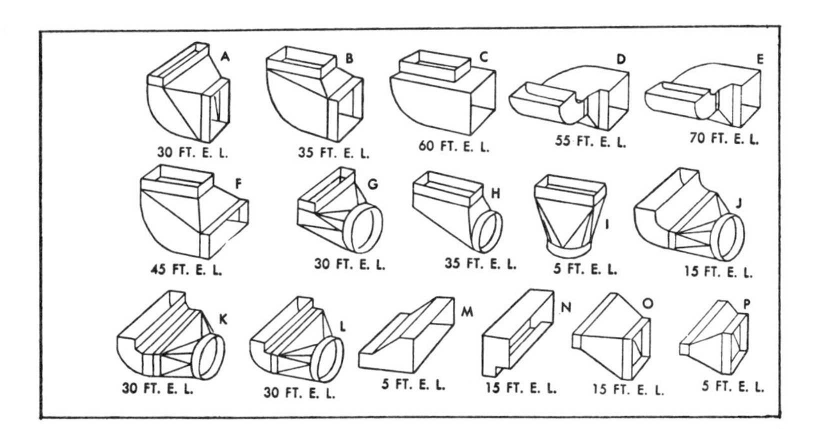

(D) BOOT FITTINGS
FROM BRANCH TO STACK

Fig. 15-12 Equivalent lengths for various types of fittings (Continued)

(E) STACK ANGLES, ELBOWS AND COMBINATIONS

(F) RETURN AIR COMBINATIONS

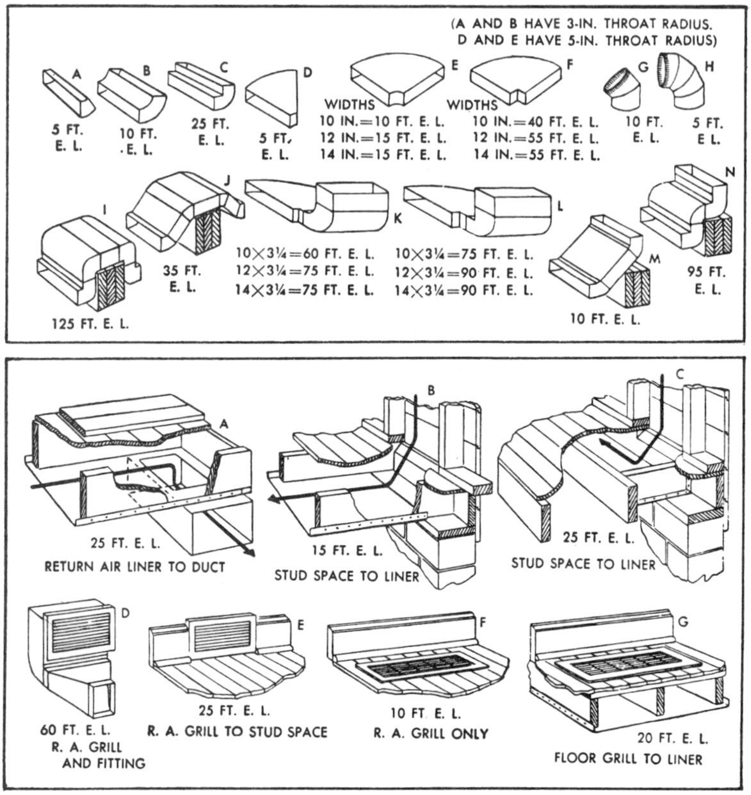

Fig. 15-12 Equivalent lengths for various types of fittings
(Courtesy National Environmental Systems Contractors Association)

example, a 6-inch diameter, three-piece 90° elbow has approximately the same friction loss as a 6-inch diameter duct that is 12 feet long.

Equivalent lengths are also determined for outlet registers and return grilles. For example, a supply register at the end of a 6-in. diameter branch duct run may have the same friction loss as 48 feet of straight 6-in. diameter duct. The equivalent lengths for registers usually vary according to the register design and construction. Tables of equivalent lengths are supplied by register manufacturers.

To determine the total friction loss of a system, all fittings, registers, and grilles must be considered in terms of their equivalent lengths of straight duct. The equivalent amount of straight duct is then added to the actual amount of straight duct in the system. The actual straight duct length in a system may be only 100 feet in the longest run. But, when the equivalent length of straight duct due to fittings and registers is added, the total length of duct may be 200 feet or more.

Therefore, if the friction loss is determined to be 0.1 inch of water per 100 feet, the total friction loss for this run of duct is:

(0.1 inch) (100 feet of actual straight duct + 100 feet of straight duct added because of fittings and registers) = 0.2 inch

Determining the Total Resistance in the System

As stated previously, the total resistance of a system results from the friction loss per 100 feet of duct and the friction loss in elbows, registers, and other fittings.

For example, if the friction loss is 0.15 inches of water per 100 feet and the duct run with the most resistance has a total duct length of 200 feet (including the equivalent lengths of fittings), the total system resistance is 0.15 in. for the first 100 feet plus 0.15 in. for the second 100 feet for a total of 0.30 in. for 200 feet of duct.

The total resistance is one factor considered in selecting the proper fan size. The fan must be able to supply enough air to counteract the friction in the duct system to maintain comfort conditions.

SUMMARY

- To size ducts it is necessary to complete each of the following steps:
  1. Review the building plans
  2. Select the type of system to be used
  3. Select the supply and return outlet locations
  4. Determine the air quantities for each outlet
  5. Supply the air outlet registers and return grilles
  6. Sketch the final supply and return outlet locations on the plans
  7. Size the supply and return duct system
- Ducts can be sized using either the equal friction or the static regain method.
- Duct friction is usually measured in inches of water per 100 feet of straight duct.
- For duct fittings, the friction loss is determined by considering the fittings in terms of an equivalent length of straight duct.
- The total duct friction must be determined before the proper fan size can be selected.

The application of this unit will be made in the assignments which follow Units 16 and 17.

SECTION 5
APPLIED LOAD ESTIMATING

UNIT 16
RESIDENTIAL AIR CONDITIONING

OBJECTIVES

After completing the study of this unit, the student will be able to

- complete a survey checklist.
- estimate the cooling load for a residence.
- estimate the heating load for a residence.
- based on the load estimates, select the proper equipment combination for both the heating and cooling requirements.
- size the duct system for a residence.

This unit presents typical problems in residential air conditioning, cooling, and heating load estimating, and duct system design.

Regardless of the type of survey and estimating form used, or the sequence and approach followed to make the estimate, the final estimate must present values that represent the Btu load for a structure. These values are used to determine the equipment size and to establish a satisfactory air distribution system.

The back cover envelope of this text provides the following building plans which are to be used in preparing the estimates in this unit.

- Floor plan of a seven-room residence
- Basement plan
- Elevations

SURVEY

A systematic check of the items to be considered in a survey is one of the most important steps in load estimating. To lessen the possibility that survey items are omitted, air-conditioning equipment manufacturers provide checklists. A typical example of a systematic survey checklist is included. The Carrier Air-Conditioning Company forms used in this unit are typical of the forms supplied by all equipment manufacturers. The information recorded on the checklist is used as the basis for the load estimate examples presented in this unit.

A checklist can be used for either an existing residence or for a structure that is to be built and for which only the floor plans and elevation drawings exist.

If the survey is made for an existing house, the checklist information is obtained directly by examining the structure and by taking actual measurements of window, wall, floor, and roof areas. If the survey is made for a house that is to be built, the checklist information is obtained from the drawings supplied by the architect or the builder. The load estimate examples in this unit are based on the drawings provided in the back cover envelope.

CHECKLIST ENTRIES

The following sections cover briefly each of the items on the checklist as they relate to the house plans provided with this text. (Note: in actual situations, information not included in the plans must be obtained from the architect or builder.) The shaded figures represent parts of a typical checklist and are used to illustrate the information to be recorded.

Item 1: Orientation

The house faces north as indicated on the floor plan by a symbol.

Item 2: Construction

Type. The Floor Plan and the Elevations show that the house has a single story. Figure 16-1 illustrates the three basic types of house design: single story, split level, and multiple story.

Walls. The Elevation drawings indicate that wood shingles and siding are to be used. Although insulation is not shown on the drawings, the specifications call for 4-in. insulation. This information can be shown by a sectional view through the exterior wall. The color of the exterior walls is also given in the specifications; in this case, it is to be a dark color. The ceiling height for the first floor is shown on the Elevation drawings.

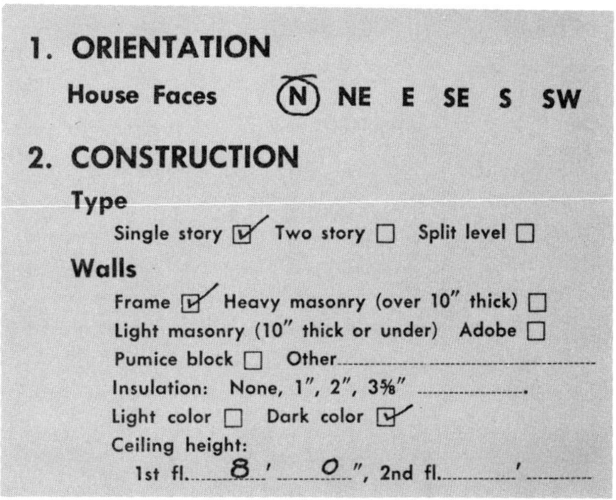
1. ORIENTATION

House Faces (N) NE E SE S SW

2. CONSTRUCTION

Type

Single story ☑ Two story ☐ Split level ☐

Walls

Frame ☑ Heavy masonry (over 10" thick) ☐
Light masonry (10" thick or under) Adobe ☐
Pumice block ☐ Other..........
Insulation: None, 1", 2", 3⅝" ....
Light color ☐ Dark color ☑
Ceiling height:
1st fl. 8 ' 0 ", 2nd fl. ...' ...

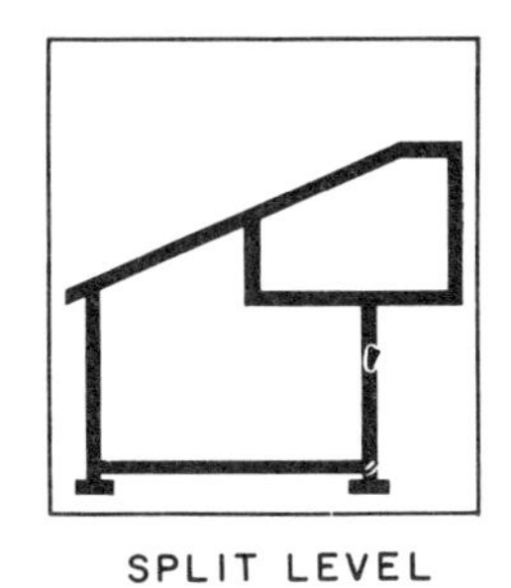

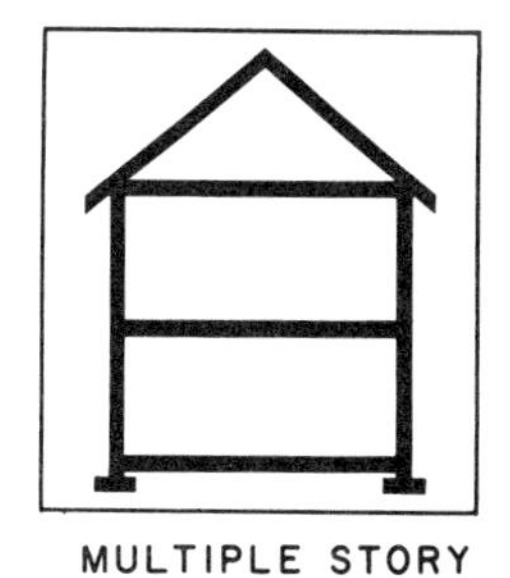

Fig. 16-1 Three basic house designs

Roof and Ceiling. The roof type is shown on the Elevation drawings. Figure 16-2 illustrates the basic roof designs which are either pitched or flat. The roof insulation is to be 4 in. thick as given in the building specifications. The color of the roof is also indicated in the specifications.

Roof & Ceiling

Pitched roof ☐ Flat roof ☐ Rafters exposed ☐
Studio ceiling ☐ Rafters covered ☐
Insulation: None, 1", 2", (4") ....
White roof ☐ Dark roof ☐

Floor. Information about the floor is provided on the Floor Plan. In this case, the floor is to be installed over a basement. The Basement Plan shows that the house has a full basement. However, there are many other combinations of floor construction as shown in figure 16-3.

Floor

Slab on ground ☐ Edge insulation: None, 1", 2"
Over Crawl space ☐
Open ☐ Closed: Vented ☐ Unvented ☐
Vapor seal on ground ☐
Insulation: Underside of floor ☐ ...."
Crawl space walls ☐ ...."
Over Basement ☑
Over garage ☐
Insulation: None, 1", 2", 4"..........
Other ....

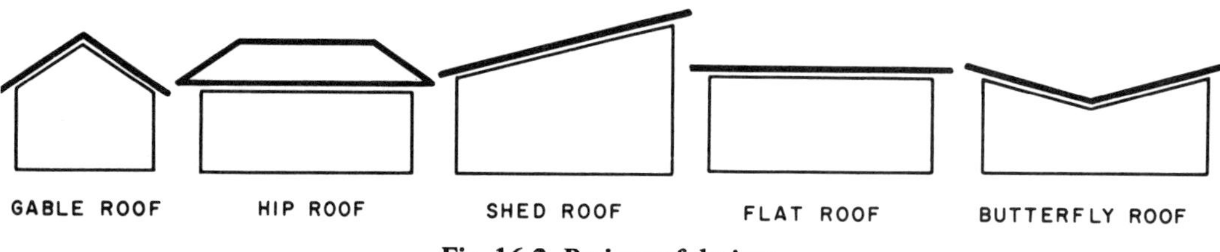

Fig. 16-2 Basic roof designs

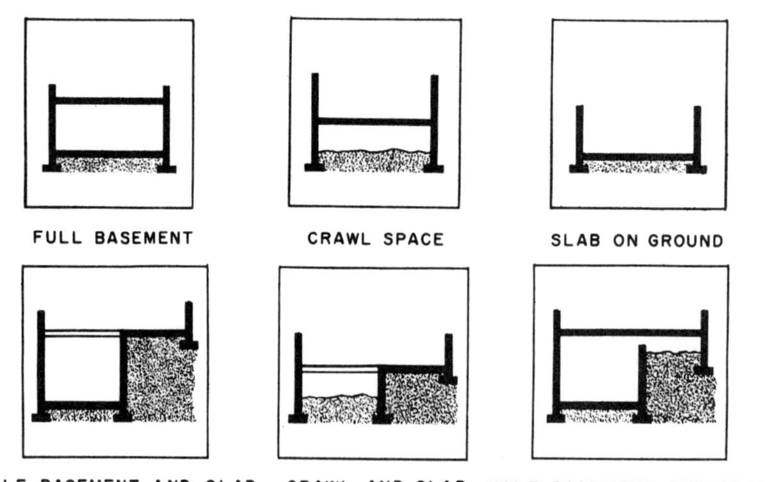

OVER GARAGE OR OVER CARPORT

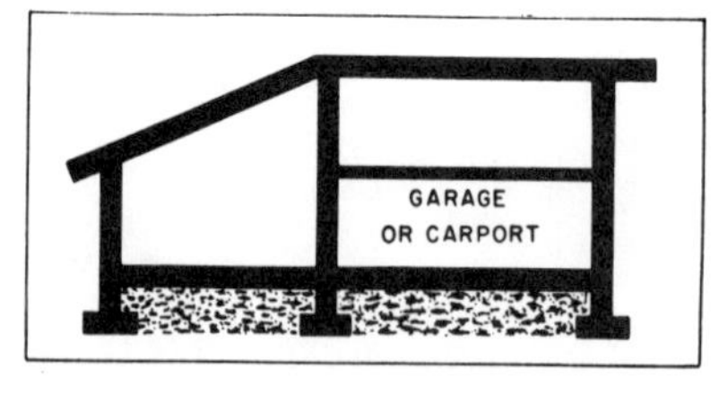

Fig. 16-3 Combinations of floor construction

Windows. The window schedule on the Floor Plan specifies the types of windows. (This information is also provided on the Elevations.) The window schedule shows eight A-type, two B-type, and three C-type windows. The specifications call for weatherstripping around the windows.

Windows

Movable ☐

Wood ☐ Metal ☐
Single pane ☐ Double pane ☐
Double hung ☑ Casement ☐
Weatherstripped ☑ Plain ☐

Fixed

Single pane ☐ Double pane or glass block ☐

Item 3: Shading

Shading is provided by a roof overhang of 24 in. on the north and south sides. (The amount of overhang is determined from the east and west elevations. The drawing scale is 1/8 in. = 1 ft. and the roof overhang on the drawing is approximately 1/4 in. Therefore, the actual roof overhang is 2 ft. or 24 in.)

Inside blinds are to be installed throughout the residence as indicated in the specifications. Therefore, the north and south exposures are to have shades and the east and west exposures have solid walls.

3. SHADING

TYPE OF SHADING	LOCATION					
	N	NE	E	SE	S	SW
Roof overhang (in.)	24				24	
Trees						
Awnings						
Trellis						
Outside shade screen						
Inside blinds or shades	✓				✓	
No inside shading						
Garage or carport						
Wings or offsets						
Other buildings						

Other shading devices..

Item 4: Exhausts and Vents

Attic louvers are specified at both ends of the attic as shown on the east and west elevations. Plans can also include sections or cornice details in which venting under the eaves is indicated. Figure 16-4, page 194, shows several common schemes for ventilating roofs.

4. EXHAUSTS AND VENTS

Kitchen exhaust fan ☐
Attic louvers ☑ Attic fan ☐
Clothes dryer: vented ☐ water condensing ☐

Item 5: Miscellaneous Load

The customer has no specific or unusual requirements.

5. MISCELLANEOUS LOAD ITEMS

Customer's specific requirements:

Frequent entertaining of large groups ☐
Special temperatures: Summer ....F. Winter ....F.
Positive outside air supply ☐
Other..

Unusual sources of heat gain or loss..

Item 6: Condensing Method

Air cooled.

6. CONDENSING METHOD

Air cooled ☑ Waste water ☐
Water cooled ☐ Cooling tower ☐
Spray pond ☐

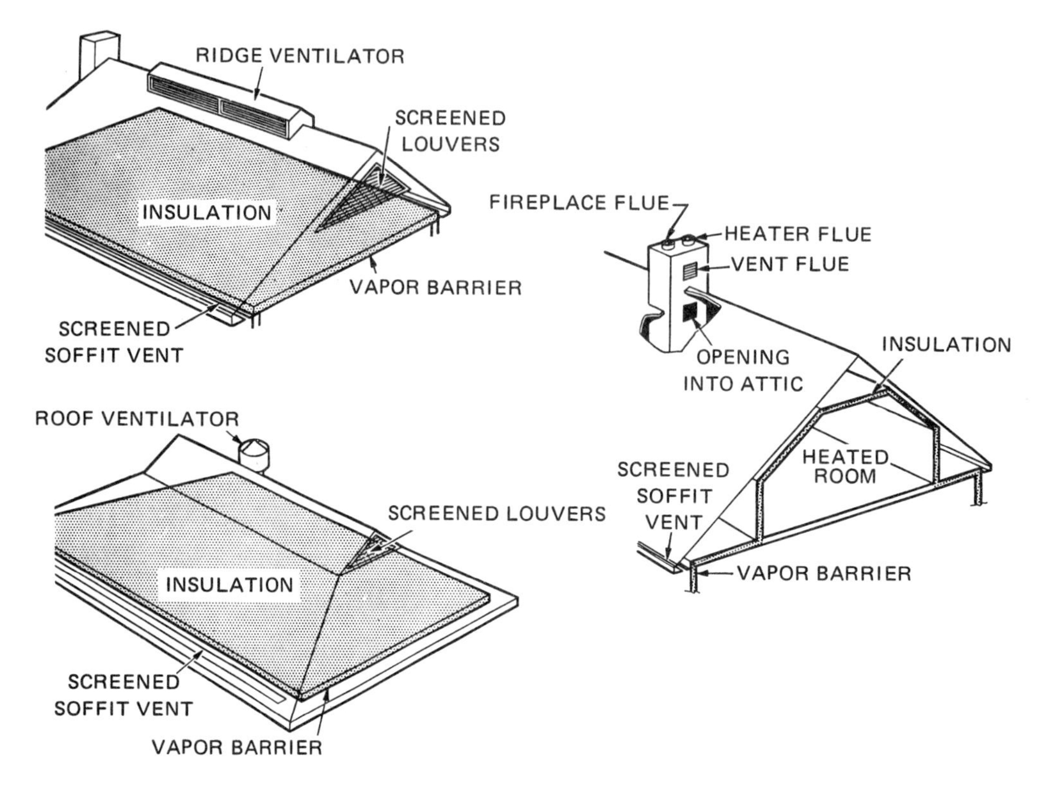

Fig. 16-4 Schemes for ventilating roofs

Item 7: Equipment Location

The unit is located in the basement since there is sufficient accessible space for the duct system. The condenser is located in the backyard.

Item 8: Entry Clearances

The Floor Plan shows that the equipment can be brought through the side kitchen door. Thus, there is direct access to the basement stairs.

Item 9: Chimney

The Floor Plan shows the chimney located in the kitchen-living room wall. Details of the fireplace are not included in these plans; thus, further information cannot be obtained.

7. EQUIPMENT LOCATION

EQUIPMENT	LOCATION							
	Basement	Utility Room	Closet	Garage	Attic	Yard	Breezeway	Other
Weathermaker	✓							
Air Cooled Condenser						✓		
Cooling Tower								
Spray Pond								
Condenser Water Pump								
Conversion Weathermaker								
Cooling Coil Section								
Condensing Unit Section								
Condensate Pump								

Notes: ....

Item 10: Combustion Air

Since the Basement Plan does not show an outside air duct to the furnace, infiltration air is used for combustion.

Item 11: Service and Utilities

The Specifications provide the following information:

a. The electrical service is 230 volts, single phase, and 60 Hz.

b. The water service is taken from the city supply at a maximum temperature of 70°F and a pressure of 60 psi.

c. The condensate from the cooling coil is drained by gravity.

d. Condenser water disposal is not necessary since the condenser is air cooled.

e. The gas service is natural gas rated at 1,000 Btu/cu. ft. of gas.

f. There is no oil service.

Item 12: Controls

Standard controls supplied by a manufacturer are used.

Item 13: Existing Air Distribution System

There is no existing system in this case. The duct, register, and grille sizes must be determined after the load estimate is completed.

8. ENTRY CLEARANCES

Check doors, halls, stairs and turns for possible passage.
Enter house through SIDE KITCHEN door.
Notes: ____________

9. CHIMNEY

Location Kitchen – Living Room Wall
Type ____________
Height above flue opening: ______ ft.
Flue size ______" x ______" or ______" dia.

10. COMBUSTION AIR

Infiltration ☑ Positive supply ☐

11. SERVICE AND UTILITIES

Electrical service

Current characteristics 230 volts, 1 phase, 60 cycles
Capacity of present service ______ amps.
Additional loads on service:
Dryer ☐ Range ☐ Water Heater ☐
Other ____________
Service entrance location ____________
Name of power company ____________

Water service (water cooled equipment only)

Source
City ☑ Well ☐ Other ______
Maximum temperature 70 F.
Pressure
Less than 30 psi ☐
30 to 70 psi ☑
Over 70 psi ☐
Meter location ____________
Meter size ______"
Service pipe size 1/2"
Available flow (wells) ______ gpm.
Name of water company ____________

Condenser water disposal (water cooled equipment only)

Storm sewer ☐ Sanitary sewer ☐ Dry well ☐ Disposal well ☐
Floor drain ☐ Sump ☐
Other ____________

Condensate disposal

Gravity ☑ Condensate pump ☐
Floor drain ☐ Sump ☐ Dry well ☐
Other ____________

Gas service

Type of gas
Natural ☑ Mixed ☐ Manufactured ☐ L.P. ☐
Heating value 1000 BTU/cu. ft.
Meter location ____________
L. P. tank location ____________
L. P. tank size ____________
Name of Gas company ____________

Oil service

Tank location ____________
Above oil pump ☐ Below oil pump ☐ Distance ______' ______"
Tank size ____________
Fill and vent location ____________
Fuel line location ____________
Name of oil company ____________

Miscellaneous Items

The specifications indicate the items that are to be supplied by other contractors. (These items are omitted here because they have no bearing on the heating and cooling loads and the duct sizing.)

Notes

The system designer is to include all special conditions or circumstances not indicated on the checklist form being used for the estimate.

12. CONTROLS

Standard ☑ Remote ☐ Automatic Seasonal Changeover ☐
Continuous fan operation ☐ Night set-back for heating ☐

13. EXISTING AIR DISTRIBUTION SYSTEMS

Type of system: Gravity ☐ Forced air ☐
Size of Furnace Fan Wheel ...."
Type of supply registers ....
Type of return grilles ....
Supply register location
Floor ☐ Baseboard ☐ Low wall ☐
High wall ☐ Ceiling ☐
Inside wall ☐ Outside wall ☐
Return grille location
Floor ☐ Baseboard ☐ Low wall ☐
High wall ☐ Ceiling ☐
Inside wall ☐ Outside wall ☐

COOLING LOAD ESTIMATE

The cooling load estimate presented in this unit is based on equipment operation over a 24-hour period. This approach makes an allowance for the storage effect in furniture and construction materials. The result is expressed as an average load per hour. When the load is averaged over a 24-hour period, the equipment requirements are smaller because the storage effect balances the peak load. As a result, the equipment can be selected for the average hourly load rather than the peak load.

The typical cooling load estimating form allows the estimator to indicate individual room load air quantity requirements (in cfm), the total residence load, and the total cfm. Using this approach, it is not necessary to proportion the total load according to the square feet of area to find individual room loads. The sample form shown is filled out according to the information presented in the following Sample Load Estimate procedure.

The estimating form provides spaces to record the area in square feet and the loads of the various rooms. In addition, the form provides a simplified survey and several tables of factors for the heat gain through windows, doors, walls, and roofs.

Table 1 provides values of heat gain per sq. ft. of window sash area for a variety of roof overhang dimensions from 0 in. to 48 in., in combination with typical outdoor design conditions and degrees of latitude.

Table 2 gives values of heat gain per linear foot through walls according to the outside design temperatures and the types of wall construction.

Table 3 gives the heat gain values per square foot through roofs for common design temperatures and daily temperature ranges.

Before any entries are made on a load estimate form, all terms used must be understood. For example, the term "Quantity" appears on the form used in this sample. This term refers to: the area in square feet of the windows, roof, and floors; the perimeter (linear feet of wall area); and the number of people. Therefore, the quantity column on the form will show measurements in terms of square feet, linear feet, and numbers with no units. All three units of measurement (square foot, linear foot, and number) for the various parts of the structure are recorded in the quantity column.

All special notes pertaining to the tables must be understood. For example, consider the third note under Table 2: "Wall adjoining attic space should be taken as roof." This means that the wall area (in sq. ft.) of the vertical wall of a room built in the attic or next to the attic (on the same floor) should be added to the roof area in the quantity column.

Sample Load Estimate

The house in this sample estimate is located in Kansas City at a latitude of 39° N. The daily temperature range is 19°F; the outside design temperature is 100°F for summer and -10°F for winter (Refer to Table 11-1). The inside design temperature is 75°F for summer and 70°F for winter. The inside design temperatures are established by agreement with the customer.

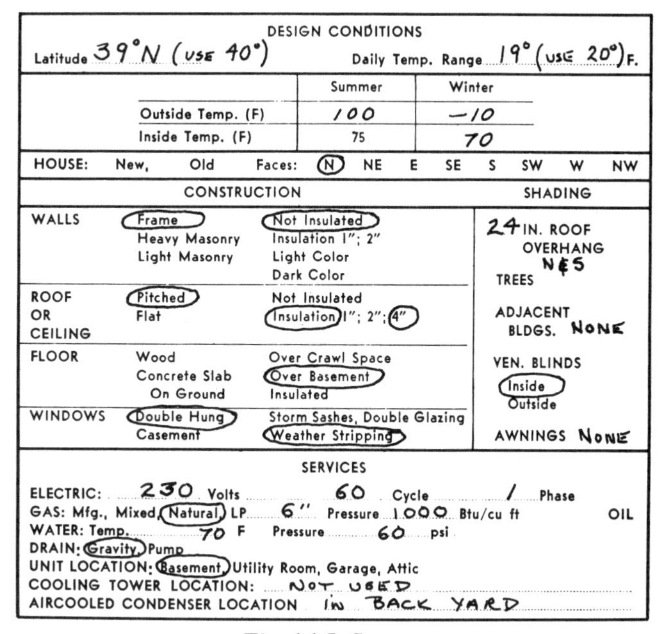

DESIGN CONDITIONS

Latitude 39°N (USE 40°) Daily Temp. Range 19° (USE 20°) F.

	Summer	Winter
Outside Temp. (F)	100	-10
Inside Temp. (F)	75	70

HOUSE: New, Old Faces: N NE E SE S SW W NW

CONSTRUCTION / SHADING

WALLS: Frame, Heavy Masonry, Light Masonry; Not Insulated, Insulation 1"; 2", Light Color, Dark Color

ROOF OR CEILING: Pitched, Flat; Not Insulated, Insulation 1"; 2"; 4"

FLOOR: Wood, Concrete Slab On Ground; Over Crawl Space, Over Basement, Insulated

WINDOWS: Double Hung, Casement; Storm Sashes, Double Glazing, Weather Stripping

24 IN. ROOF OVERHANG N & S

TREES

ADJACENT BLDGS. NONE

VEN. BLINDS Inside Outside

AWNINGS NONE

SERVICES

ELECTRIC: 230 Volts 60 Cycle 1 Phase

GAS: Mfg., Mixed, Natural, LP 6" Pressure 1000 Btu/cu ft OIL

WATER: Temp. 70 F Pressure 60 psi

DRAIN: Gravity, Pump

UNIT LOCATION: Basement, Utility Room, Garage, Attic

COOLING TOWER LOCATION: NOT USED

AIRCOOLED CONDENSER LOCATION IN BACK YARD

Fig. 16-5 Survey

Step 1

Enter the given design conditions at the top of the survey block, figure 16-5.

Step 2

In Table 1, circle the information that applies to the house in this sample estimate (refer to the survey information).

a. Zero roof overhang for the east and west roof edges.
b. 24-in. roof overhang for the north and south roof edges.
c. 100°F outside design temperature for both roof overhang entries.
d. 40° N. latitude.
e. Windows on the north side of the house have a 24-in. roof overhang and 100°F outside design temperature. Therefore, in the 40° N. latitude section, move horizontally to the right along the N row to the 100°F column for a 24-in. overhang. Circle the number 22.

Windows Facing	ROOF OVERHANG (Inches) 0				12				24				36			
	Outside Design Dry Bulb Temperature (F)															
	90	95	100	105	90	95	100	105	90	95	100	105	90	95	100	105
30° N. LATITUDE																
N	11	17	23	29	11	17	23	28	11	16	22	28	11	16	22	28
NE NW	24	30	36	41	23	29	35	40	21	27	33	39	19	25	31	37
E W	33	38	44	50	32	37	43	49	29	34	40	46	25	31	37	43
SE SW	26	32	38	44	25	30	36	42	21	26	32	38	17	23	29	35
S	14	20	25	31	11	17	23	29	11	17	22	28	11	17	22	28
35° N. LATITUDE																
N	11	17	23	28	11	17	22	28	11	16	22	28	10	16	22	28
NE NW	23	29	35	40	22	28	34	40	21	26	32	38	19	24	30	36
E W	33	39	45	50	32	38	44	49	29	35	41	46	26	32	38	43
SE SW	29	35	40	46	27	33	38	44	23	28	34	40	19	25	31	36
S	18	23	29	35	14	20	25	31	11	17	23	28	11	17	23	28
40° N. LATITUDE																
N	11	17	22	28	11	17	22	28	11	16	22	28	10	16	22	28
NE NW	23	28	34	40	22	28	33	39	20	26	32	38	19	25	30	36
E W	34	39	45	51	33	38	44	50	30	36	41	47	27	33	38	44
SE SW	31	37	43	49	30	36	41	47	25	31	37	42	21	27	32	38
S	21	27	33	39	18	23	29	35	12	18	23	29	11	17	23	28

NOTES:
Based on venetian blinds or roll shades full drawn.
For omission of venetian blinds (no inside shading) increase values in table 1
For awnings use values under 48" roof overhang.
For double glazing use 0.80 times values in Table.
Shading due to roof overhang applies to single story house. For two story home, figure shading for second floor, none for first floor.
Windows shaded all day by trees or buildings should be included as north windows.
Consider outside solid doors as outside wall.

Table 1 Solar gain and heat transmission through windows or glass doors (Btu/hr. per sq. ft. sash area)

f. Windows on the east and west sides of the house have no overhang (see the Elevation drawings). The outside design temperature is 100°F. Therefore, in the east-west row, circle the number 45.

g. The windows on the south side of the house have a 24-in. overhang. The design temperature is 100°F. In the south row, circle 23.

h. The house has venetian blinds; circle the note that describes the blinds as covering the entire window surface.

Step 3

In Table 2, circle all of the information as listed that applies to the house in the sample (refer to the survey).

a. 100°F outside design temperature
b. Dark color exterior finish
c. 20°F daily temperature range
d. Frame construction
e. Factor 54 for the dark wall and frame construction
f. Since the house has a north wall, circle the note that gives special instructions for this wall (use half of the given value).
g. The roof overhang is 24 in. and the factor is 0.81. This factor is used for walls facing the south (the sunlit side). The factor allows for the partial shading of bedrooms 1 and 2 and the living room walls by the overhang. The factor for sunlit walls with a 24-in. roof overhang is 54 x 0.81 = 43.7.

DAILY TEMP. RANGE (F)	WALL CONSTRUCTION	U	OUTSIDE DESIGN TEMPERATURE (F) 90		95		100		105	
			Light Color	Dark Color	Light Color	Dark Color	Light Color	Dark Color	Light Color	Dark Color
15	Insulated Frame (1" Ins.)	.14	13	18	20	25	26	31	32	38
	(2" Ins.)	.097	9	13	14	17	18	22	22	26
	Frame or Heavy Masonry	.27	25	35	38	48	50	60	63	72
	Light Masonry	.46	43	60	64	81	85	102	107	124
20	Insulated Frame (1" Ins.)	.14	10	15	16	21	23	28	29	34
	(2" Ins.)	.097	7	10	11	15	16	19	20	24
	Frame or Heavy Masonry	.27	19	29	31	41	44	54	56	66
	Light Masonry	.46	33	49	54	70	75	92	96	113
25	Insulated Frame (1" Ins.)	.14	7	12	13	18	20	25	26	31
	(2" Ins.)	.097	5	8	9	13	14	17	18	22
	Frame or Heavy Masonry	.27	13	23	25	35	38	48	50	60
	Light Masonry	.46	22	39	43	60	64	81	85	102

NOTES:
Based on 8 ft. wall height.
For north or shaded walls, use ½ value in table.
Wall adjoining attic space should be taken as roof.
Wall adjoining unconditioned space should be included as shaded wall.
Wall construction listed as ordinarily encountered in residences. The heat gain factor is directly proportional to the heat transmission coefficient (U).

WALL SHADE FACTORS FOR VARIOUS LENGTH ROOF OVERHANG
(Multiply Values in Table 2 by Factors Below)

Roof Overhang (in.)	12	24	36	48
Factor	.90	.81	.72	.68

Shading due to roof overhang applies to single story house. For two story home, figure shading for second floor, none for first floor.

Table 2 Heat gain through sunlit and shaded walls (Btu/hr. per linear ft. of exposed perimeter)

Step 4

Circle the items that apply in Table 3.

RANGE DAILY TEMP. (F)	ROOF Insulation	U	OUTSIDE DESIGN TEMP. (F) 90	95	100	105
15	None	.32	8.5	10.3	12.1	14.0
	1"	.15	4.0	4.8	5.7	6.6
	4"	.069	1.8	2.2	2.6	3.0
20	None	.32	7.5	9.4	11.2	13.1
	1"	.15	3.5	4.4	5.3	6.1
	4"	.069	1.6	2.0	2.4	2.8
25	None	.32	6.6	8.5	10.3	12.1
	1"	.15	3.1	4.0	4.8	5.7
	4"	.069	1.4	1.8	2.2	2.6

NOTES:
Pitched roof area is taken as the area projected on a horizontal plane, i.e. ceiling area.
No positive ventilation in attic. (With positive attic ventilation and insulation over ceiling, use 0.75 times values in Table.)
For white marble chip roof use 0.75 times values in Table.
Roof construction listed as ordinarily encountered in residences. The heat gain factor is directly proportional to the heat transmission coefficient (U).

For FLOOR over unconditioned space, 2 Btu per hour per sq ft.
No heat gain through FLOOR over cool basement, crawl space or concrete slab on ground.
For FLOOR with underside exposed to outside air 4 Btu per hour per sq ft.

Table 3 Heat gain through sunlit pitched and flat roofs (Btu/hr. per sq. ft.)

Step 5

To complete this step, it is necessary to use a drawing showing the outside house dimensions and the overall size of each room or area in the house. The Floor Plan will be used in this case to determine the following items:

a. Overall size of house: width = 24.7 ft. and length = 46 ft.

b. Total area of house = 24.7 x 46 = 1,136 sq. ft.

c. Room height = 8 ft. (from the Elevation views)

d. Total volume of house = 1,136 sq. ft. x 8 ft. = 9,088 cu. ft.

Enter these values in the space provided on the sample estimate form, figure 16-6.

Step 6

In the Factor column in figure 16-6, record the window factors circled in Tables 1, 2, and 3. For example, Table 1 shows these factors for the indicated directions: N - 22, E and W - 45, and S - 23. Repeat this process for the walls (Table 2), and the roof and floors (Table 3).

TWENTY-FOUR HOUR RESIDENTIAL COOLING LOAD ESTIMATE

Overall Size 24.7 ' x 46.0 ' = 1136 Sq Ft x 8.0 ' = 9088 cu.ft.

		FACTOR	ROOM							
			LIVING ROOM		DINING ROOM		KITCHEN		BEDROOM	
			Quantity	Btu/hr.	Quantity	Btu/hr.	Quantity	Btu/hr.	Quantity	Btu
WINDOWS (Note 1.)	N	22								
	NE NW									
	E W	45								
	SE SW									
	S	23	71.7	1649						
WALLS (Note 2.)	Sunlit	54	12.5	675						
		43.7	20.9	913						
	Shaded	27								
ROOF (Note 3)		2.4	261	626						
FLOOR (Note 4)		0								
PEOPLE (Note 5)		300	4	1200						
COOKING		1200						1200		
ROOM SENSIBLE HEAT ■				5063						
CFM = Room Sensible Heat ÷ 20										
Supply Register	No. and Size									
Return Grille	No. and Size									

Fig. 16-6 Portion of form: steps 1 through 12

Step 7

Compute and record the values for the Living Room Windows in the Quantity and Btu/hr. columns. The window schedule on the Floor Plan indicates that the living room has three C-type windows, each 3 ft. 8 in. x 6 ft. 6 in.

a. Determine the area of the windows in square feet: 3.67 ft. x 6.5 ft. = 23.9 x 3 = 71.7 sq. ft.

b. Enter the value of 71.7 sq. ft. for the living room windows (southern exposure) in the Quantity column.

c. Determine the heat gain (in Btu/hr.) through windows: 71.7 sq. ft. x 23 (factor for the windows facing south) = 1,649 Btu/hr.

Step 8

Compute and record the values for the living room walls in the Quantity and Btu/hr. columns. The quantity value for the walls is the length of the wall:

a. The length of the sunlit east wall in the living room is 12.5 ft. Enter 12.5 ft. in the Quantity column next to factor 54.
b. The length of the sunlit south living room wall is 20.9 ft. Enter 20.9 ft. in the Quantity column next to factor 43. Factor 43 is used instead of 54 because the roof overhang shades the top part of the sunlit wall.
c. Determine the Btu/hr. value for both walls:
   Btu/hr. (east wall, no overhang) = 54 x 12.5 = 675 Btu/hr.
   Btu/hr. (south wall, 24-in. overhang) = 43 x 20.9 = 913 Btu/hr.
d. Record these Btu values in the appropriate spaces in the Btu/hr. column.

Step 9

Record the Quantity and Btu/hr. values for the living room roof:

a. Determine the roof area in sq. ft. over the living room. (Use the overall outside dimensions shown on the outer edges of the floor plan.)
   Roof area = 12.5 x 20.9 = 261 sq. ft.
b. Enter 261 sq. ft. in the Quantity column for the roof.
c. Determine the heat loss in Btu/hr. through the roof:
   Btu/hr. = factor x sq. ft. = 2.4 x 261 = 626 Btu/hr.
d. Record this value of heat loss in the Btu/hr. column for the roof.
   (Note: The heat loss through the floor is not considered since this house has a basement; see the note below Table 16-3.)

Step 10

Compute and record the Quantity and Btu/hr. values for people (for the living room only). The heat gain from people in the house is always listed as a heat gain in the living room.

a. Record the number of people in the Quantity column; in this case, the number is 4. If this figure is unknown, count the number of bedrooms and add 1.
b. Multiply the number of people by 300 to determine the value of Btu/hr.:
   Btu/hr. (people) = 4 x 300 = 1,200 Btu/hr.
   Record 1,200 in the Btu/hr. column for people.

Step 11

Find the total of the values in the Btu/hr. column for the living room.

Windows	1,649
Walls	675
	913
Roof	626
People	1,200
Living Room Total	5,063 Btu/hr.

Record this value in the Btu/hr. column for the heading Room Sensible Heat.

Step 12

A dining room is not shown on the plans; hence, there are no calculations to be made.

Step 13

Determine and record the Quantity and Btu/hr. values for the kitchen window:

a. Determine the total window area (see the Window Schedule on the Floor Plan)
   11.7 sq. ft. x 2 (windows) = 23.4 sq. ft.
b. Determine the heat gain through the windows.
   Area x factor = 23.4 x 22 = 515 Btu/hr.
   Record the values 23.4 in the Quantity column and 515 in the Btu/hr. column opposite N.

			ROOM							
		FACTOR	LIVING ROOM		DINING ROOM		KITCHEN		BEDROOM	
			Quantity	Btu/hr.	Quantity	Btu/hr.	Quantity	Btu/hr.	Quantity	Bt
WINDOWS (Note 1.)	N	22					23.4	515		
	NE NW									
	E W	45								
	SE SW									
	S	23	71.7	1649						
WALLS (Note 2.)	Sunlit	54	12.5	675			12.2	659		
		43.7	20.9	913						
	Shaded	27					16.5	446		
ROOF (Note 3)		2.4	261	626			201	482		
FLOOR (Note 4)		0								
PEOPLE (Note 5)		300	4	1200						
COOKING		1200						1200		
ROOM SENSIBLE HEAT ■				5063				3302		
CFM = Room Sensible Heat ÷ 20										
Supply Register	No. and Size									
Return Grille	No. and Size									

Fig. 16-7 Portion of form: steps 13-16

Step 14

Determine and record the values for the quantity and Btu/hr. for the kitchen walls:

a. The sunlit east wall is 12.22 ft. long and the factor is 54. The heat gain through the east wall = 12.22 x 54 = 659 Btu/hr.
b. Record the values: 12.22 ft. in the quantity column and 659 in the Btu/hr. column across from the heading Sunlit Wall.
c. The shaded north wall is 16.5 ft. long. The factor for the shaded wall is 27. The heat gain through the north wall = 16.5 x 27 = 446 Btu/hr.
d. Record the value of 16.5 ft. in the Quantity column and 446 in the Btu/hr. column across from the heading Shaded Wall.

Step 15

Compute and record the values for the Quantity and Btu/hr. for the kitchen roof:

a. Kitchen roof area = 16.5 x 12.2 = 201 sq. ft. The factor for the roof is 2.4. The heat gain through the roof = 201 x 2.4 = 482 Btu/hr.
b. Record 201 sq. ft. and 482 Btu/hr. in the spaces provided.
c. Add 1,200 Btu/hr. for cooking.

Step 16

Find the total of all the values in the Btu/hr. column under Kitchen:

Window 515 + Walls 659 + 446 + Roof 482 + Cooking 1,200 = 3,302 Btu/hr.

Step 17

Find the heat gain in the remainder of the rooms, according to the procedure used for the living room and kitchen, figure 16-8. Calculate the total Btu/hr. for each remaining room.

ROOM NAMES

BEDROOM #1		BEDROOM #2		BEDROOM #3		BATH		LAV.		ENTRANCE	
Quantity	Btu/hr.	Quantity	Btu/hr.	Quantity	Btu/hr.	Quantity	Btu/hr.	Quantity	Btu/hr.	Quantity	Btu/hr.
				23.4	515	7.0	154	7.0	154		
23.4	538	23.4	538								
13.0	702			11.7	632						
12.8	559	12.3	538								
				12.8	346	7.8	211	4.5	122	4.4	119
16.6	398	191	458	150	360	7.2	173	4.1	98	5.4	130
	2198		1534		1853		538		374		249

Fig. 16-8

Step 18

The total room sensible heat is to be determined.

Living room	5,063
Kitchen	3,302
Bedroom 1	2,198
Bedroom 2	1,534
Bedroom 3	1,853
Bath	538
Lavatory	374
Entrance	249
Total Room Sensible Heat	15,110 Btu/hr.

Record this total in the Summary block at the bottom of the form, figure 16-9.

SUMMARY		
Subtotal Room Sensible Heat	= 15110	Btu/hr
Heat Gain to Ducts, Fan Heat		
Uninsulated duct in slab floor, closed crawl space, basement or furred ceiling 10%	= 1511	Btu/hr
Insulated duct in attic		
1" insulation 15%	=	Btu/hr
2" insulation 10%	=	Btu/hr
Grand Sensible Heat	= 16621	Btu/hr
Grand Total Heat = GSH/0.75	= 22161	Btu/hr

Fig. 16-9 Summary

Step 19

An allowance is to be made of 10% of the total room sensible heat for heat losses in the ducts and from the fan motor:

a. 15,110 Btu (room sensible heat) x 0.10 = 1,511 Btu/hr.
b. Record the value 1,511 Btu/hr. in the appropriate space in the Summary.
c. Determine the Grand Sensible Heat.
   Grand Sensible Heat = 15,110 + 1,511 = 16,621 Btu/hr.
   Record the value 16,621 in the appropriate space.
d. Calculate the Grand Total Heat by dividing the Grand Sensible Heat by 0.75
   Grand Total Heat = 16,621 ÷ 0.75 = 22,161 Btu/hr.
   Record this value in the appropriate space.

Step 20

Determine the total cfm by dividing the grand total heat by 30.

$$\text{Total cfm} = \frac{22{,}161}{30} = 739 \text{ cfm}$$

HEATING LOAD ESTIMATE

The heating load estimate is determined in much the same manner as the cooling load estimate. A sample estimate is shown in the following paragraphs for the same house used for the cooling load estimate.

Most of the Survey Block is the same as that for the cooling load estimate. However, the room temperature is established as 70°F.

Step 1

Complete the Survey Block, figure 16-10.

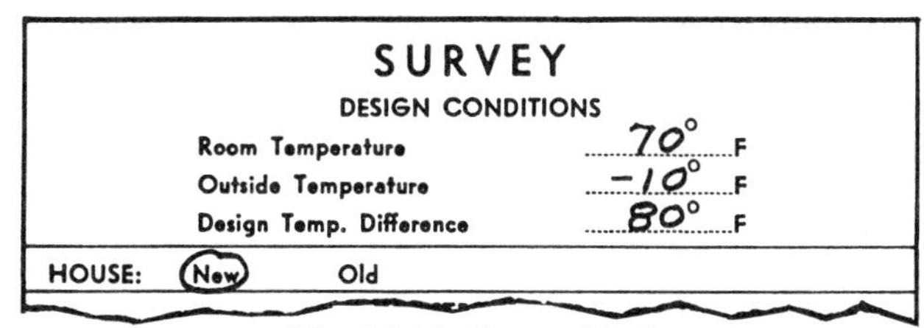
SURVEY

DESIGN CONDITIONS

Room Temperature	70° F
Outside Temperature	−10° F
Design Temp. Difference	80° F

HOUSE: (New) Old

Fig. 16-10 Survey block

Step 2

Circle all of the items that apply in Tables 1, 2, and 3, page 204. Circle and use the fifth note under Table 2 for the basement walls.

Step 3

Record the Overall Size in figure 16-11, page 205.

Step 4

Indicate the types of windows, walls, roof, and floors used and record their factors, figure 16-11.

TABLE 1
WINDOWS AND DOORS
(BTU/HR. PER SQ. FT. SASH AREA)
WINDOWS

WALL CONSTRUCTION	MOVABLE										FIXED	
	SINGLE PANE					DOUBLE PANE						
	Wood		Metal			Wood		Metal				
	Double Hung or Casement		Double Hung		Casement	Double Hung or Casement		Double Hung		Casement	Single Pane	Double Pane or Glass Blocks
	Weather Stripped	Plain	Weather Stripped	Plain	Plain	Weather Stripped	Plain	Weather Stripped	Plain	Plain		
Insulated Frame 1" to 3⅝"	96	109	102	141	128	58	72	65	104	91	72	35
Frame or Heavy Masonry	84	97	90	130	116	46	60	53	93	79	60	23
Light Masonry	79	93	86	126	112	42	56	49	88	75	56	19

NOTES:

Assume outside doors as single pane, wood windows.
For storm sash or door use 50% of values under "Plain."
Infiltration and transmission losses included.
Temperature difference between outside and inside, 70° F.

TABLE 2
EXPOSED WALLS
(BTU/HR. PER FT. OF EXPOSED PERIMETER)

WALL CONSTRUCTION		U	FACTOR
Insulated Frame	1" Insul.	.14	78
	2" Insul.	.097	54
	3⅝" Insul.	.073	41
Uninsulated Frame or Heavy Masonry		.27	151
Light Masonry		.46	258

NOTES:

Based on 8 ft wall height. For other heights, factors are directly proportional.

For partition adjoining unheated space, figure ½ linear feet of length.

Assume wall adjoining garage as outside wall.

Assume wall adjoining attic space as roof.

For basement walls under ground, use 25% of value for heavy masonry wall.

Wall construction listed as ordinarily encountered in residences. Heat loss factor directly proportional to heat transmission coefficient (U).

Temperature difference between outside and inside, 70° F.

TABLE 3
ROOFS AND FLOORS
(BTU/HR. PER SQ. FT.)

CONSTRUCTION		U	FACTOR
Pitched Roof Flat Roof or Studio Ceiling with Rafters Covered	No insulation	.32	22.0
	1" Insulation	.15	11.0
	4" Insulation	.069	4.8
Studio Ceiling with Exposed Rafters	No insulation	.50	35.0
	Insulation 1"	.15	11.0
	Insulation 2"	.10	7.0
Floor	Over unheated basement	.34	7.0
	Over closed-in crawl space	.34	7.0
	Underside exposed (uninsulated) to outside air	.40	28.0
	Basement floor on ground	.50	2.0
	Over garage (4" insulation)	.072	5.0

NOTES:

For pitched or flat roof with horizontal ceiling, use ceiling area.

For studio type ceiling, use actual roof area.

For slab-floor on ground at grade level with 1" edge insulation use wall perimeter x 50 Btu/hr.

Figure no loss through floor over heated basement.

Roof construction listed as ordinarily encountered in residences. Heat loss factor directly proportional to heat transmission coefficient (U).

Temperature difference between outside and inside, 70° F.

Step 5

Record the values for the Quantity and Btu/hr. for each room. The values of several of the quantities require special consideration. For example, the windows in the living room have one movable light (pane) in each of three windows. The area of these panes is equal to the area of one complete window, or 23.9 sq. ft. The two fixed lights (panes) in each of the three windows have an area equal to 47.8 sq. ft.

RESIDENTIAL

Overall Size 24.7 ' x 46.0 ' = 1136 Sq Ft x 8.0 = 9088 Cu Ft

SURFACE	TYPE	FACTOR	LIVING ROOM		DINING ROOM		KITCHEN		BEDROOM 1		BEDROOM 2	
			Quantity	Btu/hr	Quantity	Btu/hr	Quantity	Btu/hr	Quantity	Btu/hr	Quantity	Btu/hr
WINDOWS (Note 1)	MOV. WOOD	84	23.9	2008			40.1	3368	23.4	1966	23.4	1966
	FIXED WOOD	60	47.8	2868								
	MOV. METAL	112										
WALLS (Note 2)	UNINS. FR.	151	33.4	5043			28.7	4334	25.8	3896	12.3	1857
	BASEMENT	151 x .25 / 258										
ROOF (Note 3)	INSULATED	4.8	261	1253			201	965	166	797	191	917
FLOOR (Note 4)	BASEMENT	2.0										
ROOM HEAT LOSS (For 70° F Diff.)				11173				8667		6659		4740
ROOM HEAT LOSS (For Design Diff.) (Note 5)				12770				9900		7610		5420
20% RESERVE FOR SPECIAL ROOMS (Note 6)												
MAXIMUM ROOM HEAT LOSS ■				12770				9900		7610		5420
APPROX. CFM (Note 7)												
Supply Register	Number and Size											
Return Grille	Number and Size											

NOTES:

Note 1. **WINDOW** heat loss = Factor (Table 1) X Quantity (Area in sq.ft.) = Btu/hr.

Note 2. **WALL** heat loss = Factor (Table 2) X Quantity (Lin ft of perimeter) = Btu/hr.

Note 3. **ROOF** heat loss = Factor (Table 3) X Quantity (Area in sq.ft.) X Btu/hr.

Note 4. **FLOOR** heat loss = Factor (Table 3) X Quantity (Area in sq.ft.; Lin ft. of perimeter for slab floor) = Btu/hr.

Note 5. **ROOM** Heat Loss (for Design Diff) = Loss for 70° Diff X $\frac{\text{(Design Temp Diff in Deg F)}}{70°}$ = Btu/hr.

Note 6. **SPECIAL ROOMS**—Increase heat loss 20% for baths, sun porches, basement rooms, rooms over garages & isolated rooms.

Note 7. **APPROX. CFM** (at 135°F Register Temperature) = $\frac{\text{Maximum Room Heat Loss}}{63}$

Fig. 16-11

As another example of the special conditions to be considered, refer to the outside kitchen door. The total area of the door is regarded as movable window area. Therefore, the value for the Quantity column for movable windows in the kitchen is the sum of 23.4 sq. ft. for the windows and 16.7 sq. ft. for the door, or a total of 40.1 sq. ft. The walls in the basement also require special consideration. The wall factors are based on a height of 8 ft. Seven feet of the 8-ft. height are underground (7/8) and 1 foot is above ground (1/8). Therefore, the perimeter of the basement (141.4 ft.) is multiplied by 7/8 (below ground) and 1/8 (above ground). The resulting values are then multiplied by their respective factors to obtain the heat loss in Btu/hr.:

Basement wall (under ground) = (151 x 0.25) x (7/8 x 141.4)
= 37.8 x 123.7 = 4,670 Btu/hr.

Basement wall (above ground) = 258 x (1/8 x 141.4)
= 258 x 17.7 = 4,560 Btu/hr.

Step 6

Add the heat losses for each room and enter the results in the spaces provided in the row headed Room Heat Loss (for 70°F Difference).

Step 7

Tables 1, 2, 3 are based on a temperature difference of 70°F; however, since the actual temperature difference is 80°F (70°F inside and -10°F outside), an adjustment must be made. This adjustment is explained in Note 5 at the bottom of the estimate form. An example of how the adjustment is made is given for the living room:

$$\text{Living room heat loss (70°F diff.)} = 11{,}173$$

$$\text{Living room heat loss (80°F diff.)} = 11{,}173 \times \frac{\text{Design temp. difference}}{70°\text{F}}$$

$$= 11{,}173 \times \frac{80}{70} = 12{,}770 \text{ Btu/hr.}$$

The value for the adjusted heat loss is entered on the form as the Maximum Room Heat Loss. This adjustment must be made for all of the rooms. It is necessary to add a reserve value in the amount of 20% for the Bath and Lavatory in addition to the adjustment.

Step 8

Complete the Summary block, figure 16-12.

Add the maximum Room Heat Losses and enter this value as the Sum of Maximum Room Heat Losses (66,650 Btu/hr.).

Add 20% for duct heat losses: 66,650 x 0.20 = 13,330 Btu/hr.

Thus, the Maximum Heat Loss for the house is: 66,650 + 13,330 = 79,980 Btu/hr. This value is used to select the heating capacity of the equipment.

HEATING ESTIMATE

BEDROOM 3		BATH		LAV.		ENTRANCE		BASEMENT					
Quantity	Btu/hr.	Quantity	Btu/hr.	Quantity	Btu/hr.	Quantity	Btu/hr.	Quantity	Btu/hr.	Quantity	Btu/hr.	Quantity	Btu/hr.
23.4	1966	7.0	588	7.0	588	20.0	1680						
								21.4	2397				
24.5	3700	7.8	1178	4.5	680	4.4	664						
								7/8 x 144.4 1/8 x 141.4	4670 4560				
150	720	72	346	41	197	54	259						
								1135	2270				
	6386		2112		1465		2603		13897				
	7300		2420		1680		2860		15870				
			480		340								
	7300		2900		2020		2860		15870				

SUMMARY

Sum of Max. Room Heat Losses =	66650	Btu/hr.
Duct Loss (+20%) =	13330	Btu/hr.
Maximum Heat Loss ■ =	79980	Btu/hr.

EQUIPMENT SELECTION

Quan.	
Size	
Capacity	Btu/hr.

CFM (Std. Air) = $\frac{\text{Max. Heat Loss}}{108}$ = 740

Fig. 16-12

Step 9

$$\text{Total cfm} = \frac{\text{Maximum Heat Loss}}{108}$$

$$\text{Total cfm} = \frac{79{,}980}{108} = 740 \text{ cfm}$$

This value agrees very closely with the cfm value determined in the Cooling Load Estimate (739 cfm). Therefore, the fan size selected for the cooling requirement is also satisfactory for the heating requirement. Incidentally, the fan can be sized to deliver approximately 800 cfm, instead of the 740 cfm indicated.

EQUIPMENT SELECTION

The load figures obtained in the Cooling Load Estimate (22,161 Btu/hr.) and the Heating Load Estimate (79,980 Btu/hr.) are the basis for the selection of the proper equipment combination for both the heating and cooling requirements. The calculated load values are checked against the actual equipment capacities as shown in rating tables supplied by the manufacturers of cooling and heating units.

Equipment Combinations

The information in figure 16-13 covers many of the common combinations of heating and cooling equipment.

Cooling	Heating
1. Year-round cooling-heating residential unit.	1. Year-round cooling-heating residential unit.
2. Self-contained cooling unit or fan coil unit with remote condenser and compressor.	2. Gravity air furnace. Forced air furnace. Gas fired heaters. Hot water furnace. Electric system.
3. Cooling coil in duct (remote refrigeration equipment).	3. Forced air furnace.
4. Heating-cooling room units with remote water chiller.	4. Heating-cooling room units with remote hot water boiler.
5. Duct units with remote supply of cooled air.	5. Duct units with remote supply of heated air.
6. Central cooling (any cooling type).	6. Partial heating with space heaters.
7. Room air conditioners (partial cooling)	7. Central heating (any furnace type).

Fig. 16-13 Combinations of heating and cooling equipment

Cooling Equipment Selection

The capacity of the cooling equipment selected must be as close as possible to, and always equal to or greater than, the actual cooling load and the actual air quantity (cfm) required. In other words, although the equipment capacity can be greater than the amount required, it must never be excessive. If the actual load and cfm required do not correspond exactly to the rated capacity as expressed in manufacturer's catalogs, select the rated equipment size that is just above the actual requirements.

If a combination of cooling and heating equipment is to be used, and the equipment capacity can handle the cooling load, but not the heating load, then the heating load may be reduced by the following means: (1) Thermopane® or double-glazed windows or storm sashes, (2) weatherstripping around windows and doors, or (3) wall and roof insulation.

Heating Equipment Selection

The Btu/hr. output of the heating equipment selected should be equal to the actual load requirement, or as close to this value as possible. If equipment for the required rated output is not available, the equipment selected should have a rated output that is just above the actual load requirement. The designer should avoid equipment undersizing and excessive oversizing.

If the equipment can handle the heating load but not the cooling load, it may be possible to reduce the cooling load by the following means: (1) wall and roof insulation, (2) double-glazed windows, (3) outside shades or roof overhang, or (4) light outside colors.

EQUIPMENT LOCATION

The equipment to be installed inside the house should be placed in a central location since there is less cost involved to run the ductwork than is required for flue pipes, drains, or electrical connections.

Equipment such as condensers can be placed on a cement slab in the backyard, a carport, a breezeway, or in the attic. The most important consideration for equipment placement is the sound (noise) level. All equipment should be located so that the sound level is as low as possible.

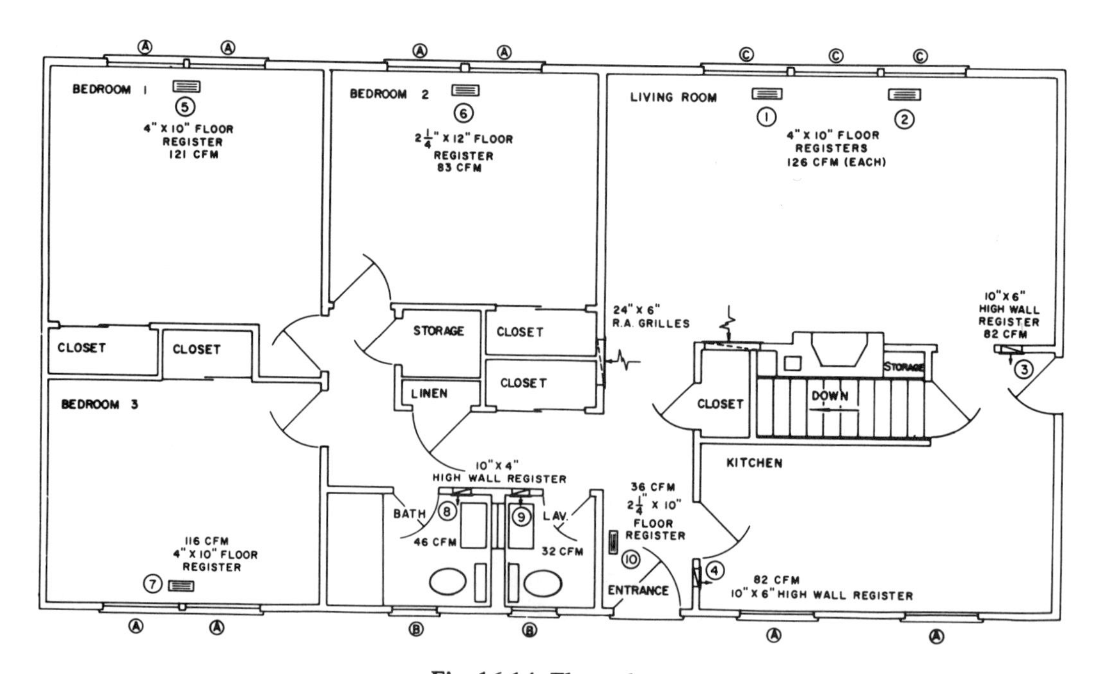

Fig. 16-14 Floor plan
(Courtesy Carrier Corporation)

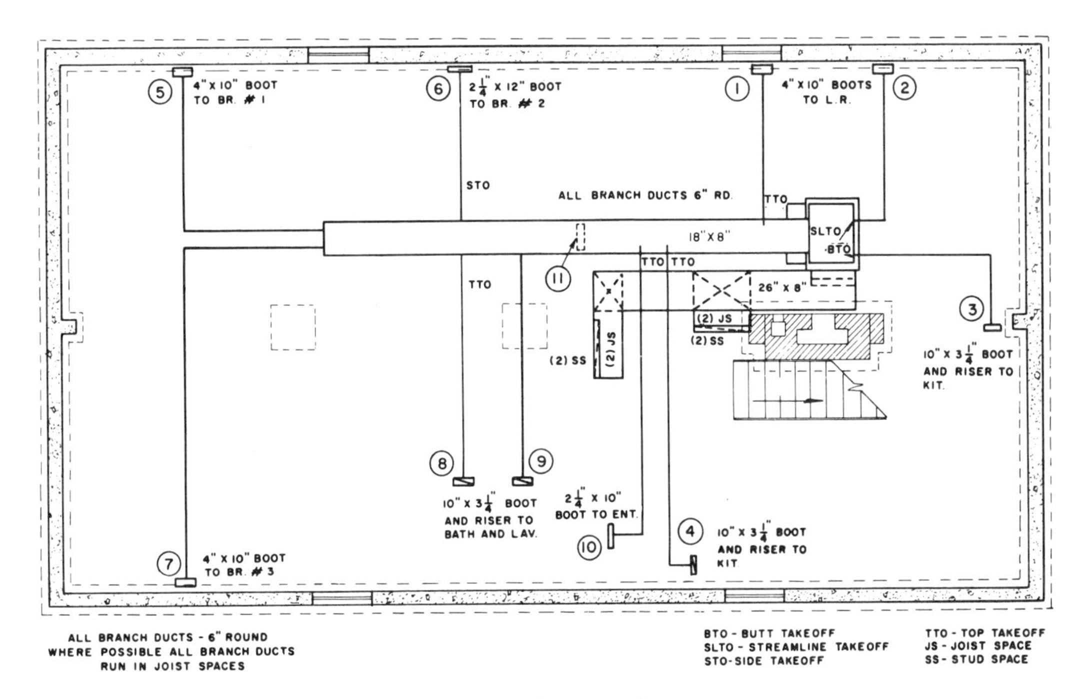

Fig. 16-15 Duct layout in basement
(Courtesy Carrier Corporation)

RESIDENTIAL DUCT SIZING

The following example of the procedure for sizing ducts is based on the residence used to make the previous cooling and heating load estimates. The first floor plan (figure 16-14), basement plan (figure 16-15), and a tabulation of the Btu/hr. values for each room are required to size the supply and return duct system and the supply registers and return grilles. Locate and number the supply outlets and locate the return grilles as shown on the floor plan, figure 16-14. Sketch the duct system as shown on the basement plan, figure 16-15.

The Btu/hr. values determined for the specific residence and listed on the cooling and heating load estimates are used to make the duct sizing data table shown in figure 16-16, page 210.

The plenum size is given in Table 5 as 18 in. x 8 in. (An additional 40 cfm added to the basic values equals a total cfm of 740. Interpolating from the table, 700 = *16* in. x 8 in. and 800 = *20* in. x 8 in. Therefore, 740 cfm equals a plenum size of *18* in. x 8 in.).

The return grille size (two grilles) is given in Table 4: 24 in. x 6 in.

The return duct size is 28 in. x 8 in. from Table 5 in Unit 14.

Duct Sizing Procedure

The procedure outlined as follows is used to size the plenum, branch supply, supply risers, supply outlets, return outlets, and return duct as tabulated in figure 16-16 and shown on the basement and floor plans.

1. List the supply register numbers and describe the register location (such as floor or wall).

Room	Outlet No.	COOLING Btu/hr. ÷ 20 = cfm		HEATING Btu/hr. ÷ 63 = cfm		Type of Register	Supply Register	Branch Supply	Supply Riser
Living Room	1 2	2531 ÷ 20 2532	126 126	6385 ÷ 63 6385	101 101	floor floor	4" x 10" 4" x 10"	6" 6"	
Kitchen	3 4	1651 1651	82 82	4950 4950	77 77	high wall high wall	10" x 6" 10" x 6"	6" 6"	10" x 3 1/4" 10" x 3 1/4"
Bedroom 1	5	2198	110	7610	121	floor	4" x 10"	6"	
Bedroom 2	6	1534	77	5420	83	floor	2 1/4" x 12"	6"	
Bedroom 3	7	1853	93	7300	116	floor	4" x 10"	6"	
Bath	8	583	29	2900	46	high wall	10" x 4"	6"	10" x 3 1/4"
Lavatory	9	374	19	2020	32	high wall	10" x 4"	6"	10" x 3 1/4"
Entrance	10	249	12	2860	45	floor	2 1/4" x 10"	6"	
Basement	11			2270	36	ceiling	8" x 6"	6"	

Fig. 16-16 Duct sizing data

2. Using the values from the cooling and heating load estimate forms, record the Btu/hr. for each room. (Use two supply registers for the living room and kitchen and split the load.)
3. The value of the cooling cfm for each register is determined by dividing the Btu/hr. by 20, as indicated in figure 16-16.
4. The value of the heating cfm for each register is determined by dividing the Btu/hr. by 63 as indicated in figure 16-16.
5. For each register, the largest cfm value is used to determine the appropriate register size from Tables 1 and 2 in Unit 15. The student is cautioned to select values from the proper table according to the type of register used (such as floor or wall).
6. Determine the supply branch duct size from Table 5 in Unit 15. Most residential supply ducts are 6 in. in diameter.
7. Determine the riser size from Table 5 (Unit 15) for the high wall outlets in the kitchen, bath, and lavatory.
8. Determine the supply air plenum size from Table 5 in Unit 15. The total cfm is 740. This value is approximately midway between 700 and 800 cfm as listed in the table. Thus, the plenum should be slightly larger than the 16 in. x 8 in. size shown for 700 cfm; use a size of 18 in. x 8 in. for the full length of the plenum. The plenum length is usually terminated at the point where there are only three branch runs remaining.

The size of the plenum can be reduced after each runout to three outlets.

a. For a total cfm of 740, the plenum size is 18 in. x 8 in. (from Table 5, Unit 15).
b. First reduction:

cfm after runouts to outlets 1, 4, and 10

$$= 740 - (126 + 82 + 12)$$
$$= 740 - 220$$
$$= 520 \text{ cfm}$$

The plenum size is 14 in. x 8 in. (from Table 5, Unit 15).

c. Second reduction:

cfm after runouts to the next three outlets (6, 8, and 9)

$$= 520 - (83 + 46 + 32)$$
$$= 520 - 161$$
$$= 349 \text{ cfm}$$

The plenum size is now reduced to 10 in. x 8 in. (from Table 5).

9. Two return grilles are used; thus, the grille size is divided into two 24 in. x 6 in. grilles. Since the 24-in. width of the grille is larger than one stud space, the grille must cover two stud spaces. Therefore, two return duct risers are used for each grille. Since joist space is used for the return duct, the two risers should be fed into one joist space duct where possible.

10. The return air plenum is sized according to Table 5 in Unit 15. For 740 cfm, the suggested plenum size is 28 in. x 8 in.

SUMMARY

- Regardless of the load estimating method used, the final value of the Btu/hr. for each room is the important value.
- Cooling Load Estimating Procedure
  1. Complete the survey information.
  2. Circle all of factors in the tables that apply to the residence.
  3. Determine the following values:
     a. the overall size of the house
     b. the floor area in sq. ft.
     c. the room height
     d. the living area in cubic feet
  4. Determine the Btu/hr. value for each room by multiplying the appropriate factor (from the tables) by the window area in sq. ft., the linear feet of walls, the roof area in sq. ft., and the floor area in sq. ft.
  5. Add the heat gain due to people to the living room value and the heat gain from cooking to the kitchen value.
  6. Find the total Btu/hr. in the rooms.
  7. Add 10% for the duct loss to the total load from the rooms. The resulting value is the total load from the rooms and ducts in Btu/hr.
  8. Multiply the total load from the rooms and ducts by 0.75 to determine the Grand Total Cooling Load.

- Heating Load Estimating Procedure
  1. Complete the survey information and circle the applicable factors in the tables.
  2. Multiply the areas or linear feet for the windows, walls, roofs, and floors by the appropriate factors to determine the heat loss per room in Btu/hr.
  3. Find the sum of the room heat losses and adjust this total for the actual temperature difference, if necessary.
  4. Add 20% for duct heat losses to the loss from the rooms. The resulting value is the total loss from the rooms and ducts.
  5. Determine the heating cfm requirement by dividing the total loss by 108.

Equipment Selection

- Select the proper combination of heating and cooling equipment to handle the cooling and heating load.
- Select the equipment so that the capacity is close to but never less than the load requirements.

Equipment Location

- Place the inside equipment at a central location.
- Equipment such as condensers can be located on a slab in the backyard, a carport area, a breezeway, or the attic.
- Factors to be considered in locating the equipment include: sound level, space clearance, duct system, vents, drains, local codes, and services.

Residential Duct Sizing

- Locate each outlet in the rooms.
- Determine the cfm requirement for each outlet.
- Size the outlet.
- Size the branch runs.
- Size the plenums.

REVIEW

Calculate the cooling and heating loads for the residence shown. Figure 16-17 and figure 16-18, page 214, provide typical survey information.

LOAD FACTORS

1. Design Conditions

Location — Birmingham, Alabama

Dry-bulb	Summer	Winter
Indoor, °F	75	70
Outdoor, °F	95	+ 10
Difference	20	60

2. General Conditions

a. Direction house faces: N NE (E) SE S SW W NW
b. House type: 1-story ✓ 2-story Split level
c. House age: New Planned ✓ Under construction Existing Approx. age
d. Direction of prevailing wind: Winter N Summer SW

3. Construction

a. Walls: Frame Brick Veneer ✓ Masonry
Insulation: (None) 1″ 2″ 3 5/8″
Interior partitions: Single Double ✓ Color: Dark ✓ Light
b. Ceiling Heights: 1st floor 8 Ft. 2nd floor Other Basement 7 Ft. Crawl space
c. Roof: Pitched ✓ Flat Vented ✓ Unvented Color: Dark ✓ Light
d. Ceiling: Attic above Yes Under occupied floor
Insulation: None 1″ 2″ (4″) Attic fan
Roof color: Light Dark ✓
e. Floors: Over basement ✓ Over garage Car-port Over crawl space
(Vented Unvented) Moisture barrier on ground
Insulation: (None) 1″ 2″ 3 5/8″
Slab on ground Edge insulation
f. Windows: Single pane ✓ Double pane or storm sash Double hung ✓ Movable
Fixed Plain Weatherstripped Casement Glass block
Fit: Tight Average ✓ Loose

4. Shading

Type	Location E	SE	S	SW	W
Roof overhang (feet)	2	2	2	2	2
Awnings None Trees None					
(Garage) Car-port, Porch	On	North	Side		
Inside shades Yes - Venetian Blinds					
No shading					

5. Miscellaneous

a. Number of people 5
b. Kitchen exhaust fan ✓
c. Clothes dryer: Vented Unvented ✓ In Storage Area
d. Special customer requirements: None Frequent entertaining Temp: Summer Winter
e. Special room treatment: None Workshop Game room Other
f. Unusual lighting or appliance loads No

Fig. 16-17 Survey information for a residence

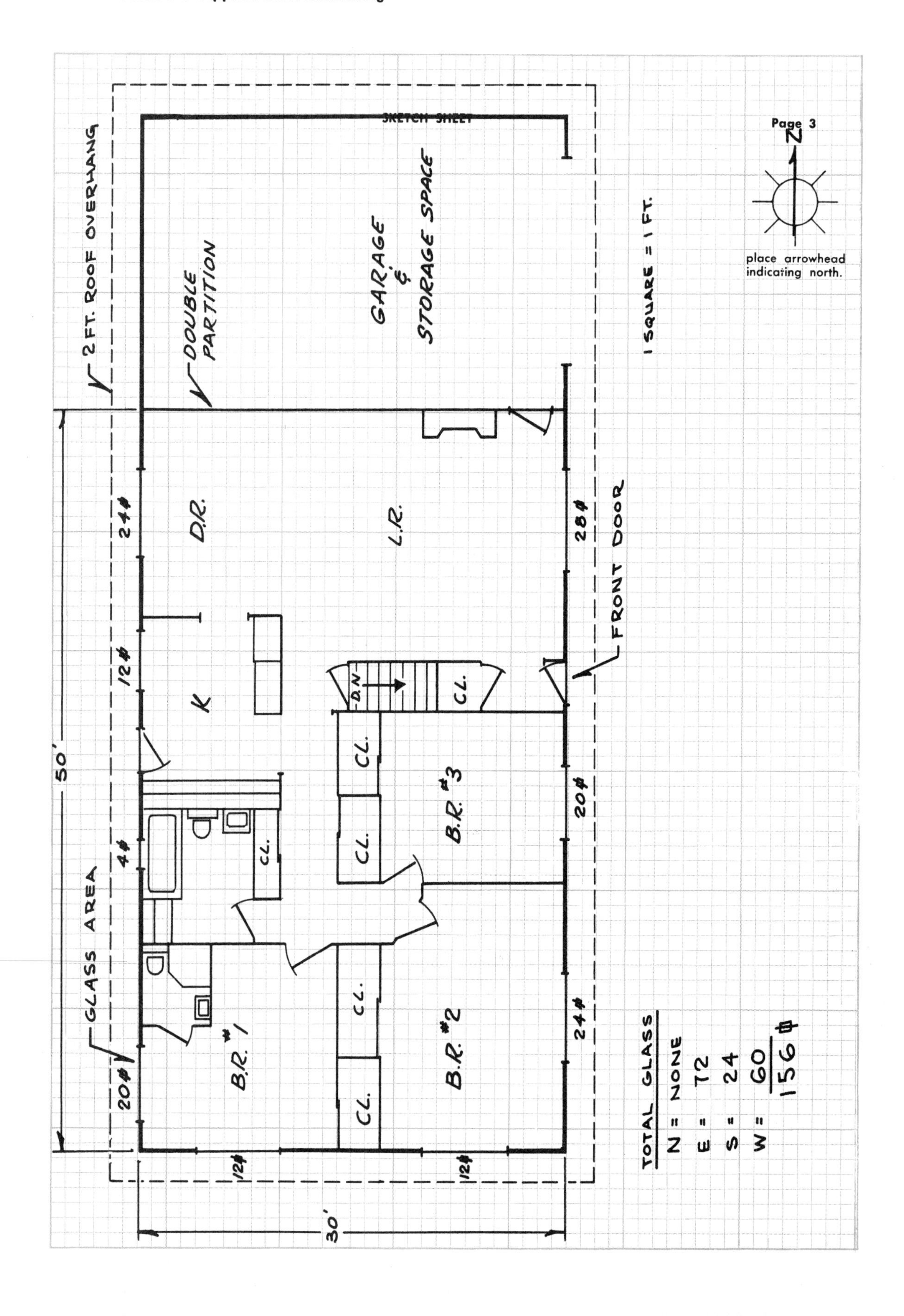
SKETCH SHEET
Page 3
N
place arrowhead indicating north.
2 FT. ROOF OVERHANG
DOUBLE PARTITION
GARAGE & STORAGE SPACE
1 SQUARE = 1 FT.
D.R.
L.R.
K
FRONT DOOR
24#
12#
4#
20#
28#
20#
24#
12#
12#
50'
30'
GLASS AREA
CL.
D.N.
B.R. #1
B.R. #2
B.R. #3
TOTAL GLASS
N = NONE
E = 72
S = 24
W = 60
156#

UNIT 17

COMMERCIAL AIR CONDITIONING, LOAD ESTIMATING, AND DUCT DESIGN

OBJECTIVES

After completing the study of this unit, the student will be able to

- describe a typical self-contained single-package air-conditioning unit, listing components.
- describe a typical split package air-conditioning unit, listing components.
- estimate the heating and cooling loads for a commercial structure.
- based on the estimated loads, select the conditioning equipment to meet all requirements and determine the best location for this equipment.

Commercial air-conditioning applications include restaurants, churches, office buildings, supermarkets, small businesses, and shopping malls.

The load estimating form described in detail in Unit 12 can be used to determine the load for commercial applications. A load estimate for each room usually is not required since most commercial structures have comparatively large open areas. For many structures, enclosed areas such as private offices are present in addition to the open areas. In general, the load is still determined on the basis of the total cooling or heating requirements for a large open space. The smaller enclosed areas are then conditioned by providing sufficient cooling or heating through separate supply and return outlets to each area. The large open space is still considered to be one area.

EQUIPMENT

The air-conditioning equipment for commercial use is available as self-contained, single-package units or split package units.

Single-Package Unit

The single-package unit contains the cooling coil, water-cooled condenser, compressor, heating coil, fan, and fan motor, figure 17-1. In other words, the self-contained single-package unit is a complete air-conditioning system in one package. This unit can be installed in the air-conditioned space either with or without a duct system.

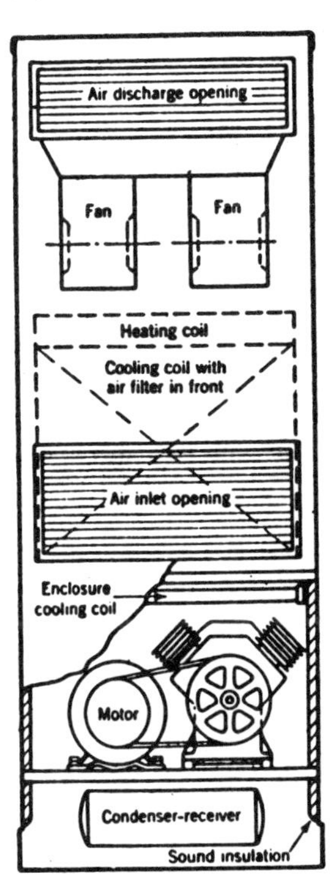

Fig. 17-1 Self-contained water-cooled air conditioner
(Courtesy ASHRAE Guide 1965, by permission)

Many single-package units are fitted with supply and return plenums, supply registers, and return grilles. If the space to be conditioned is sufficiently free from obstructions, the air is delivered to the entire space directly from the supply air register on the unit. The air throw varies from approximately 25 ft. to more than 70 ft., depending on the size of the unit and the height of the room. The spread varies to fit the space. The return air is taken in through the return grille in the unit. If obstructions are present, a standard duct system can be designed. For such a system, the unit supply and the return plenum are omitted and the fan outlet is connected to the duct system.

Split Package Unit

Split package units are designed with the fan and cooling or heating coil in one equipment section, and the condenser and compressor in another section, figure 17-2. The fan coil section is usually located in the air-conditioned space. The condenser and compressor are generally placed in a remote location either indoors or outdoors. The condenser can be air cooled or water cooled. If a water-cooled condenser is to be used, the condenser is provided with connections for either city water or well water. For an air-cooled condenser, these connections are not required.

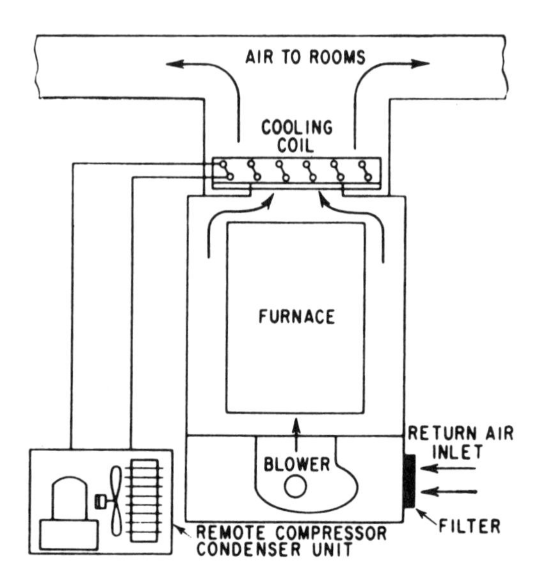

Fig. 17-2 Split package unit containing cooling coil added to warm air furnace

If a water-cooled condenser is installed, there are two ways in which the water can be used: (1) it can be run through the condenser once and then discarded, or (2) the water can be run through the condenser and then diverted to a remote cooling tower where it is cooled and returned to the condenser. The cooling tower approach is not practical if water is expensive or in short supply, or if drainage of the waste water is a problem. Because of these problems, air-cooled condensers are commonly used.

The typical estimating form in the following examples includes: a survey check-list; factors for the average window, wall, roof, and floor constructions; a method of determining cfm requirements; and a heating load estimate. This composite form provides much of the information required to estimate the load for standard, small commercial applications such as restaurants, offices, and stores. A number of types of composite forms are available for use by the estimator.

The ARI Cooling and Heating Load Estimating forms discussed in Unit 12 are also suitable for use with the examples in this unit.

RESTAURANT LOAD ESTIMATE

The Commercial Air-conditioning Estimate form provided in the back cover envelope of this text should be removed from the envelope and used with this estimating problem. The specifications for the restaurant are as follows:

Floor size	20 ft. x 40 ft.
Wall height	12 ft.
Inside conditions	80°F DB temperature, 50% RH

Outside design conditions	95°F DB temperature, 78°F WB temperature
Peak load time	1:00 to 4:00 P.M.

The outside design conditions for Houston, Texas can be obtained from the ARI outside design tables in Unit 12, the ASHRAE Guide outside design tables, and the local weather bureau.

If the city in which the installation is to be made is not listed in the design tables, one way of determining the design conditions is to use the temperatures shown for a listed city close to the desired city. The best approach, however, when outside design conditions are not listed for a specific city or town, is to contact the nearest weather station for this information.

The peak load time for restaurants generally is the time at which the greatest number of customers is present. For this problem, assume that the peak period is from 1:00 to 4:00 P.M.

The sketch on page 5 of the estimate form shows the floor plan, including window, door, wall, and floor dimensions. The sketch also indicates the wall construction and the compass direction. This simple sketch gives much key information if it is examined with the principles of load estimating in mind. For example, two important heat load factors are evident from the sketch: (1) the adjacent store is not air conditioned and (2) the west windows are key glass areas since the sun is in the west during the peak load period. As a result, a shading device for the west-facing windows may be appropriate to reduce the total load from the customers and the sun during the period between 1:00 and 4:00 P.M.

The location of the unconditioned kitchen is another important factor. Since this area usually produces large quantities of heat, the partition between the kitchen and the dining space requires special consideration. This detail is examined in the description of the load estimate portion of the form.

Survey Information

Much of the information required for a restaurant survey can be obtained from sketches or drawings and from the owner and builder. This sample survey is not complete in every detail. For example, survey items such as fluorescent lights and exhaust fans in the air-conditioned space do not apply to the restaurant in this problem.

The following survey entries are made on page 2 of the sample estimate form.

Space Dimensions. The floor plan sketch shows the basic dimensions of the restaurant. If a floor plan is not available, make a sketch similar to the one shown. The dimensions for the restaurant are 20 ft. x 40 ft. x 12 ft.

Wall Construction. All restaurant walls are of heavy masonry construction. Indicate the type of construction for each wall in the spaces provided on the survey.

Gross Wall Area. For each wall, find the gross area by multiplying the length by the height.

North Wall	40 ft. x 12 ft. = 480 sq. ft.
East Wall	20 ft. x 12 ft. = 240 sq. ft.
South Wall	40 ft. x 12 ft. = 480 sq. ft.
West Wall	20 ft. x 12 ft. = 240 sq. ft.

Record the computed areas in the spaces provided.

Windows. Determine the window areas facing each compass direction.

North	4 ft. x 2.5 ft. x 5 windows	=	50 sq. ft.
East	None		
South	None		
West	6 ft. x 4 ft. x 2 windows	=	48 sq. ft.
	3 ft. x 6.67 ft. (door)	=	20 sq. ft.
			68 sq. ft.

(Note: the door is considered to be a part of the window area for the west wall.)

Record the window area in sq. ft. in the spaces provided. For each compass direction, subtract the window area from the gross wall area. Enter the result for the Net Wall Area in the spaces provided:

North	=	430 sq. ft.	South	=	480 sq. ft.
East	=	240 sq. ft.	West	=	172 sq. ft.

Partitions. The east wall of the conditioned space is the partition between the dining area and the kitchen. The area of this wall is 240 sq. ft. and is entered in the space indicated.

Since a restaurant kitchen produces more heat than the average unconditioned space, the partition area is entered on the second line under Partitions rather than on the line for an area adjacent to an unconditioned space. Note the large difference in the factors used on the form for unconditioned areas and for kitchen areas.

Roof. The information that medium construction is used is obtained from the builder or is determined by observation. Two inches of insulation are used.

Ceiling. A flat roof is used; thus, the ceiling is considered to be part of the roof.

Floor. Since none of the items on the survey under this heading apply, add the word slab and enter the dimensions: 20 ft. x 40 ft. = 800 sq. ft.

Electrical. A value of lighting only in use at the time of the peak load is to be recorded. The value of 1,000 watts incandescent is obtained from the electrical contractor or is determined by observation.

Appliances. The amount of heat gain from appliances is obtained from Table 3 of the estimate form, page 5. The total heat gain is determined in Table 3 and then transferred to the appropriate space on the form. The restaurant has one coffee maker (1/2 gallon) and one toaster (two slices).

People. The restaurant seats 24 people which is considered to be the maximum load.

Outside Air for Ventilation. Follow the instructions given in Note 6 on the estimate form. Enter the result in the appropriate space on the estimate form.

Power Supply. The power supplied is 230 V, three phase, 60 hertz. This information is obtained from the electrical contractor or the local power company.

Condensing Method. The equipment supplied contains a water-cooled condenser because water in the locality is plentiful and inexpensive.

Water Service. The water connection is within five feet of the unit through the kitchen partition. The pipe size used and the water pressure available is standard for the locality.

Drain Service. The water and condensate are drained within five feet of the unit in the kitchen area.

Heating. A hot water heating coil in the unit is supplied with 140°F water from a boiler in a remote location.

Equipment Layout

If a duct system is used, the equipment location is selected to minimize the length of the duct required. For the restaurant, the air is distributed from the register in the unit plenum directly into the space to be conditioned. Therefore, the unit is located at one end of the space to insure that the air blow and spread pattern covers the entire area.

COOLING ESTIMATE

The first column of the cooling estimate (page 3 of the Commercial Air Conditioning Estimate) shows that the major sources of heat are the windows, walls, roof, ceiling, floor, electrical devices, appliances, people, and outside air. The factors for each of these heat sources are determined by the specific methods of identification or description. For example, the correct factor for windows is determined by the type of shading present at the time of the peak load for a given compass direction.

The cooling load for each of the items listed in the restaurant estimate form is determined as shown in the following procedures.

Windows

Step 1. The restaurant has 68 sq. ft. of sunlit window area (including the door) with awnings, on the west exposure. The north exposure has 50 sq. ft. of window area without awnings. Note 2 (of the Cooling Load Estimate form notes on page 4) indicates that for windows on a sunlit exposure (west, for the restaurant), a factor is selected according to the time of the peak load. The peak load on the west side occurs between 1:00 PM to 4:00 PM. Since the west windows are shaded on the outside with awnings, circle Outside Shades, 1:30 P.M. to 4:30 P.M., and factor 68.

Step 2. Note 2 also indicates that for the windows on the north side of the restaurant, the factor should be selected from the Other Facings column. Since the north windows are not shaded, circle factor 33.

Step 3. Opposite the appropriate factor selected for each window area, record the area in the sq. ft. column: north facing windows = 50 sq. ft. and west facing windows = 68 sq. ft.

Step 4. Multiply the factor by the area in sq. ft.

North windows 33 x 50 = 1,650

West windows 68 x 68 = 4,624

Step 5. Enter the result obtained in Step 4 in the Btu/Hr. column, as shown.

Walls

Step 1. Select the factor for the walls using a procedure similar to that for the window factors.

a. Find the appropriate Time of Peak Load column (1:30 PM to 4:30 PM) and follow it down the type of wall construction (the restaurant has heavy masonry walls).
b. Select factor 3 for south, north, and west walls.

Step 2. Record in the Sq. Ft. column the total area in sq. ft. of the west, south, and north walls.

Step 3. To find the Btu/hr. value, multiply the total area recorded in Step 2 by the factor 3:

West, south, and north walls 1,082 x 3 = 3,246 Btu/hr. Record this value.

Step 4. Since the east wall of the air-conditioned space is a partition between the kitchen and the dining area, select the factor for the kitchen partition and multiply the east wall area by this value:

240 x 14 = 3,360 Btu/hr.

Record this value in the Btu/Hr. column.

Roof

Step 1. Follow the Time of Peak Load column (1:30 PM to 4:30 PM) to the appropriate type of roof construction. For the restaurant, the roof is of medium construction. In this case, the ceiling is considered to be part of the roof; therefore, select factor 5.

Step 2. Record the roof area in sq. ft. Multiply this value by the selected factor:

800 x 5 = 4,000 Btu/hr.

Record this value in the Btu/Hr. column.

Ceiling

The ceiling is considered to be part of the roof.

Electrical

The restaurant has a total of 1,000 watts of incandescent lighting. Multiply the total number of watts by the appropriate factor:

1,000 x 3.4 = 3,400 Btu/hr.

Record this value in the Btu/Hr. column.

Appliances

Add the total Btu/hr. established from Table 3 of the estimate form and record the value in the Btu/Hr. column:

1,700 + 2,900 = 4,600

People

Under the heading Persons, record the maximum capacity of the restaurant (24 people).

$$24 \times 450 = 10{,}800 \text{ Btu/hr.}$$

Record the result in the Btu/Hr. column.

Outside Air

Step 1. Refer to Note 6 on the estimate form, page 4. Since the air-conditioned area does not have an exhaust fan, select the recommended cfm per person (for restaurants) from Table 2, page 4:

Air Quantity = 15 cfm per person

Step 2. Multiply the maximum number of people in the restaurant by 15 cfm:

$$24 \times 15 = 360 \text{ cfm}$$

Step 3. Circle the established design conditions:

Inside: 80°F DB and 50% RH
Outside: 78°F WB

The appropriate factor for these conditions is 46; circle this value.

Step 4. Record the quantity 360 in the Cfm column and multiply this value by the factor:

$$360 \times 46 = 16{,}560 \text{ Btu/hr.}$$

Record this value in the Btu/Hr. column.

For applications where an exhaust fan is used or where there is no outside air, refer to the instructions in Note 6(b), page 4 of the form.

Step 5. Add all of the values entered in the Btu/Hr. column:

Windows	1,650
	4,624
Walls	3,246
Roof	3,360
	4,000
Electrical	3,400
Appliance	4,600
People	10,800
Outside air	16,560
Total Cooling Load	52,240 Btu/hr.

Step 6. Determine the cooling load in terms of tons of refrigeration:

$$52{,}240 \div 12{,}000 \text{ Btu/ton} = 4.3 \text{ tons}$$

Total Cfm Requirements

The outside air quantity (360 cfm) is only part of the total air quantity that must be supplied by the air-conditioning equipment fan.

The total cfm per ton for the restaurant is determined from Table 1 of the estimate form (page 4). Before the cfm per ton can be selected, it is necessary to determine the percentage of the outside air used as compared to the total cfm required for the restaurant. This percentage is used to select the appropriate Percent Outside Air column in Table 1. A formula for determining the percent outside air is given in Note (b) of Table 1.

$$\text{Percent Outside Air} = \frac{360 \text{ (cfm outside air)} \times 100}{400 \times 4.3 \text{ tons}} = 20\%$$

Find the 20% column in Table 1 and follow this column down to the inside room conditions for the restaurant: 80°F and 50% RH. The required cfm per ton is 310.

Since the cooling load is 4.3 tons, the total cfm requirement for the restaurant is:

$$4.3 \times 310 = 1{,}333 \text{ cfm}$$

Once the total load and the cfm requirements of the restaurant are known, it is possible to select the cooling equipment. However, since the equipment selected must be able to heat the area as well, the heating load must also be determined.

HEATING LOAD ESTIMATE

The heating load estimate appears on page 6 of the estimate form. The load estimate is based on the heat loss in Btu/hr. through the items listed for each degree of temperature difference between the outside winter design temperature (0°F) and the temperature to be maintained inside the restaurant during the winter (70°F). As shown on the form, the heat loss through 118 sq. ft. of window area is 189 Btu/hr. for each degree of temperature difference. At an outside temperature of 0°F and an inside temperature of 70°F, the temperature difference (at design conditions) is 70°F. The amount of heat lost by the windows is as follows:

$$189 \times 70 = 13{,}230 \text{ Btu/hr.}$$

For each item (1-7) of the heating estimate, it is necessary to find the value of the Btu per hour per degree temperature difference.

Item 1: Windows

The restaurant has 118 sq. ft. of single-pane glass. The glass area in sq. ft. is multiplied by the factor for single-pane glass to obtain the Btu/hr. value.

$$118 \times 1.6 = 189 \text{ Btu/hr.}$$

Record this value in the Btu/Hr./Degree Temperature Difference column.

Items 2 Through 7

Find the Btu/hr./degree amount for each item using the procedure in Item 1. Then find the sum of all of the values in the Btu/Hr./Degree Temperature Difference column.

118 sq. ft. x 1.6	=	189 Btu/hr.
20 sq. ft. x 4.0	=	80 Btu/hr.
60 lin. ft. x 4.2	=	252 Btu/hr.
800 sq. ft. x 0.13	=	104 Btu/hr.
60 lin ft. x 0.69	=	41 Btu/hr.
360 x 1.10	=	396 Btu/hr.
Total Btu/Hr./Deg. F Temp. Diff.	=	1,062 Btu/hr.
Multiply 1,062 by 70°	=	74,340 Btu/hr.
Total Heat Load	=	74,340 Btu/hr.

SELECTING EQUIPMENT FOR THE RESTAURANT

The estimates completed for the restaurant have resulted in the following information:

Cooling load requirement	=	52,240 Btu/hr. (4.3 tons)
Heating load requirement	=	74,340 Btu/hr.
Air quantity requirement	=	1,333 cfm

It is now possible to select the required cooling and heating equipment.

Manufacturers of combination cooling and heating equipment normally provide rating tables that show the cooling capacity, heating capacity, and air quantity (cfm) capacity of the equipment. For the loads determined in the estimates, the equipment selected is in the 4 to 5 ton range. By applying the calculated loads to the manufacturers' rating tables, it is possible to select a package unit that meets the requirements. For the restaurant, the cooling and heating unit selected has a water-cooled condenser and the following capacity:

Cooling	54,000 Btu per hr.
Heating	75,000 Btu per hr. with 140°F water
Air Quantity	1,200 to 1,600 cfm

Although the equipment selected in this case from the particular manufacturers' ratings has a capacity larger than the actual load requirements, the listed equipment ratings are often exactly the same as the actual load.

For the restaurant, the water supply is plentiful and inexpensive; thus, a water-cooled compressor was selected. However, if the water supply is a problem, other equipment can be selected. For example, an alternative choice is an air-cooled condenser located on the roof with a coil and fan section placed in the attic space.

EQUIPMENT LOCATION FOR RESTAURANT

The equipment selected consists of a floor standing single package unit with a heating and cooling coil. Suitable free floor and wall area is available at the partition wall between the dining area and the kitchen. The location of the equipment is shown in figure 17-3, page 224. To obtain the best air distribution pattern, the unit should be located in the middle of the front (west) wall. In this position, the air blows toward the kitchen. However, the location of the front door prevents the location of a floor unit at this point. In addition, longer water supply pipes and water drain lines would be required in this location since the drain and the city water supply are in the kitchen.

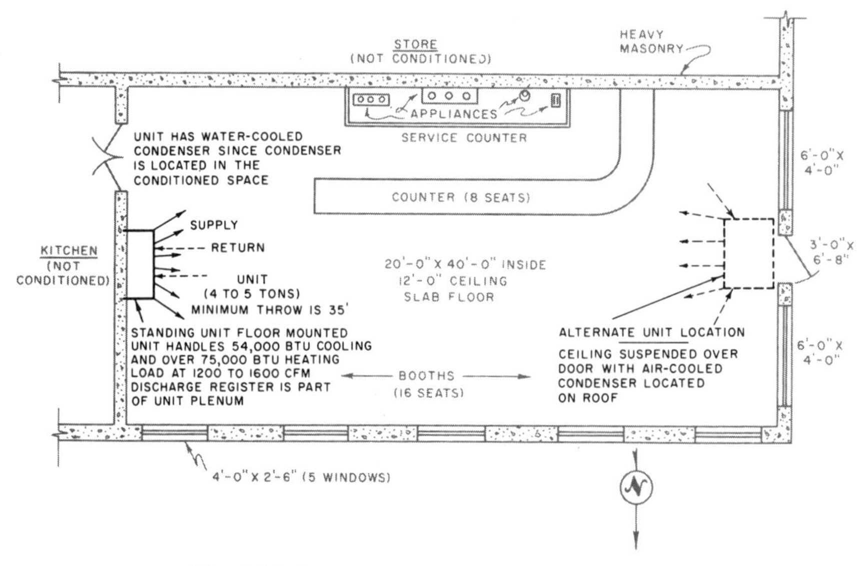

Fig. 17-3 Locating equipment for a restaurant

AIR DISTRIBUTION

The equipment selected for the restaurant can handle a slightly larger air quantity than is required. This situation is preferred in a restaurant application since more and stronger odors are likely to occur than in other types of comfort applications.

A duct system is not used in the restaurant because the unit can distribute air up to 65 ft. horizontally. As a result, the distribution system is quite simple. All that is required is an adjustment of the register directional vanes to insure that the length of blow is satisfactory. The equipment selected has a blow of 35 ft. when the vanes are set at a 45° angle.

If the restaurant contains obstructions to the airflow, a simple duct system can be furred into the ceiling. A standard plenum (without register) can be connected to the duct run and two ceiling diffusers installed.

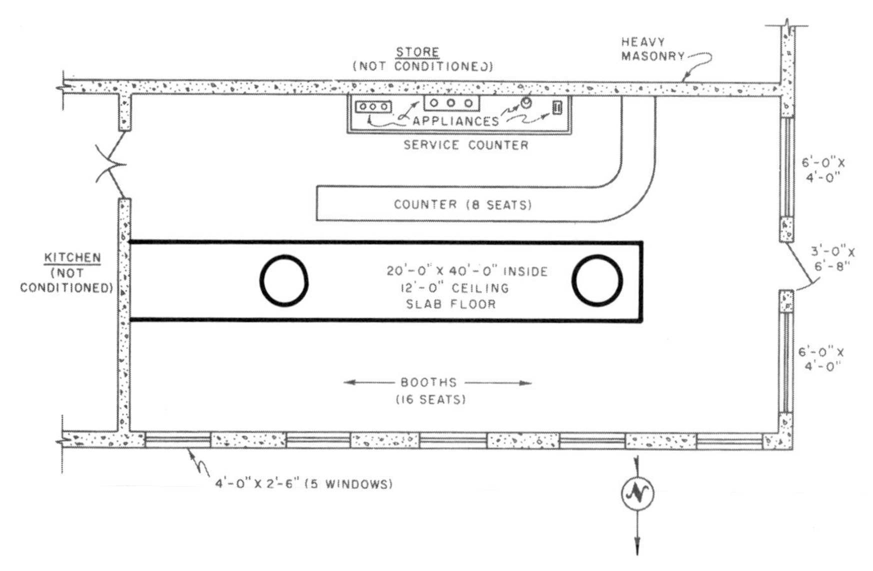

Fig. 17-4 Air distribution for a restaurant

SUMMARY

- Commercial air-conditioning applications include restaurants, churches, offices, supermarkets, small businesses, and other structures which can be air conditioned by self-contained, single-package units or split package units.
- Air distribution in commercial applications using self-contained equipment is accomplished directly from a supply register in the unit plenum, or through a duct system. The direct approach is possible only if the space does not contain obstructions to the airflow.
- If a water-cooled condenser is used, the water supply must be plentiful and economical. If water is expensive, consideration should be given to a cooling tower installation. The tower cools the condenser water so that it can be reused. If water is not plentiful, an air-cooled condenser should be used.
- Commercial equipment is selected according to the cooling load; heating load, and air quantity requirements. Once these items are known, manufacturers' ratings can be used to select the required equipment size.
- The location of air-conditioning equipment in commercial applications is determined by the available space, location of services, appearance, type of air distribution method, type of equipment used, and sometimes by owner preference.

REVIEW

Figures 17-5 and 17-6, page 226, provide survey information for an office located on the second floor. Figure 17-6 also includes a suggested duct system layout. Determine the cooling load in Btu and the cfm requirements. Size the duct system according to the layout shown.

COOLING ONLY

DESIGN TEMP. 93°F DB; 75°F WB OUTSIDE
80°F DB AND 50% RH INSIDE

Record all information essential to the cooling and heating estimates, the air distribution system, and the equipment selection, location, and installation.

DIMENSIONS OF SPACE

Sketch the floor plan in figure 17-6 or obtain the building plans. Indicate scale. Record all dimensions including ceiling heights, heights under beams, and floor to floor heights.

WALLS

Direction Facing	N-NE	E-SE	S-SW	W-NW
Construction		MEDIUM	MEDIUM	MEDIUM
Insulation thickness				
Shaded				
Gross Area (Sq. Ft.)		575	256	575

WINDOWS

Type Shading		INSIDE	INSIDE	INSIDE
Area (Sq. Ft.)		126	72	126
Net Wall (Sq. Ft.)		449	184	449

PARTITIONS

Area (adjacent) to ~~un~~ (conditioned space) — sq. ft.
Area adjacent to kitchen, boiler rm., etc.______ sq. ft.

ROOF

Construction: (light) ((medium)) (heavy)
Ceiling ((Yes)) (No)
Insulation thickness 1 in.
Attic Space: Ventilated by fan (Yes) (No).

CEILING

Unconditioned space above – area______ sq. ft.
Insulation thickness______ inches

FLOOR

Unconditioned space below – area______ sq. ft.
Kitchen, boiler room, etc., below – area______ sq. ft.
Insulation thickness______ inches

ELECTRICAL

LIGHTS IN USE AT TIME OF PEAK LOAD

Incandescent:______ watts or______ watts per sq. ft.
Fluorescent: 4690 watts or______ watts per sq. ft.

MACHINES OR MOTORS

Rating in Watts or Horsepower______

APPLIANCES

PEOPLE

Number in space at time of peak load: 10

OUTSIDE AIR FOR VENTILATION

EXHAUST FANS

TYPE	H.P.	DIAMETER (in.)	SPEED

POWER SUPPLY

Volts,______ phase,______ hertz,______
Panel______ ft. from unit.
Main switch capacity______ amperes

CONDENSING METHOD

(City water.) (Cooling tower.) ((Air Cooled.))
(Evaporative condenser.)

WATER SERVICE

Connection______ ft. from unit. Size______ in.
Water pressure______ lb. New service, new meter.
Pressure reducing valve required? (Yes) (No)

CONDENSER WATER AND/OR CONDENSATE DRAIN

Distance from unit______ ft.
Low enough for gravity flow of condensate? (Yes) (No).
New drain or condensate pump required? (Yes) (No).

Fig. 17-5 Survey information

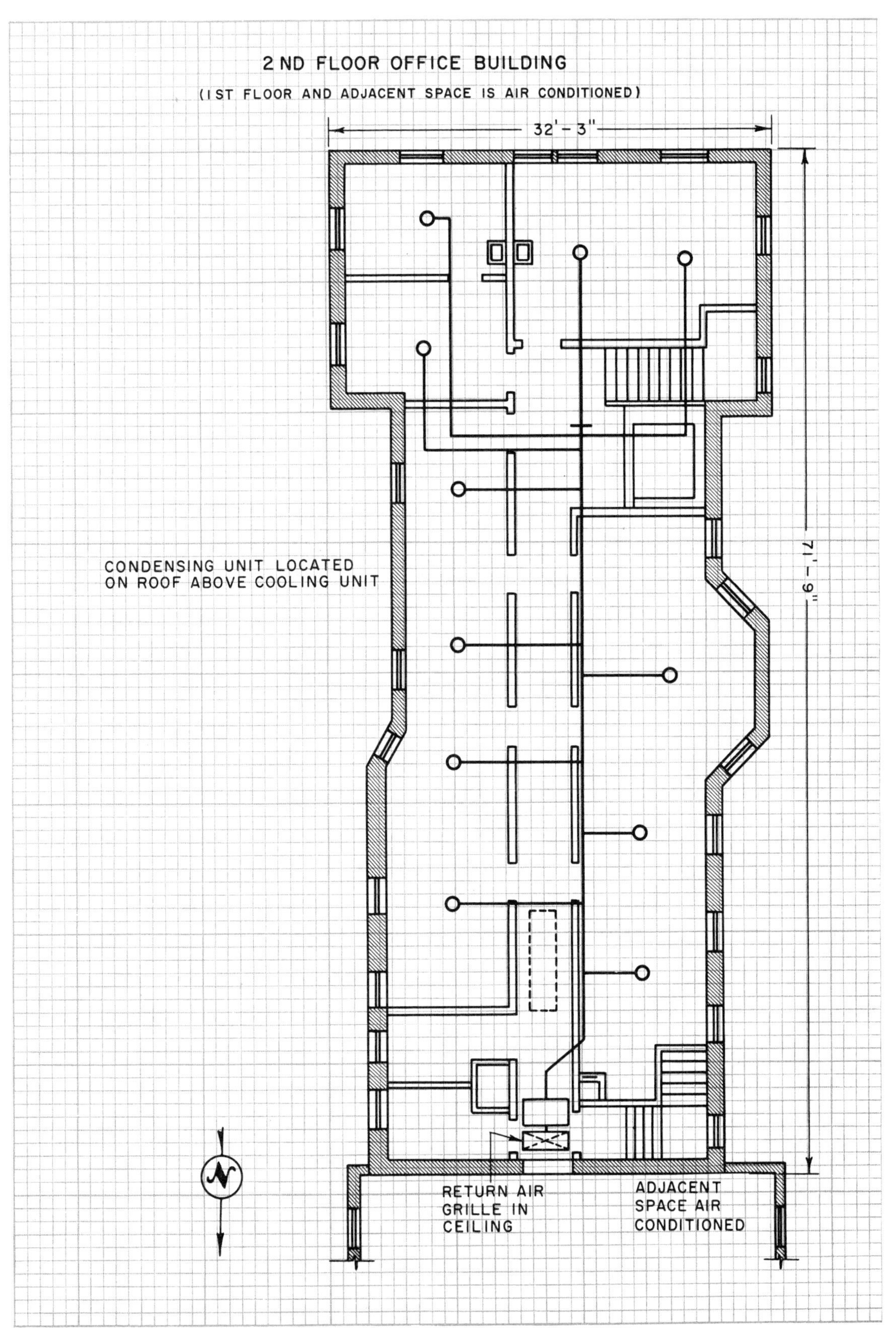

Fig. 17-6 Layout for office building

SECTION 6
RESIDENTIAL AND COMMERCIAL EQUIPMENT

UNIT 18
AIR-CONDITIONING EQUIPMENT

OBJECTIVES

After completing the study of this unit, the student will be able to

- state the advantages of window-type air-conditioning units.
- state the advantages of central air-conditioning units.
- describe the cooling equipment combinations that can be installed in a one-story residence and in a two-story residence for heating systems of the following types –
  1. radiator and panel heating system
  2. forced warm air heating
- list the components of the following residential systems and state the function of each component –
  1. warm air winter air-conditioning system
  2. summer air-conditioning system
- describe the operation of two types of cooling towers.
- list the advantages of commercial self-contained air-conditioning equipment.
- define what is meant by a commercial built-up air-conditioning system.

RESIDENTIAL EQUIPMENT

Residential air conditioning is accomplished by two general types of equipment: window units and central systems.

WINDOW UNIT

Window units (also known as room units), figure 18-1, are designed to air condition a portion of the residence. The capacity of such a unit is designed so that a given unit size is adequate to condition one room. Thus, a unit in a larger size is satisfactory for more than one room. If the unit is to be used for more than one room, however, the arrangement of the rooms must insure good airflow. Although a unit

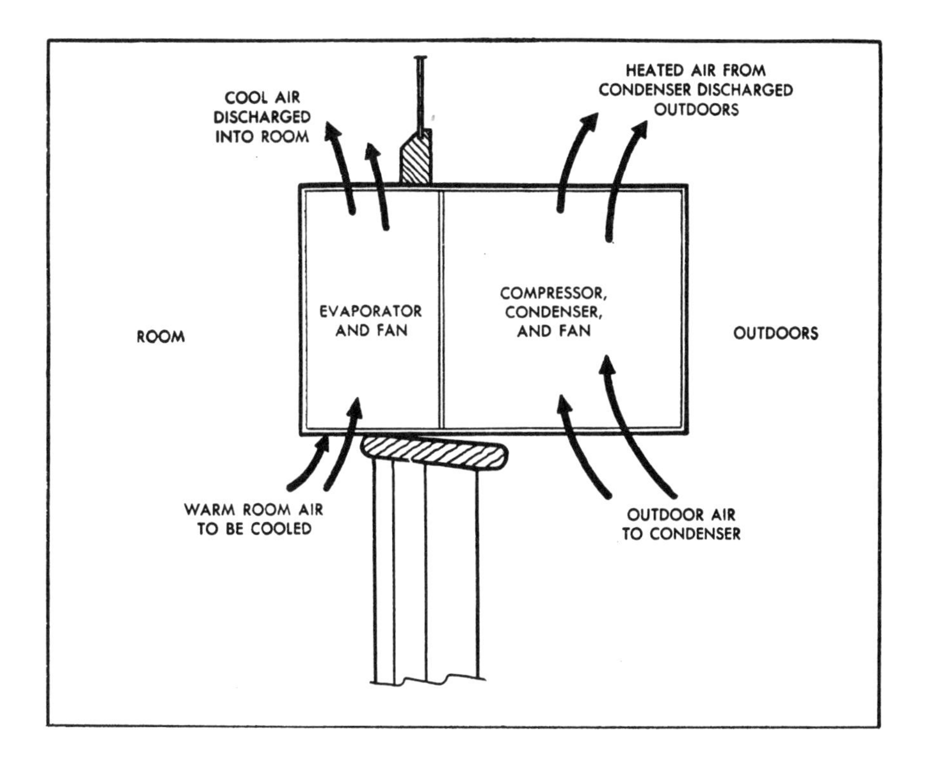

Fig. 18-1 The window-type room cooling unit

may be designed to air condition only one room, several window units can be used to air condition an entire residence. In fact, it is possible to air condition an entire multi-room commercial building using window units.

Advantages of the window unit include the following:

- A separate temperature control is provided in each room or area in which a unit is installed.
- Ducts are not required. This advantage is especially noticeable in residences in mild climates where central heating systems are not required. The advantage may not apply in cooler regions where duct systems are necessary for winter heating, regardless of the summer requirements.
- Plumbing is not required.
- The installation of the unit is simple; therefore, changes in the construction of the residence usually are not required.
- Some types of window units are fitted with heating coils and can be used as supplementary sources of heat.

Disadvantages of window units include the following:

- The unit requires space in a window.
- In general, the unit has a fixed air quantity.
- The installation must be made on an outside wall.
- Appearance may be a factor.

CENTRAL SYSTEMS

There are two categories of central air-conditioning equipment for residences. The first category is cooling equipment that is added to or combined with radiator and radiant panel heating systems. In the second category, the cooling equipment is added to or combined with warm air heating systems.

In either case, the type of equipment selected depends (1) on the style of the residence (single story or two story), and (2) for an existing structure, the type of heating system installed.

Central systems offer the following advantages:

- Comfort conditions are maintained in every room in the house.
- A central automatic control point is provided.
- Since the cooling and heating equipment is installed in one location rather than several locations, the maintenance of the system is easier.
- Better air distribution.

One of the major differences between central systems and window units is the higher equipment and installation cost of the central system. Several typical configurations of central systems are shown in figures 18-2, 18-3, and 18-4.

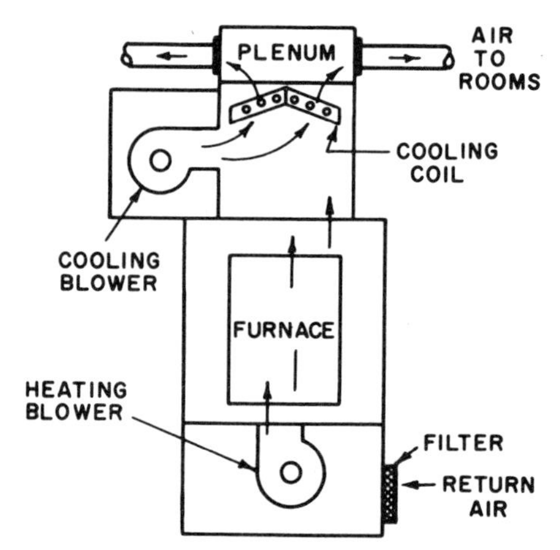

Fig. 18-2 Heating and cooling with a central system; the cooling unit is added to a warm air system

RADIATOR AND PANEL HEATING

The heating systems of many residences use steam or hot water radiators or radiant panels. For a ranch or single-story residence containing this type of heating system, cooling can be achieved with the following equipment combinations.

- A fan coil unit with a remotely located water- or air-cooled condensing unit. The fan coil section can be installed in the basement, attic, or crawl space and is connected

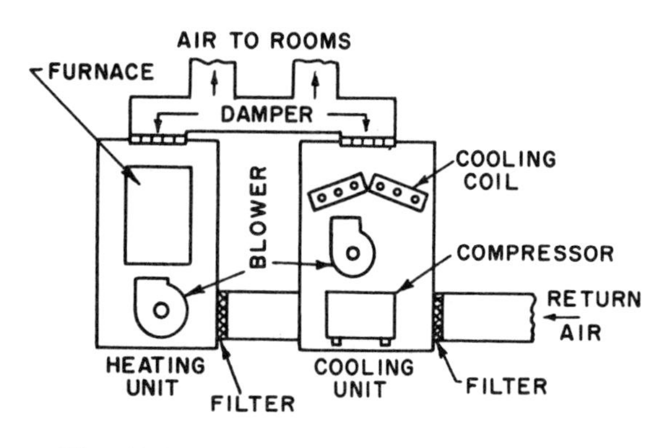

Fig. 18-3 Separate heating and cooling units using a common duct system

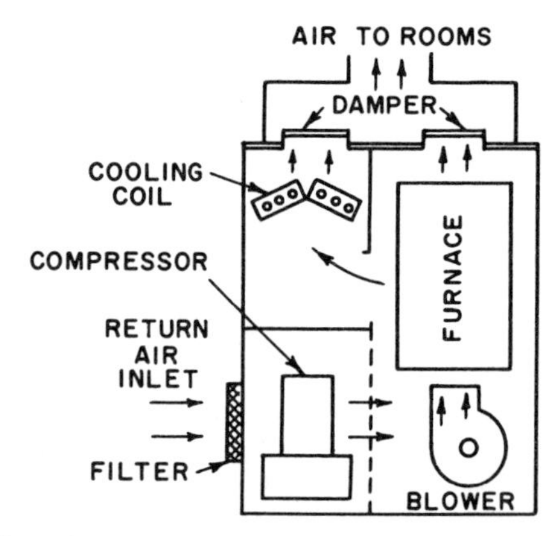

Fig. 18-4 A combination heating and cooling unit

to its own duct system. If the house is very large, two fan coil units can be used with each unit having its own duct system.

- A self-contained unit with fan, coil, condenser, and compressor in one package can be installed inside or outside of the house and connected to its own duct system.

A two-story house with steam or hot water radiators or radiant panel heating can be cooled using the following equipment combinations:

- One fan coil unit located in the basement, attic, or crawl space, and one condensing unit placed in a remote location. This arrangement usually requires an extensive duct system.
- One fan coil unit in the basement or crawl space to supply a separate first floor duct system, with a second fan coil unit in the attic to supply a separate second floor duct system. One remotely located condensing unit can be used for both fan coil units or a separate condenser can be installed for each unit.
- One or two self-contained packages may be used. The duct systems will be the same as those described for the previous equipment combinations.

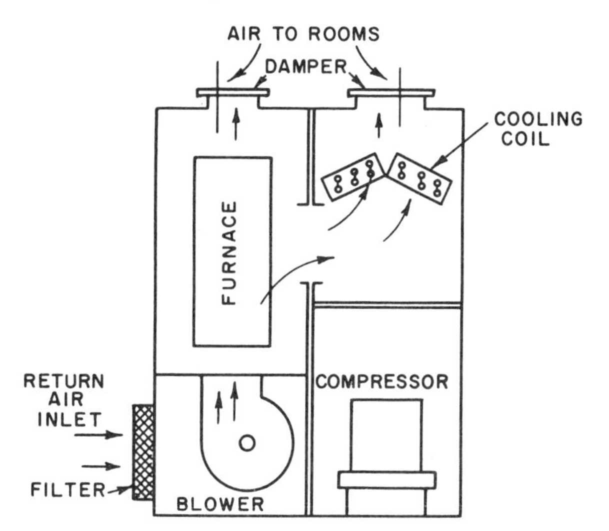

Fig. 18-5 A dual-unit system with one blower

FORCED WARM AIR HEATING

Single-story or two-story houses having forced warm air heating systems can be cooled using any one of the following equipment combinations:

- a cooling coil located in the discharge plenum of the duct system and a remotely located condensing unit. The capacity of the unit should be checked to determine if the same fan can be used for both the heating and cooling applications.
- a single package combining the cooling and heating equipment with the condensing unit (for cooling) in a remote location. The combination unit can be placed in the basement, attic, or crawl space.
- a single package containing the cooling and heating equipment as well as the condensing unit for cooling.

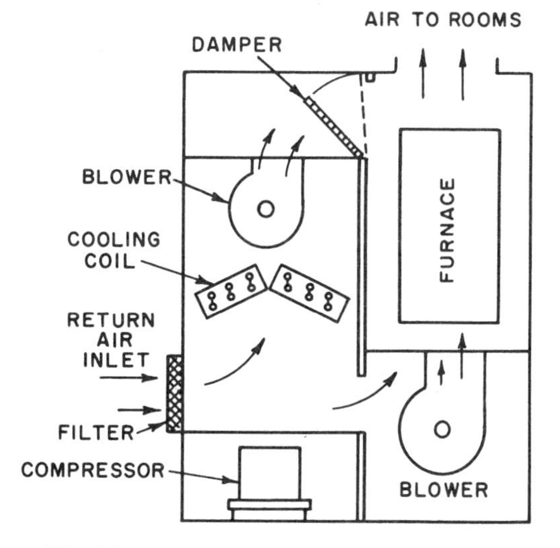

Fig. 18-6 A dual-unit system with two blowers

If the residence has a gravity warm air heating system, it is usually a good procedure to replace the furnace with a forced air system. Such a change may also require a change in the duct system.

WARM AIR RESIDENTIAL SYSTEM EQUIPMENT

(The information presented on pages 232 through 235 was adapted by permission from the National Environmental Systems Contractors Association, NESCA.)

Winter Conditioning System

A complete warm air, winter air-conditioning system is shown in figure 18-7. This air-conditioning system consists of the blower, burner, controls, humidifier (where required), and air filter. The ducts, dampers, return air intakes, and warm air outlets make up the duct system. A winter air conditioner burns fuel and provides heat. It may be hand fired or provided with a burner using a solid, liquid, or gaseous fuel. The main parts of the system are described in the following sections.

- A room thermostat, figure 18-7A, is a temperature actuated electrical switch. It maintains the room air temperature by turning the burner of the system on or off. In general, the room thermostat does not directly control the operation of the blower.
- Return air intakes, figure 18-7B, are grilles through which the air passes on its way from the rooms to be reconditioned.

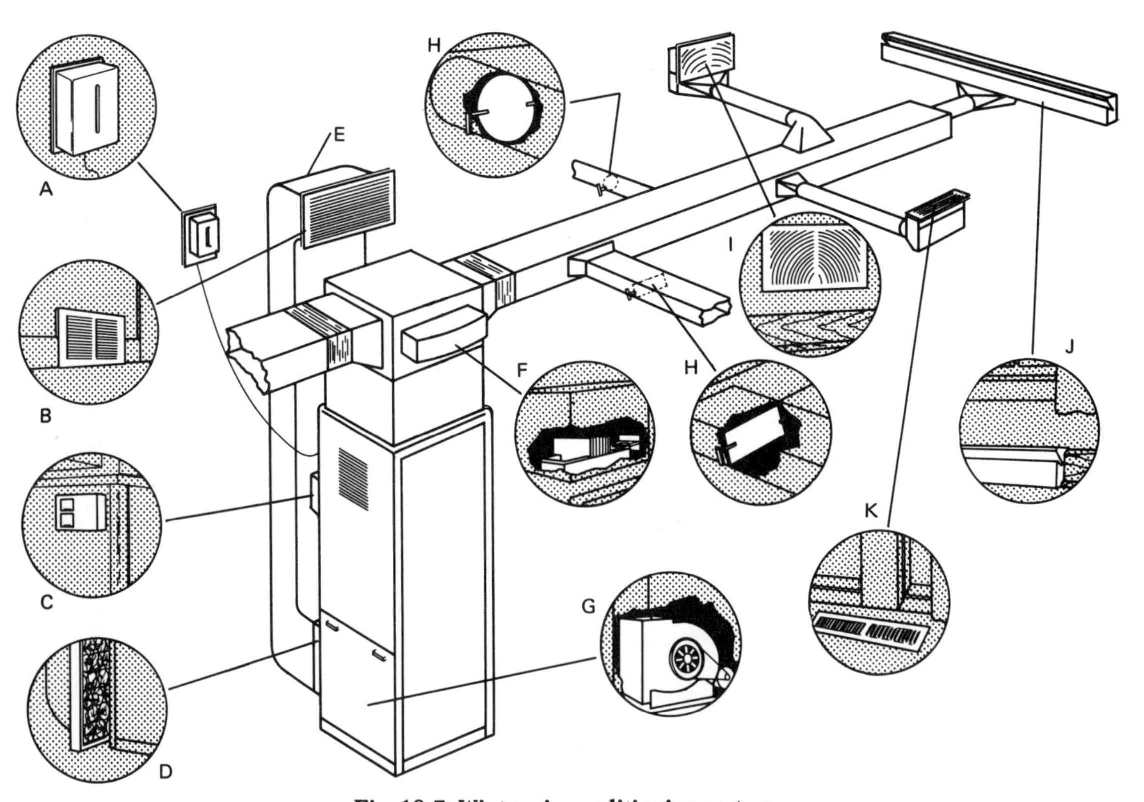

Fig. 18-7 Winter air-conditioning system

- Bonnet thermostats, figure 18-7C, are temperature actuated switches which provide automatic control of the blower and limit the maximum bonnet air temperature.
- An air filter, figure 18-7D, removes lint and dust particles from the air circulated through the unit and the duct system.
- A duct system, figure 18-7E, returns the room air from the return air intakes to the unit. The duct system also supplies conditioned air from the unit to the rooms. Ventilation air can be introduced into the building through the duct system as well.
- A humidifier, figure 18-7F, if required, adds moisture to the circulating air in the system.
- A blower (also known as a fan), figure 18-7G, circulates the air through the winter air conditioner and duct system.
- Volume dampers, figure 18-7H, are located in the supply ducts, stackheads, or outlets. In systems with more than one return air inlet, volume dampers may be installed in the return ductwork. These dampers regulate the quantity of air flowing through the various ducts and are used to balance the system.
- Supply outlets are diffusers or registers through which the conditioned air is introduced into the rooms. Certain types of outlets are designed for use in specific locations in a room. Perimeter diffusers, figure 18-7I are usually placed in the outside walls. This type of diffuser directs the air upward in a fan-shaped pattern so that it blankets the outside wall. Ceiling diffusers usually direct the air parallel with the ceiling. Registers are normally located either on the inside wall, figure 18-7J, where they direct the air toward the outside wall, or in the floor, figure 18-7K, where they direct the air upward.

Summer Conditioning System

A complete summer air-conditioning system, figure 18-8, page 234, includes an air distribution system, air handling unit, and refrigeration system. The refrigeration system or cooling unit consists of a compressor, condenser, expansion valve or capillary tube assembly, cooling coil, and interconnecting tubing. The air handling unit, figure 18-8A, includes the cooling coil, a blower and motor, filters, and a condensate drain pan. All of these components are contained in a suitable insulated enclosure. (The cooling coil actually is part of both the air handling unit and the refrigeration system.) The ducts, dampers, return air intakes, and supply air outlets form the duct system.

- The room thermostat, figure 18-8B, maintains the desired room air temperatures by turning the summer air conditioner on or off during the cooling season. For year-round operation, it is necessary to provide either two thermostats (one for heating and one for cooling) or a single thermostat for both applications.
- The air filter, figure 18-8C, removes lint and dust particles from the air circulated through the conditioning unit and duct system.
- The blower fan, figure 18-8D, circulates air through the summer or year-round air conditioner and duct system.

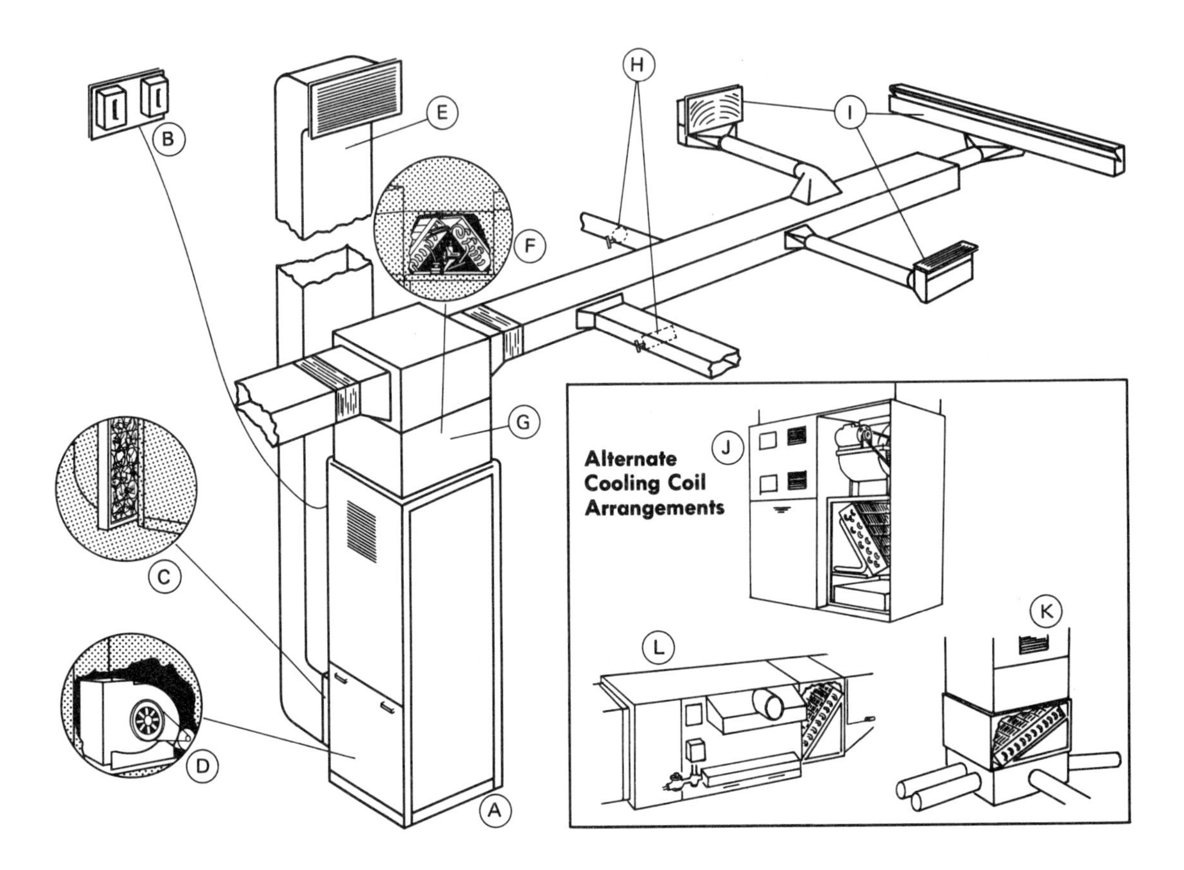

Fig. 18-8 Summer air-conditioning system

- The duct system, figure 18-8E, brings the room air from the return air intakes back to the unit and supplies conditioned air from the unit to the rooms. In addition, ventilation air can be introduced into the building through the duct system.
- The cooling coil, figure 18-8F, removes heat and moisture from the air passing through it. The type of cooling coil used and its location depend on the type of summer air-conditioning unit provided. Heat removed from the air is carried by the cold refrigerant in the cooling coil. Moisture removed from the air is drained from the coil. The cooling coil may be located in the supply plenum, figure 18-8G, of the winter air conditioner or it can be located in the supply duct system. Other components of the summer air conditioner usually are placed in a remote location when a plenum-mounted cooling coil is used.
- Volume dampers, figure 18-8H, are placed in supply ducts, stackheads, or outlets. In systems with more than one return air inlet, volume dampers may also be installed in the return ductwork. The dampers regulate the quantity of air that is to be removed through the various ducts. In addition, the dampers are used to balance the system.
- The supply outlets, figure 18-8I, consist of diffusers or registers through which conditioned air is introduced into rooms. Certain types of outlets are designed for use in specific locations in a room.

Alternate Cooling Coil Arrangements

Figure 18-8 also indicates that the cooling coil may be located in a twin or companion summer air conditioner. Thus, during the cooling season, air passes through the summer air conditioner; during the heating season, air passes through the winter air conditioner. The twin unit, such as the one shown in figure 18-8J, contains a blower. Other components of the cooling cycle may be mounted in the cabinet of the twin unit (self-contained unit) or they may be mounted remotely. The plenum-mounted cooling coil is used with a down-flow winter air conditioner, figure 18-8K. The cooling coil and housing are placed below the supply outlet of the winter air conditioner. The cooling coil is mounted in the supply outlet or duct system of a horizontal-type winter air conditioner, figure 18-8L. In this system, as in other systems using plenum-mounted coils, the air passes through the winter air conditioner before it passes through the cooling coil.

TYPES OF HUMIDIFIERS

One of the most common methods of adding humidity to a conditioned space is to supply air moistened by the evaporation of water. This moisture is provided by a humidifier, figure 18-9. The humidifier should have a water supply control valve with an overflow outlet and should be easily accessible for inspection and cleaning. The humidifier may contain porous plates to increase the surface area from which evaporation can take place. These plates should be inspected periodically. Replacement of the plates is required if they are coated with minerals from the water or other materials with the result that their absorbent qualities are reduced. Spray-type humidifiers can be used as well.

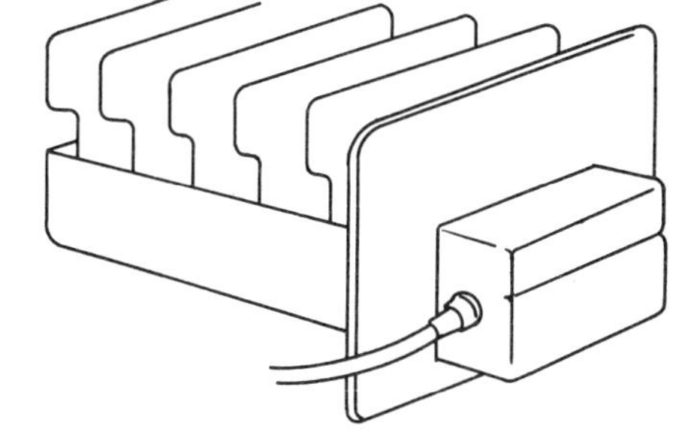

Fig. 18-9 Humidifier

MECHANICAL FILTERS

Mechanical or dry-type filters, figure 18-10, are provided as an integral part of the winter or year-round air-conditioning equipment. Such filters trap or entrain most large airborne dirt, dust, and lint particles. However, the filters have little effect on small particles. The filtering medium is tightly packed between a mesh-type or woven material. The medium is usually some type of fibrous material such as bonded glass fibers, aluminum, hair, or treated paper.

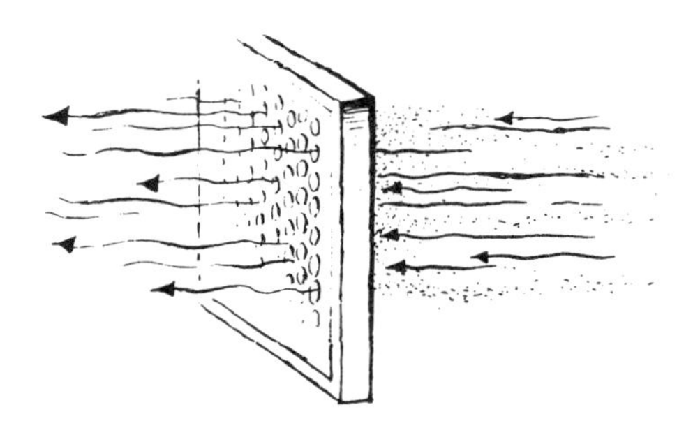

Fig. 18-10 Mechanical filter

Filter Maintenance

Mechanical filters should be inspected periodically (monthly), particulary when the conditioning system is in use in areas near industrial plants or other sources of airborne dirt. Disposable-type filters should be replaced as soon as their effectiveness is reduced. Washable-type filters should be cleaned in accordance with the manufacturer's recommendations. Dirty filters limit the flow of circulation air. As a result, the performance of the winter or year-round air-conditioning unit can be adversely affected.

SELF-CHARGING FILTERS

Another type of filter is similar in appearance to a mechanical filter but uses a plastic filtering medium. Such a filter combines a mechanical and an electrostatic filtering action. The electrostatic charge is generated by air passing through the filtering medium. Dirt particles are attracted to the surfaces of the material by this electrostatic charge. Self-charging filters must be inspected and cleaned frequently in accordance with the manufacturers' recommendations. Once the filter is cleaned, it is reusable.

CHARGED-MEDIA ELECTRONIC FILTERS

Charged-media electronic filters, figure 18-11, operate on the principle that unlike electrical charges attract one another. The airstream passes through a series of charging plates so that the dirt particles in the air acquire a strong charge. The airstream then passes through a bank of collector plates which have an electrical charge opposite to that on the particles. The charged dirt particles are attracted to the collector plates. When these plates are loaded, they can be removed for cleaning.

This type of filter requires a source of high voltage to supply the charge for the particles in the airstream and the collector plates.

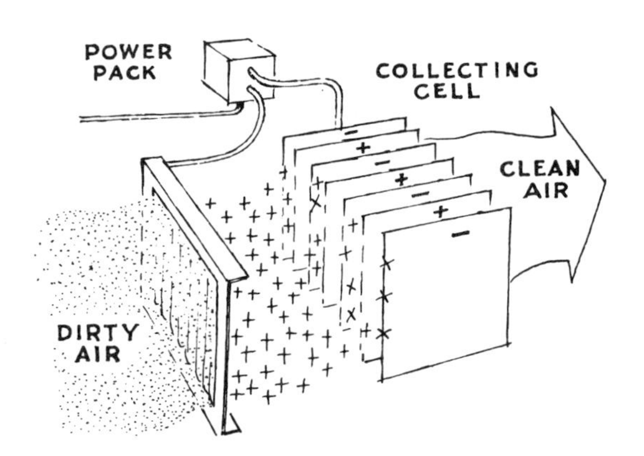

Fig. 18-11 Electronic filter

COOLING TOWERS

There are two types of cooling towers as shown in figures 18-12 and 18-13. The atmospheric cooling tower, figure 18-12, depends upon wind action to bring the air into contact with a water spray. Warm water enters the top of the tower through

Fig. 18-12 Atmospheric cooling tower

Fig. 18-13 Forced draft cooling tower

a distribution header. The water is then broken into a fine spray by nozzles. As the water spray drops through the tower, it comes in contact with the air. The water is cooled by evaporation and is collected in the sump in the base of the tower. Makeup water, entering the tower at the base, and recirculated water are drawn into the circulating pump from the sump.

A forced draft cooling tower, figure 18-13, is similar to the atmospheric tower in many respects. However, the forced draft tower contains a fan to provide positive air circulation. Wood fill or louvers are used to increase the area of contact between the air and the water. The water is introduced from the distribution header in the form of droplets rather than as a fine spray. In general, the water losses in a forced draft cooling tower are less than those in an atmospheric cooling tower having the same capacity.

COMMERCIAL AIR-CONDITIONING EQUIPMENT

Offices, restaurants, churches, supermarkets, bowling alleys, and buildings having similar or related uses can be air conditioned using single- or split package self-contained units or built-up systems.

Single- and Split Package Self-Contained Equipment

Self-contained equipment is designed to be used with or without a duct system. Such equipment can condition comparatively large open areas, individual small areas, or combinations of both types of areas. In most commercial applications, self-contained equipment offers great flexibility as shown by the following possible equipment combinations:

- Single-package, floor-mounted unit with a water-cooled condensing unit, no duct system, and with or without a cooling tower.
- Single-package, floor-mounted unit with a water-cooled condensing unit, a duct system, and with or without a cooling system.
- Split package, floor-mounted unit with a remote air-cooled condensing unit with or without a duct system.

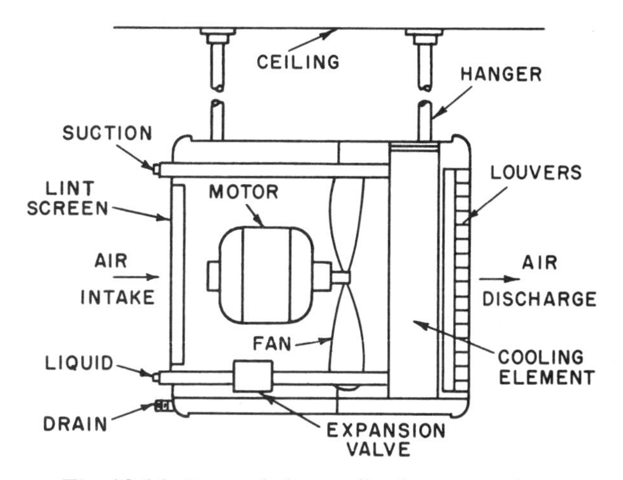

Fig. 18-14 Suspended propeller fan-type unit

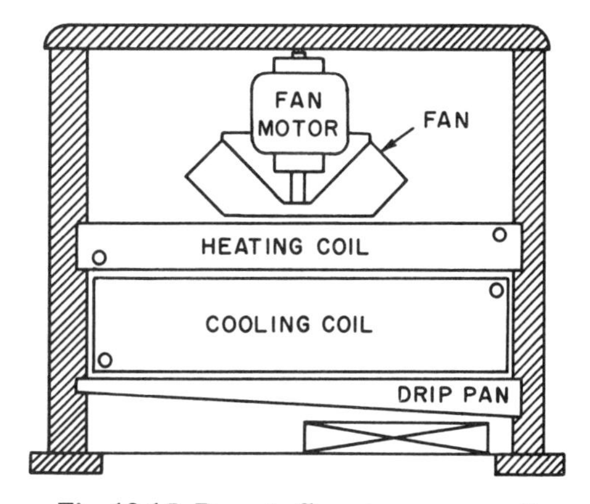

Fig. 18-15 Remote floor-type room unit

- Split package, ceiling-mounted unit with a remote air-cooled condensing unit with or without a duct system.
- Single- or split package unit with a cooling coil only and in the combinations listed here.
- Single- or split package unit with a heating and cooling coil and in the combinations listed here.
- Any of the previous combinations where all of the equipment is located outside of the air-conditioned space. Suitable ducting is provided to connect the fan discharge outlet to the conditioned space.

Self-contained air-conditioning equipment offers the following advantages:

- Complete flexibility provides a solution to space and water supply problems.
- Individual space temperature control.
- Low cost air distribution if the area is free of obstructions.
- Extremely simple operating controls.
- Simple installation which does not require special foundation construction.
- Relocation requires a minimum of effort in the event of business expansion.

Although self-contained equipment has numerous advantages, the following details may offset the advantages and should be considered before such equipment is installed:

- Water-cooled condensing units require water piping, drain piping, and possibly cooling towers.
- Air-cooled condensing units require refrigerant piping and electrical wiring to the remote location.
- Floor-mounted equipment located in the conditioned space uses valuable floor area.
- Air-cooled condensing units may be prohibitive if a reasonably close location is not possible. Long runs of refrigerant piping may not be desirable.
- Cooling towers may be difficult to locate without considerable plumbing, foundation, and rigging costs.

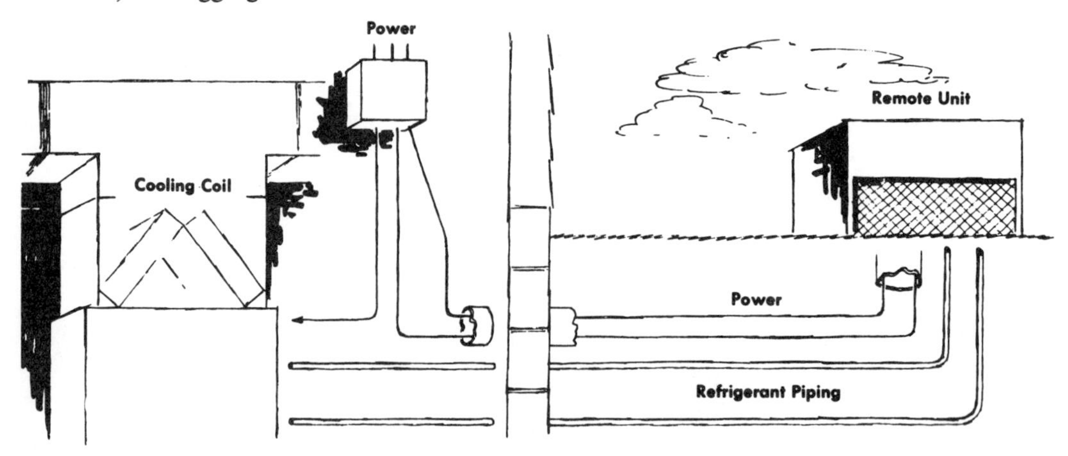

Fig. 18-16 Remote component arrangements

Built-Up System

A built-up air-conditioning system, figure 18-17, offers almost unlimited flexibility for commercial applications where there are many individual private areas or several zones. Equipment for built-up systems is available in a variety of sizes and combinations which can be selected and arranged to maintain exacting comfort conditions. Large or small areas with different comfort requirements can be handled simultaneously. A built-up system can be designed to supply simultaneous cooling to one area and heating to another area. There are two basic types of built-up systems. The first type uses a ceiling or high wall air distribution pattern with manual dampers for each large or small space. The fan, cooling coil, and condensing unit or water chiller are located outside the air-conditioned space.

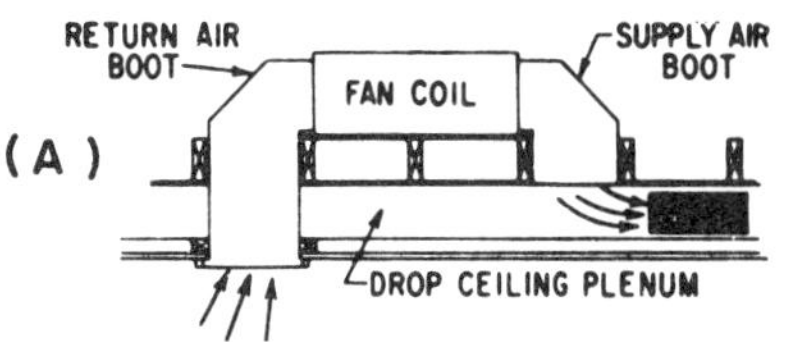

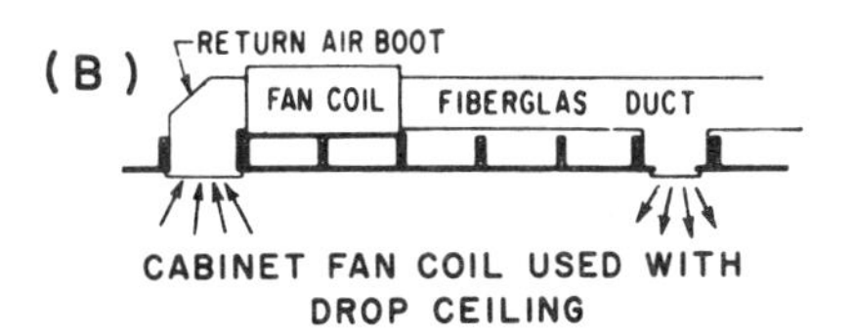

Fig. 18-17 Built-up system

Such a system requires an equipment room which is usually located in the basement or in a special enclosure on the roof. The equipment room contains the fan, cooling and heating coil, filter, condensing unit or evaporative condensing coil, water cooling equipment, and a heater.

The second type of built-up system uses fan coil units located under windows or suspended from the ceilings in closets or similar available spaces. When a ceiling unit is concealed in the closet, only the supply and return outlet face of the unit is visible in the space.

This type of built-up system has a fan coil in each window or closet installation. Cold or hot water is supplied to the coil from a remote equipment room containing condensers, water chillers, and a hot water heating source.

If ventilation air is supplied by an overhead duct system from a central source, heating and cooling can be supplied simultaneously. That is, hot or cold water can be supplied to the coil and heated or cooled air can be supplied for ventilation. This dual arrangement is particularly useful during marginal weather when the outside temperature may change daily from cold to warm or from warm to cold. Ventilation air can also be supplied through openings in the outside wall for each unit. In this case, hot or cold water is available to the unit installed under the window and the unit coil handles the total load.

Still another type of system uses an induction assembly only, figure 18-18, page 240. This assembly is designed to bring room air across the coil. The induction-type conditioner has a central fan that delivers conditioned air through a duct system to each induction unit. The coil in the induction unit is supplied with hot or cold water. Because there is a combination of warm or cold air and hot or cold water available at the unit, almost any temperature and humidity condition can be maintained, regardless of outdoor conditions.

Built-up systems have a number of advantages. For example, such a system can be designed with individual room control. Cooling or heating capacity not needed in one zone can be directed to another zone. The equipment is remotely located. As a result, equipment noise usually is not a problem. Since this type of equipment can be designed to supply conditioned air continuously, humidity control is constant.

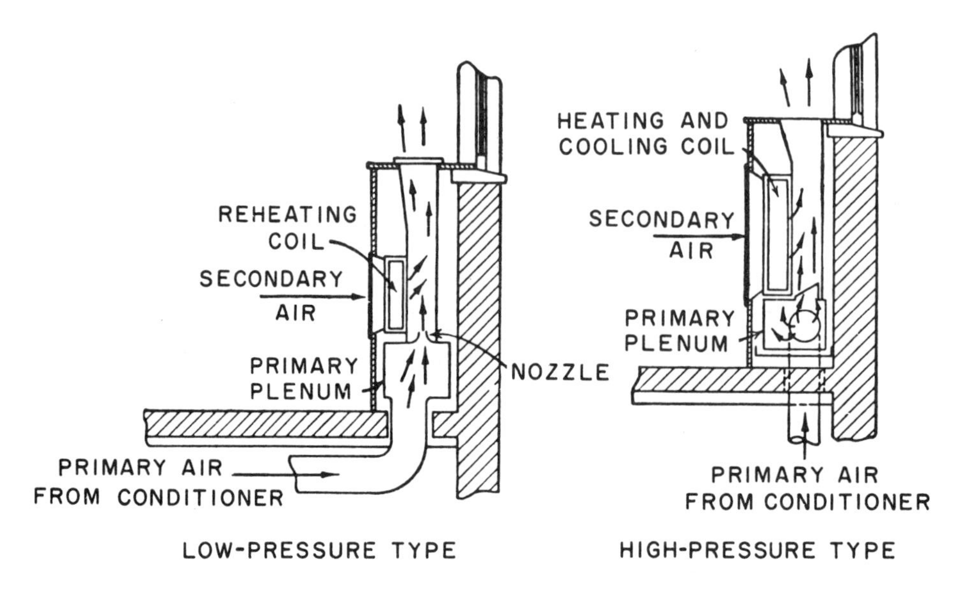

Fig. 18-18 Induction-type room air conditioners

The refrigeration and air handling equipment can be located in one area; therefore, maintenance is simplfied.

While built-up equipment systems have the advantages listed, there are several factors which may offset some of the advantages. For example, the complete system must be used to supply air conditioning for overtime work even though only a small portion of the building is occupied. In addition, valuable space may be required for equipment rooms. Additional controls and ductwork may be required to provide individual room control.

SPECIAL CONSIDERATIONS

Each buyer of air conditioning equipment may have preferences or requirements that apply only to a particular system. In addition, the physical arrangement and use of the structure to be air conditioned may require special installation techniques. These two considerations have a direct influence on the selection of the air-conditioning equipment for a given application. Therefore, the system designer should investigate each of the following questions to insure that the equipment selected is the best choice for the application.

- Are there special areas requiring unusual heating or cooling treatments?
- Is the owner aware of the cost of one system as compared to another system (where both systems will do a similar job but each system has special features)?
- Is the owner aware that the cost of the equipment and the cost of operating the system for a given application can vary considerably? For example, it is usually more economical to install a water-cooled system without a water tower. However, the cost over the years of the additional water required when the system is operating is usually greater than the initial cost of the tower. In other words, although the system costs more in the beginning because of the addition of a tower, the cost of the water saved by using the tower actually pays for the tower. Thus, the addition of the tower actually becomes a long-run cost saving.

- Is the sound level of equipment installed in the space as compared to that of equipment installed in a remote equipment room significant to the occupants of the building?
- Are there unusual problems in locating the equipment?
- Is future expansion planned?
- Is the future use of the space likely to change?
- Can the installation be made with the occupants in the building?
- Are changes to the building structure required?
- Is central shutdown required, or is building zone shutdown more appropriate?
- Is appearance a major factor?

SUMMARY

- Room or window units are designed to condition a single room; but, several units may be used to completely air condition a house or larger building. Window units offer individual room temperature control, air distribution without ducts, and heating and cooling. However, these units require space in the room, outside wall installations, and usually have a fixed air quantity.
- Residential central air-conditioning equipment offers cooling equipment that is added to or combined with radiator or radiant panel heating systems, or warm air heating systems. The cooling equipment requires a separate duct system unless it is to be combined with warm air systems. A central system supplies comfort conditions to every room, has better air distribution, and is easier to maintain than room units. A central system, however, has a higher initial equipment and installation cost.
- For single- or two-story residences, single-package fan coil units with a water- or air-cooled condenser are available as are units with the condenser remotely located.
- Mechanical (washable or disposable) and electronic filters are available for the air-conditioning equipment, as are natural draft and forced draft cooling towers.
- Self-contained commercial equipment is available in single- or split package units that can be used with or without duct systems. Such equipment can be used for heating or cooling applications, or both, in restaurants, offices, stores, supermarkets, churches, and similar structures.
- Single- or split-package units can be floor- or ceiling-mounted in or outside of the conditioned space. Self-contained equipment is flexible, gives individual space temperature control, is simple to operate, and can be relocated easily.
- Water-cooled condensing units require plumbing; air-cooled condensing units require refrigerant and electrical lines to the condenser.

- Applied or built-up system equipment offers many combinations of fan, cooling coil, heating coils, condensing units, evaporative condensers, filters, water coolers, and water heaters. Built-up systems can use fan coil units installed under windows or induction units. The air can be supplied to the unit from a central point, or, in the case of the fan coil unit, from a central point through a separate overhead duct system. Built-up systems can supply heating and cooling simultaneously in different areas. Such systems usually require a separate equipment room.

REVIEW

1. List three advantages of window air-conditioning units.
2. List three advantages of central air-conditioning systems.
3. Identify the basic parts of window or room units.
4. If a home has gravity warm air heating, should the furnace be replaced for air conditioning? Why?
5. What is an important requirement for the use of electronic filters?
6. Describe the two common types of cooling towers and their basic differences.
7. Discuss three advantages and three disadvantages of self-contained equipment.
8. Identify the parts of the self-contained air-cooled unit in figure 18-19.

A. ____________

B. ____________

C. ____________

D. ____________

E. ____________

F. ____________

G. ____________

H. ____________

I. ____________

J. ____________

K. ____________

L. ____________

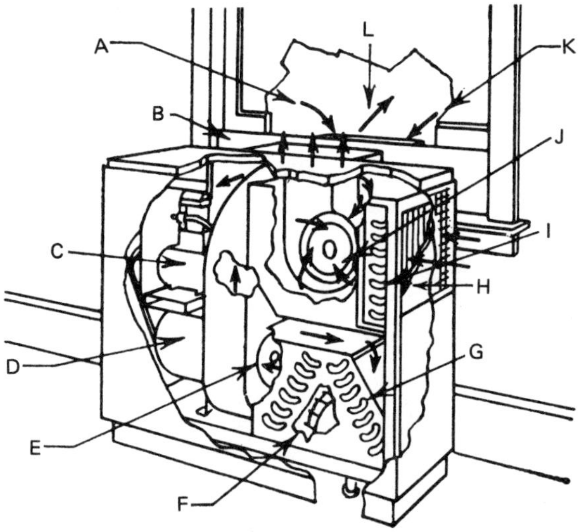

Fig. 18-19 A self-contained air-cooled unit

UNIT 19

INSTALLING RESIDENTIAL AND SMALL COMMERCIAL EQUIPMENT

OBJECTIVES

After completing the study of this unit, the student will be able to

- uncrate, install, and check the operation of a window-type room air conditioner.
- list the basic items in a preventive maintenance program for a window air conditioner.

The popularity of residential and commercial air conditioning has increased enormously in the past 25 years. To fill the demand for inexpensive and dependable equipment designed to meet all requirements, manufacturers have developed a wide variety of units in most price ranges. The two types of equipment generally used are window units (room air conditioners) and package units (self-contained).

A limitation in the amount of space available for the installation is often the factor which determines the selection of the best type of unit for a particular requirement. Each job must be considered on an individual basis.

Earlier units of this text outlined the procedures used in estimating loads and selecting equipment. It is the responsibility of the job estimator to decide the best location for the air-conditioning equipment. This equipment may be located on the roof, in the attic, in the basement, in a closet, suspended on outside walls, or placed in the yard outside of the building.

Units 19 through 23 describe the installation, startup, maintenance, and service of residential and commercial air-conditioning equipment.

INSTALLING AND SERVICING ROOM AIR CONDITIONERS

Room air conditioners are small self-contained units whose basic function is to cool a single room or space. These units vary in capacity from 1/2 to 2 tons. The majority of room units are air cooled. They are designed to be installed on a window sill or through a hole cut in a wall. Floor-mounted models called consoles are also available.

Room units must be located near a source of outdoor air in sufficient quantities to permit the proper air-cooled condenser heat rejection. (Note: the manufacturer's installation instructions shipped with each unit are to be studied carefully before the installation is started.)

Window Sill Installation (Typical Unit)

The following instructions describe the installation of a typical window unit.

1. Remove the shipping crate and inspect the unit for shipping damage, figure 19-1, page 244.
2. Tip the unit on its back and remove the bolts holding the shipping skid.

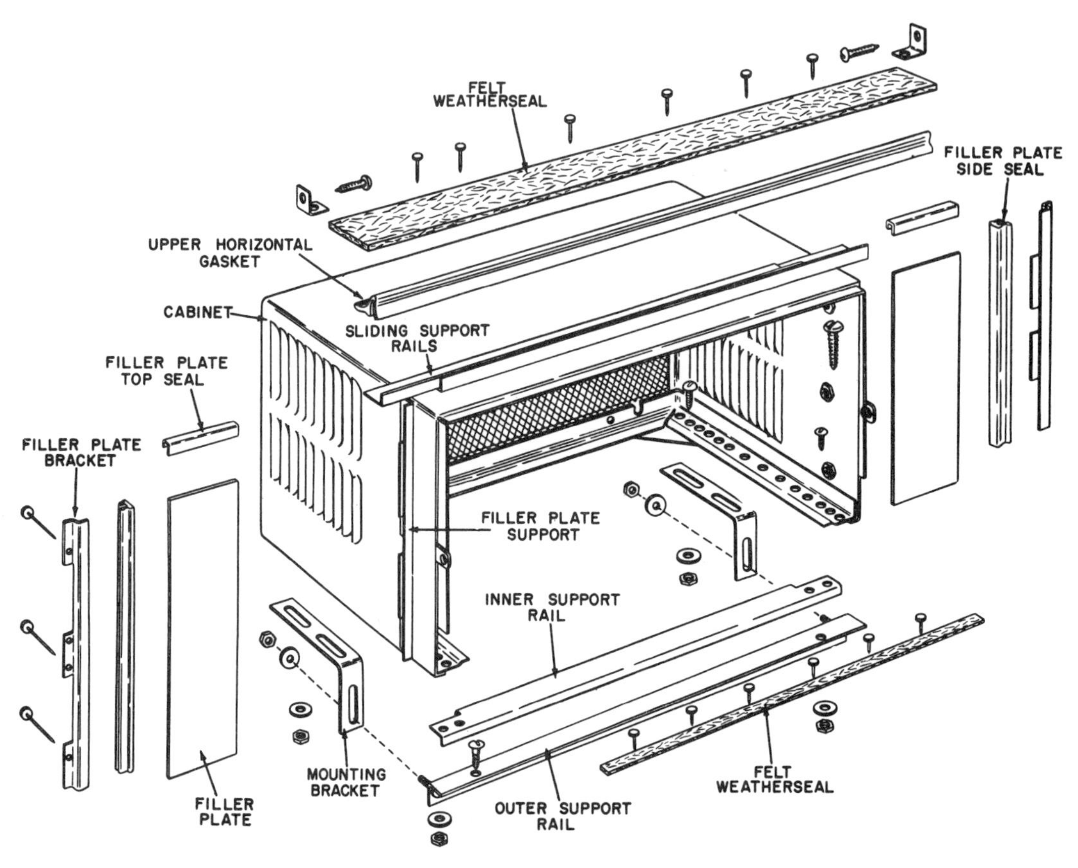

Fig. 19-1 Exploded view of mounting parts

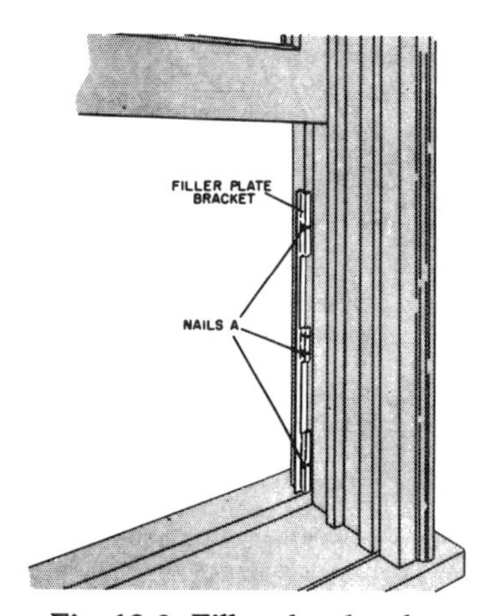

Fig. 19-2 Filler plate bracket

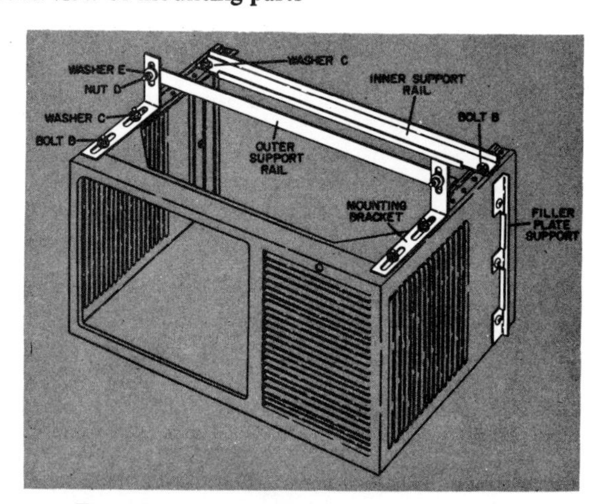

Fig. 19-3 Attaching the mounting bracket

3. Slide the chassis from the cabinet.
4. Rotate the fan motor shaft several times to insure that the fan blades do not strike any part of the chassis.
5. Attach the filler plate brackets as shown in figure 19-2. Use three nails in each bracket.
6. Place the cabinet upside down on the floor and bolt the inner support rail to the cabinet, figure 19-3. The bolt holes in the support rail are to line up with the

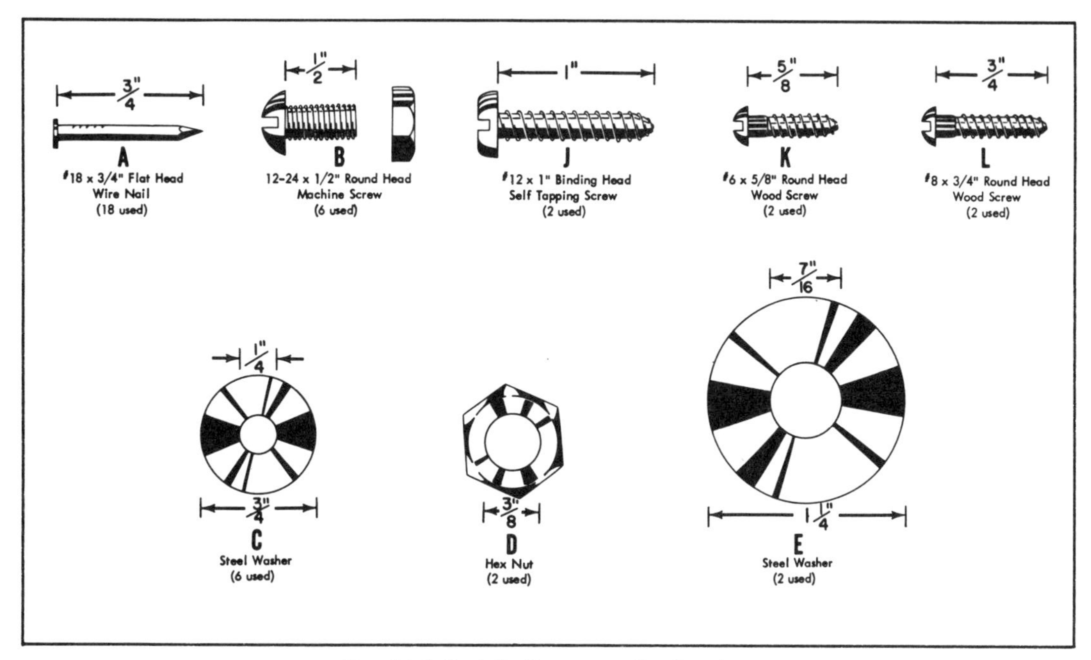

Fig. 19-4 Installation mounting hardware

third hole from the front of the cabinet, on both sides. Secure the support rail with bolts (B) and washers (C), figure 19-4.

7. Measure the window sill to obtain dimension A as shown in figure 19-5.

8. Loosely install the mounting brackets at a distance equal to dimension A from the inner support rail using bolts (B) and washers (C), figure 19-4.

9. Measure from the mounting bracket to obtain dimension B, figure 19-5. Loosely install the outer support rail at this height using nuts (D) and washers (E), figure 19-4.

10. Place the cabinet in the center of the window opening and lower the sash to hold the cabinet in position, figure 19-6.

11. Attach the cabinet to the window stool with screws (J). Attach the small felt weatherseal to the stool with nails (A).

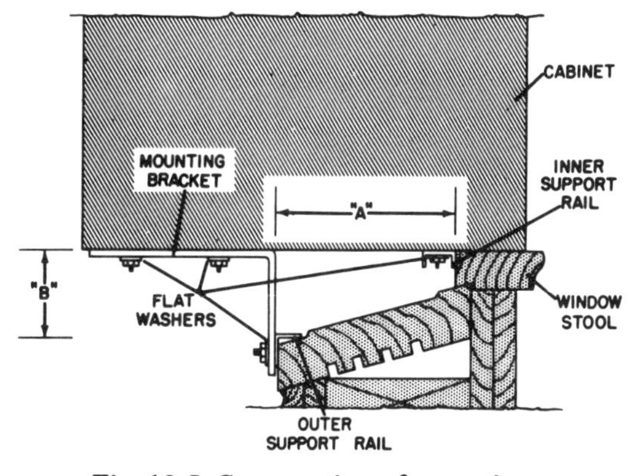

Fig. 19-5 Cross section of mounting

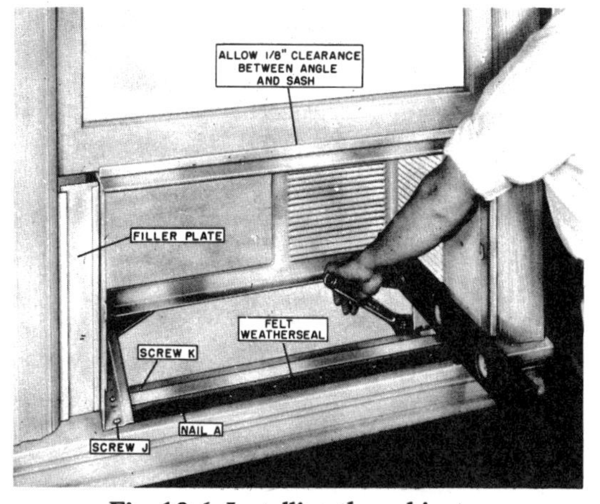

Fig. 19-6 Installing the cabinet

12. Use a level to position the cabinet at a slope so that the outside edge is approximately 1/8 in. lower than the inside edge.
13. Hold the cabinet in this position and tighten the loosely installed bolts (Steps 8 and 9) on the mounting bracket and the outer support.
14. Attach the outer support rail to the sill with screws (K), figure 19-4.
15. Measure the distance between the cabinet and the window frame. Cut the filler plate to this dimension.
16. Raise the window sash and install the filler plate so that the side seals with the beaded edge face toward the room. Slide the filler plate into position, figure 19-7.
17. Loosen the four mounting bolts and extend the sliding support rails to the window frames. Then retighten the mounting bolts.
18. Cut and install the upper filler plate top seal.
19. Cut the upper horizontal gasket to the correct length. Install the gasket.
20. Lower the window sash and install the felt weatherseal, figure 19-8.
21. Loosen the compressor shipping bolts so that the compressor floats freely.
22. Slide the chassis into the cabinet.

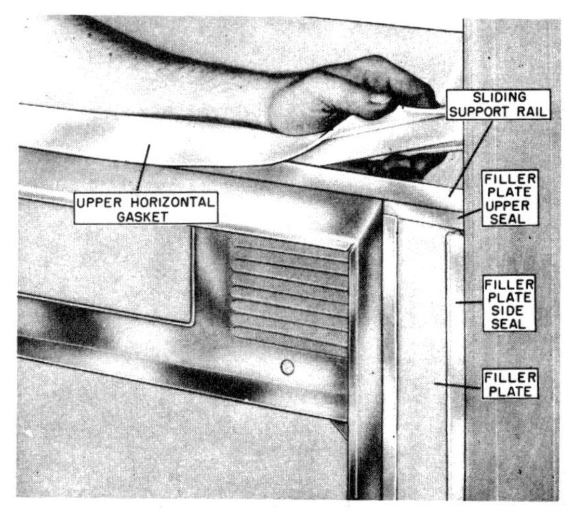

Fig. 19-7 Installing weatherseals

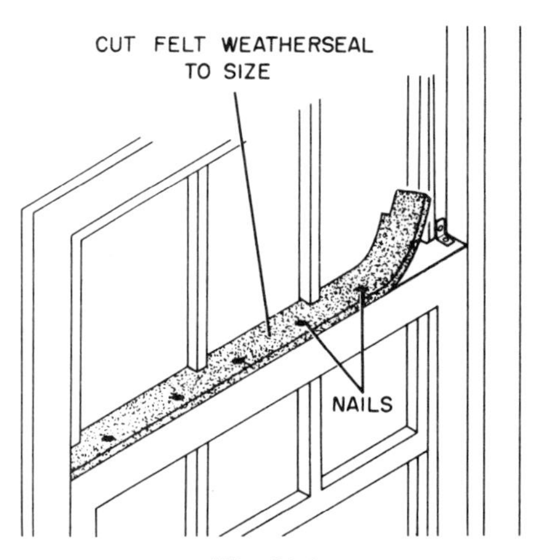

Fig. 19-8

Servicing Window Units

Preventive maintenance is just as important for window units as it is for larger units. Most air-conditioning units are hermetically sealed so that there is no access to the refrigerant circuit. As a result, testing gauges cannot be used without cutting into the refrigerant circuit. Before any attempt is made to cut into the refrigerant circuit, the technician must study the manufacturer's maintenance instructions for the particular unit and model being serviced.

The condenser and evaporator fan motors generally are permanently lubricated; thus, they require no further lubrication in the field. However, if oil cups are provided on any motor, it should be oiled every three months with a high grade oil (SAE 20 viscosity).

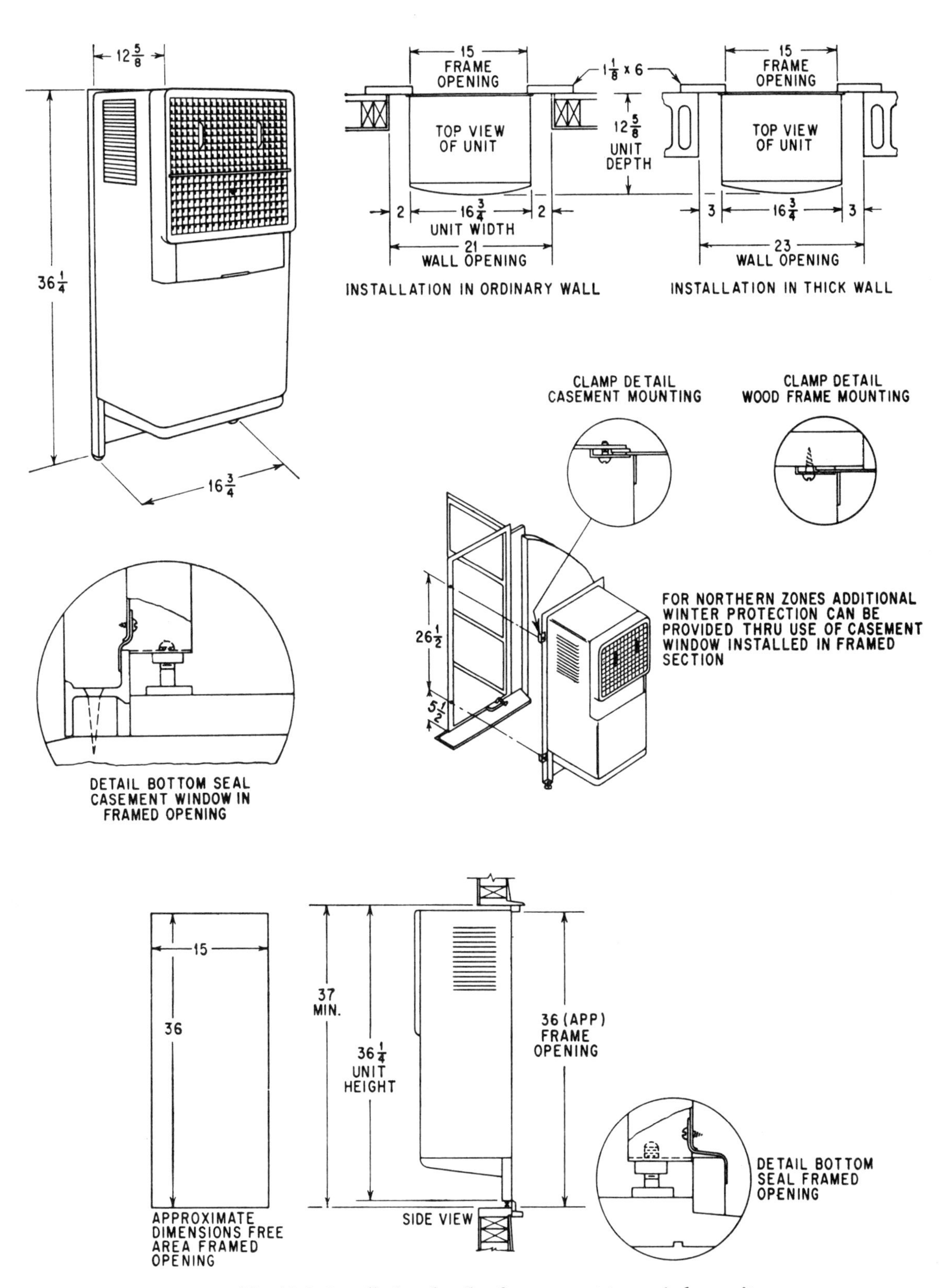

Fig. 19-9 Installation details of a casement-type window unit
(Courtesy Chrysler Airtemp.)

Filters must be changed whenever they get dirty. The amount of time required for the filter to become soiled enough for replacement varies from two weeks to three months, depending upon the location of the installation. Regardless of their appearance, filters should be changed every two months during the operation season.

Window unit design varies considerably between manufacturers and between models from the same manufacturer. As a result, it is necessary to obtain detailed service instructions for the particular model being serviced when major repairs are necessary. Such major repairs include replacing fan motors and charging the refrigerant circuit.

An annual cleaning of the unit is recommended. Both the evaporator coil and the condenser coil and any rusty surfaces must be cleaned. After the rust is removed from the surfaces, they must be repainted.

SUMMARY

- Room air conditioners are designed to be installed in a window or in a hole in the wall.
- Room air conditioners have air-cooled condensers; therefore, these units require a plentiful supply of outdoor air.
- Before installing room air conditioners, it is essential to study the manufacturer's installation instructions provided with the model being installed.
- Preventive maintenance is just as important for small window units as it is for larger commercial equipment. Periodic and thorough unit cleaning and motor lubrication are important.
- The frequency with which the filter is replaced or cleaned depends upon the amount of dust and dirt in the air.

UNIT 20
INSTALLING A WATER-COOLED, SELF-CONTAINED UNIT

OBJECTIVES

After completing the study of this unit, the student will be able to

- list the items of an inspection of the site selected for the installation of a water-cooled, self-contained air-conditioning unit.
- use the proper procedure in uncrating an air-conditioning unit.
- explain the selection and use of isolation pads under an air-conditioning unit when it is placed on a foundation.
- make the following piping and wiring connections –

  piping for the water-cooled condenser
  installing water regulating valves
  cooling tower
  condensate drain piping
  wiring to power supply

- adjust the fan speed of the air-conditioning unit.

INSPECTING THE SITE

The site selected for the equipment installation is to be inspected before the unit is unloaded. A thorough inspection will include the following points:

- Doors or stairways must be wide enough to admit the unit.
- Floors or roofs must be strong enough to support the weight of the unit.
- The unit foundation must be level.
- Enough room is provided around the foundation to install the water piping, electrical service, and ductwork.

Always read the manufacturer's installation instructions carefully before the actual installation is started. These instructions normally are attached to the outside of the shipping crate.

MOVING THE UNIT TO THE SITE AND UNCRATING

The unit is moved to its final location in an upright position. If it is necessary to put the unit on its side to move it through doorways or up and down stairs, place long planks under the unit and gently lower it on these planks. Do not dent the unit casing. At no time should the unit be dropped from a truck, platform, or shipping dock. When the unit is raised or lowered, the base is to be kept lower as the weight of the unit is concentrated at the lower end.

After the unit is moved to the site, remove the shipping crate. Place blocks under one edge of the base skid and remove two of the shipping bolts. Then remove the blocks, place them under the opposite edge of the unit, and remove the two remaining shipping bolts.

Inspect the unit thoroughly for shipping damage before sliding it from the base skid. Claims for shipping damage should be submitted to the transportation company.

PLACING UNIT ON FOUNDATION

The unit is to be installed on a level foundation having sufficient strength and rigidity to support the unit weight. An isolation pad is placed under the *entire* base

(A) FAN SECTION
(B) PANELS
(C) FILTERS
(D) COOLING COIL
(E) DRAIN PAN
(F) ELECTRICAL CONTROLS
(G) CONDENSER (WATER COOLED)
(H) REFRIGERANT DRIER/STRAINER
(I) LIQUID SIGHT GLASS
(J) RECEIVER (AIR COOLED MODEL)
(K) COMPRESSOR

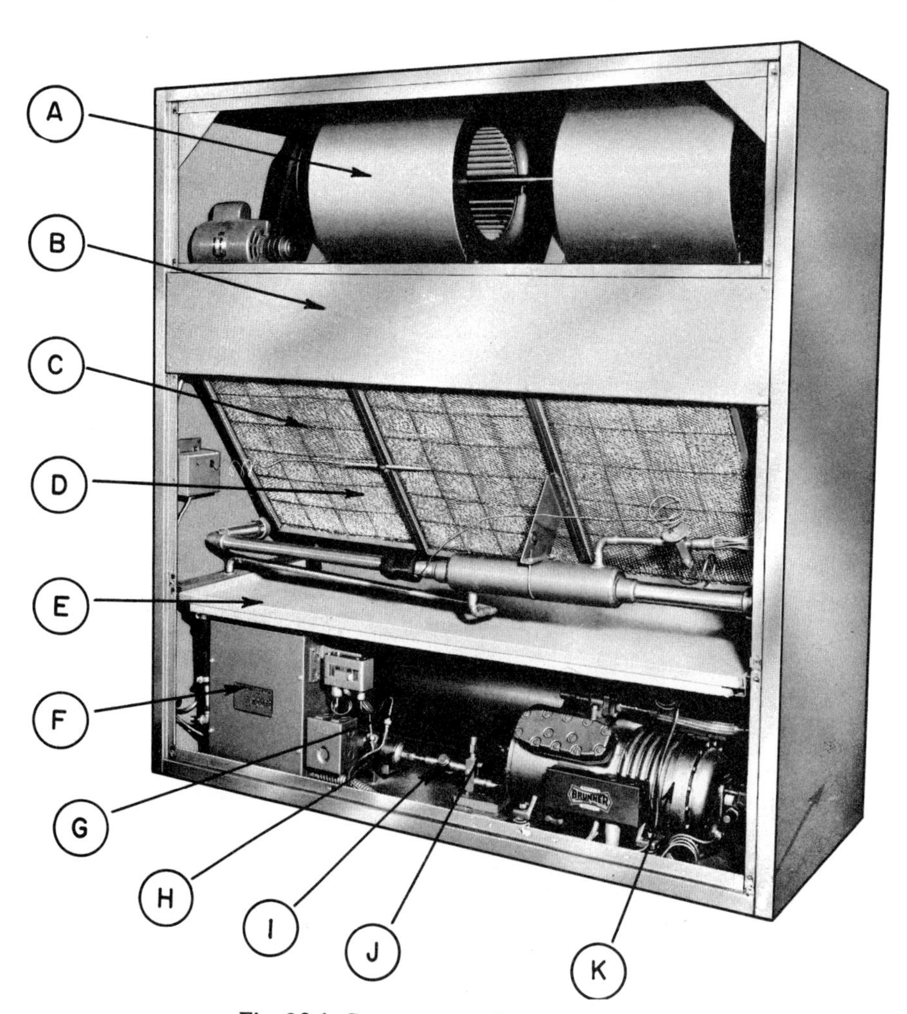

Fig. 20-1 Components of water-cooled unit

of the unit. When isolation pads are placed only at the corners of the unit base, the bottom sheet metal of the unit acts as a sounding board with the result that more noise is generated than when the unit is placed directly on the floor. In order of preference, recommended isolation pad materials are sponge rubber (1/4 in.) and fiberglass roof insulation (1/2 in.)

Structural insulation such as Celotex®, Temlock®, Insulite®, and Lockaire® are good acoustical insulators, but deteriorate and give off odors when wet.

If the unit is placed on a concrete foundation on the ground, an isolation pad is not required normally. In this case, the bottom of the unit is coated with a liquid or semiliquid waterproofing compound such as roofing cement or hot tar.

CONNECTING PIPING FOR THE WATER-COOLED CONDENSER

A water-cooled condenser installed in a package unit usually is of the shell and coil design. The coils are arranged so that there are no joints inside the shell. This design eliminates the possibility of inaccessible leaks. The water is circulated in a parallel flow pattern which is suitable for either city water or cooling tower operation.

Holes for water piping are provided in the condenser paneling; these holes are usually located in the left end panel and the back panel of the unit. Figure 20-2 shows two methods of handling condenser waste water.

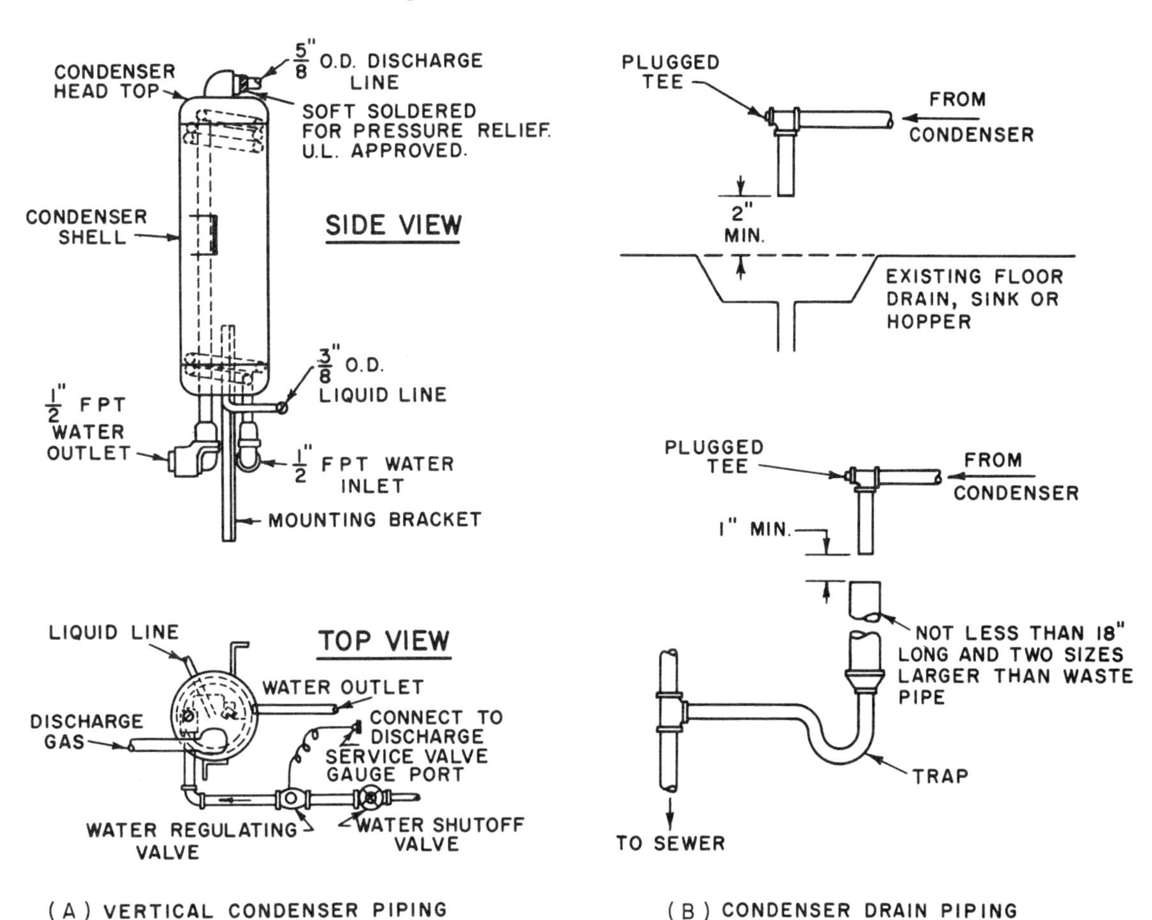

Fig. 20-2 Types of condenser piping

INSTALLING WATER REGULATING VALVES

The water regulating valve, figure 20-3, is selected on the basis of the refrigerant used and the required size to maintain an economical and safe condensing temperature under the design conditions (high load, high water temperature, and low water pressure). The condensing temperature and the valve size (in most cases) are determined in the survey.

The water regulating valve should be installed with the capillary end down. The water flow is to be in the same direction as the arrow on the valve body. The capillary tube is connected to the high-pressure side of the compressor. The gauge port on the compressor discharge shutoff valve is to be used for this connection when possible. The valve to which the capillary tube of the regulating valve is to be connected must be backseated before the connection is attempted.

Once the connection is made, the valve is opened approximately one turn from the back-seated position to allow the refrigerant pressure to reach the water regulating valve and still leave the line open.

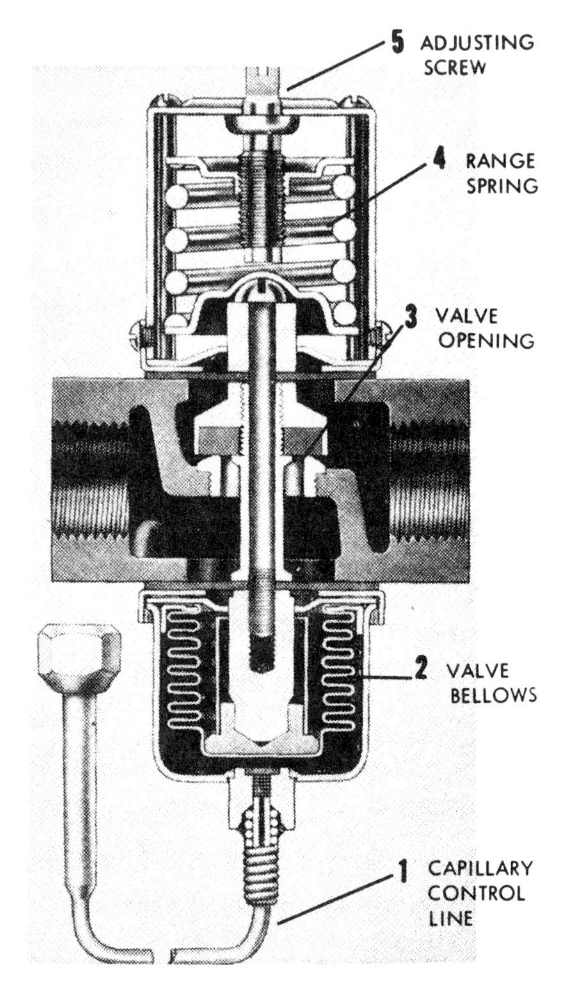

Fig. 20-3 Cutaway view of pressure-operated water regulating valve

CONNECTING THE COOLING TOWER

In those installations requiring a water-saving device, a cooling tower is selected. The cooling tower must have a sufficient capacity to maintain the condensing temperatures within safe operating limits.

When a cooling tower is used, it is good practice to interlock the compressor control circuit with the cooling tower. As a result, the compressor runs only when the tower is operating. If this interlock is not made, compressor abuse and shortened life can occur with a resulting higher maintenance cost. A water regulating valve is not recommended when a water cooling tower is installed.

CONNECTING THE CONDENSATE DRAIN PIPING

The condensate drain piping must be run full size to an open drain or sump. The piping must slope downward from the unit the full distance to the drain. Provisions are made for cleaning the piping by installing plugged tees at all turns in place of elbows.

INSTALLING THE WIRING AND CHECKING THE POWER SUPPLY

All wiring should comply with local codes and with the National Electric Code requirements. The units are wired at the factory for the electrical characteristics shown on the nameplate. The nameplate characteristics of the unit must be checked with the available power supply. This supply is run to the unit starter. The power at the unit must be within 10% of the rated voltage during normal operation and on startup. For polyphase units, the phases must be balanced to within 3%.

An adequate fused disconnect switch is to be provided in the power supply to handle the starting current. This switch is to be located within sight of the unit. If a starter is used with the fan motor, a separate disconnect switch is required.

It is essential that the proper wire size and dual element fuse sizes be installed to prevent nuisance tripping, accentuated voltage unbalance, generally unsatisfactory operation, and shortened unit life.

Figure 20-4 shows a typical wiring diagram for self-contained equipment.

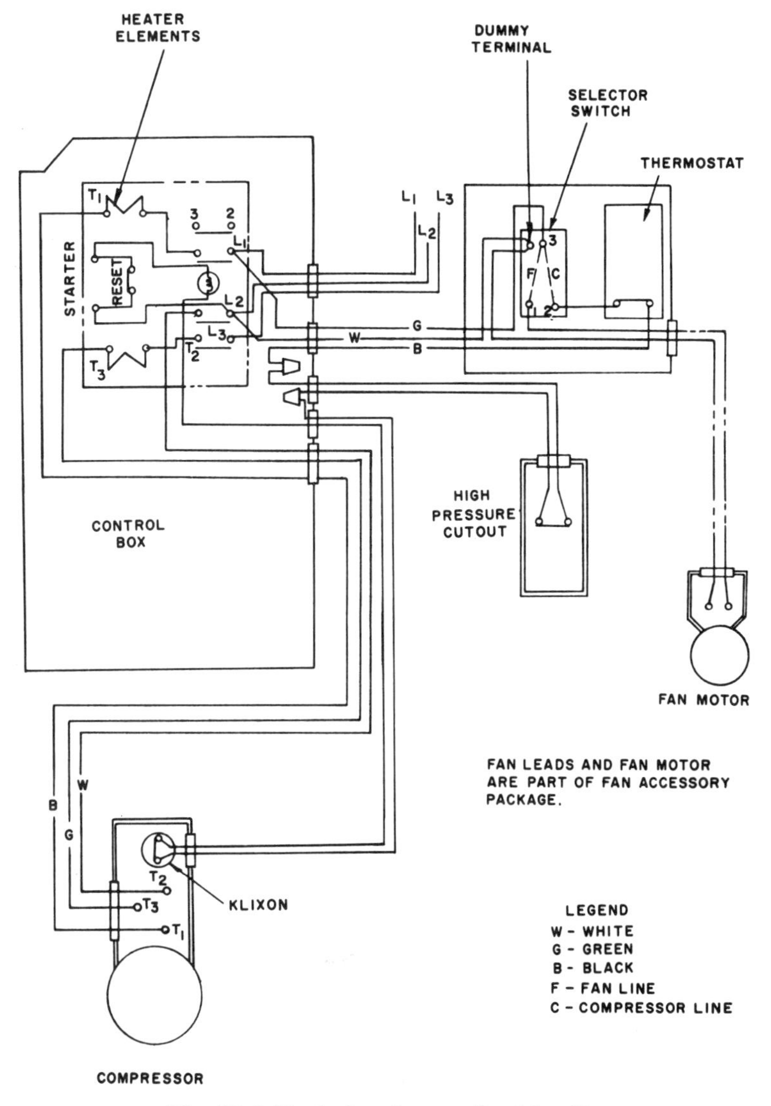

Fig. 20-4 Typical package unit wiring diagram

REMOVING THE COMPRESSOR SHIPPING BOLTS

A hermetic motor-compressor is usually spring mounted. For shipment, the motor-compressor is held rigidly in place by hex head machine bolts. These bolts are usually located at the corners of the compressor. When the bolts are removed, the compressor should float freely on its springs. To save the bolts for future use, place them in a cloth bag and tie the bag inside the unit. These bolts should be used whenever the unit is moved from one location to another.

ADJUSTING THE FAN SPEED

The fan motor is usually equipped with an adjustable pitch pulley, figure 20-5. This pulley can be adjusted to vary the fan speed to obtain a range of air quantities. To obtain the desired fan speed, adjust the fan motor pulley according to the following procedure.

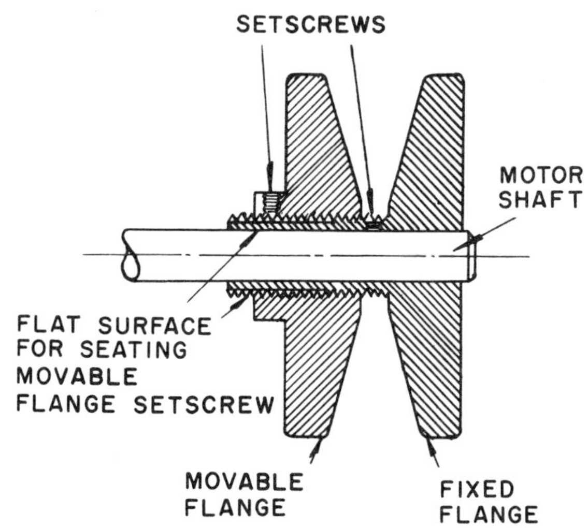

Fig. 20-5 Adjustable speed motor pulley

1. Remove the belt from the fan motor pulley after loosening the motor (as outlined in the section *Adjusting Fan Belt Tension*).
2. Loosen the setscrew in the movable flange of the pulley. Screw in the movable flange of the pulley. Screw the movable flange in toward the fixed flange as far as possible. In this position, the pulley gives the maximum fan speed.
3. Each half-turn of the adjustable flange away from the fixed flange reduces the fan speed.

   CAUTION: Under no circumstances should the flange be screwed more than six turns from the fixed flange. Before tightening the setscrew, insure that the setscrew is over the nearest flat surface of the pulley hub.

4. Replace the belt and adjust the belt tension as outlined in the following section.

It is not possible to measure the fan speed accurately while the front panel is removed from the unit. Such fan speed measurements should not be necessary if the preceding instructions have been followed.

ADJUSTING FAN BELT TENSION

The following procedure is to be used to adjust the fan belt tension.

1. Remove the cover plate and loosen the motor mounting bolts above the top plate, figure 20-6.
2. Slide the motor to a position such that the bolt can be depressed 3/4 in. at a point midway between the two pulleys.
3. Tighten the mounting bolts.

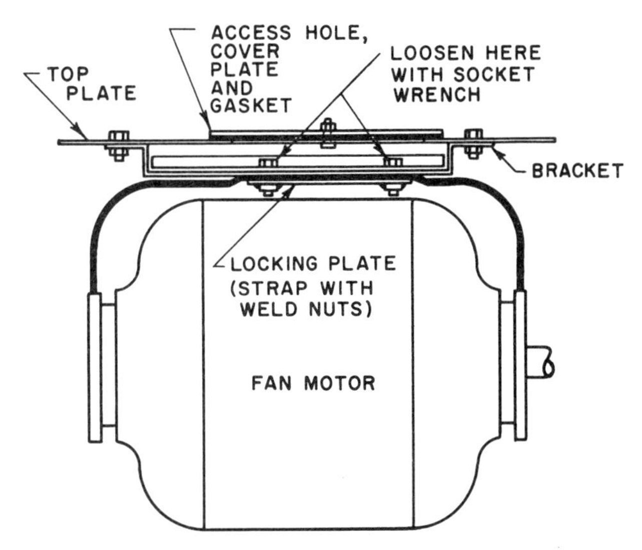

Fig. 20-6 Fan motor mounting

SUMMARY

- The major steps in installing self-contained water-cooled equipment are:
  1. Inspect the site to determine the sizes of the door, hall, or stairway and the strength and levelness of the floor.
  2. Move the unit to the site and uncrate it.
  3. Place the unit on the foundation.
  4. Connect the condenser water piping.
  5. Install the water regulating valves.
  6. Connect the cooling tower (if used).
  7. Connect the condensate drain piping.
  8. Install the wiring and check the power supply.
  9. Remove the compressor shipping bolts.
  10. Adjust the fan speed and the fan belt tension.

UNIT 21
INSTALLING AN AIR-COOLED, SELF-CONTAINED UNIT

OBJECTIVES

After completing the study of this unit, the student will be able to

- prepare and install a self-contained, air-cooled unit in any one of three locations –
  inside the building
  projecting through the wall of the building.
  outside the building
- install an air-cooled condenser indoors or outdoors including –
  the installation of the refrigerant piping
  the completion of the wiring of the fan motor and condenser.

Air-cooled package units are popular for central residential and small commercial units since they are compact and eliminate the need for cooling towers and water piping. These units are very popular in areas where hard water and the resulting condenser scaling are problems.

Air-cooled package units are designed to satisfy many different residential and commercial requirements. For example, the units can be installed in the attic or crawl space, through the wall or transom, suspended from the ceiling, placed on a roof, or set on a concrete base next to the basement wall. These units are completely

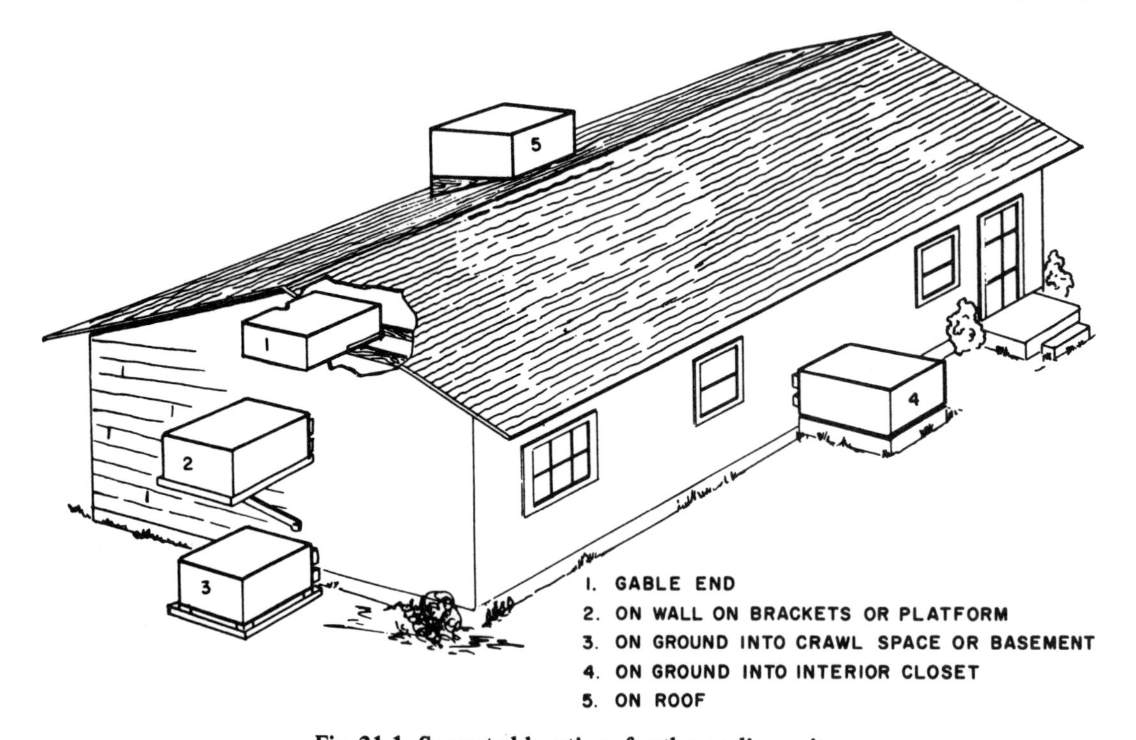

Fig. 21-1 Suggested locations for the cooling unit

weatherproofed for outside or roof installation. Air-cooled units can be connected to an existing warm air duct system or to a specially designed duct system of their own. Plenums are available for free blow distribution into the conditioned space.

Most package units supply cool air to the conditioned space through the top section of the front face of the unit. Warm air is removed through the bottom section. The condenser intake occurs through the sides and discharge takes place through the rear face of the unit.

PREPARING THE UNIT FOR INSTALLATION

The unit is uncrated and the clearance and rotation of the evaporator and condenser fans are examined. All fans must be checked to insure that they are properly adjusted.

The hermetic motor-compressor is spring-mounted. During shipment, the motor-compressor is rigidly held by hex head bolts. These bolts are usually located at the four corners of the compressor. A 9/16-in. socket wrench and extension must be used to remove the compressor holddown bolts. The compressor must float freely on its springs when the bolts are removed. Place the bolts in a cloth bag and tie the bag inside the unit so they are available for future use. The compressor must be bolted down whenever the unit is moved.

INSTALLING THE UNIT

Base Unit Inside the Building

Condenser Air. The following procedure is recommended to insure that an adequate air supply is available to the condenser. These steps will insure that there are no excessive losses in capacity.

1. Locate the air intake at a remote site from the unit, figure 21-2.
2. Remove condenser side panels if the rotating fan parts do not constitute a hazard.
3. If the air intake must be located next to the unit, the following precautions must be followed:
   a. Box in and seal the intakes.
   b. Remove the condenser side panels to decrease the air restriction.
   c. Make any necessary provisions to minimize the recirculation.

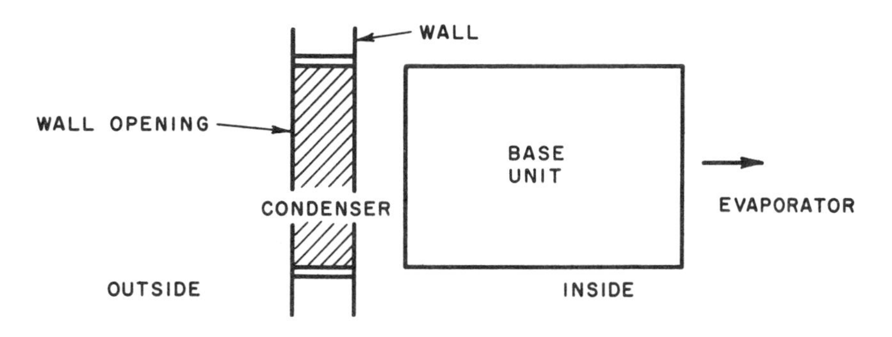

Fig. 21-2 Base unit inside building

Construction of Wall Opening. The location of the opening for the condenser supply and discharge air is determined in the preliminary survey. An existing window, vent, transom or other opening can be used if it is adequate in size and strength. The gap around the unit must be closed to insure a good weatherseal.

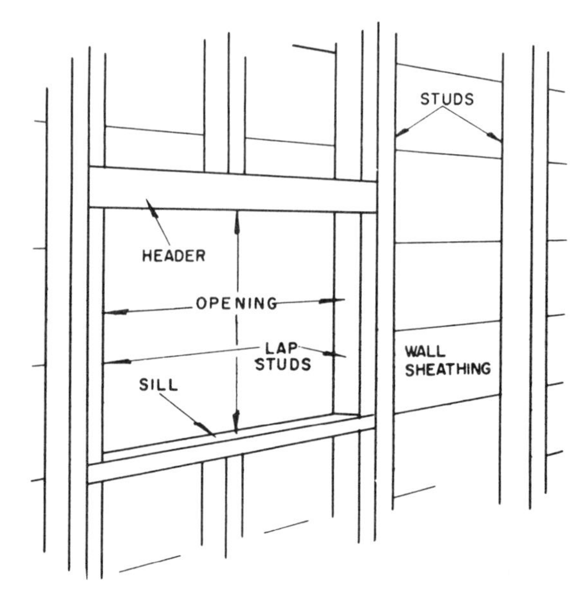

Fig. 21-3 Structural details of wall opening

Construction of Platform. The weight of the unit requires substantial support when installed. In attic installations, the unit is to be placed directly over, or as near as possible, to those points where the center partition or other walls meet the side walls of the building. The unit can be placed directly on the joists or on a platform constructed on the joists. The joists or platform must provide a level base for the unit.

Disposal of Condensation. The condensate flows through 1/2-in. FPT fittings in the base pan of the unit. An auxiliary drip pan is to be used on all inside installations even though FHA and local codes do not require it. The drip pan must be constructed of a corrosion resistant material such as galvanized steel. The pan must include a drain line to the outside of the building. The drip pan is placed on the installation platform so that it collects all of the condensate from the unit.

Sound Isolation. To prevent the transmission of vibration when the unit is operating, an isolation pad or several strips of 1/4-in. rubber is placed under the entire width of the base unit.

Base Unit Projecting Through the Wall (Figure 21-5)

Construction of the Wall Opening. The location of the opening is usually determined in the survey. A 3/4-in. clearance is provided around the unit. This clearance is the

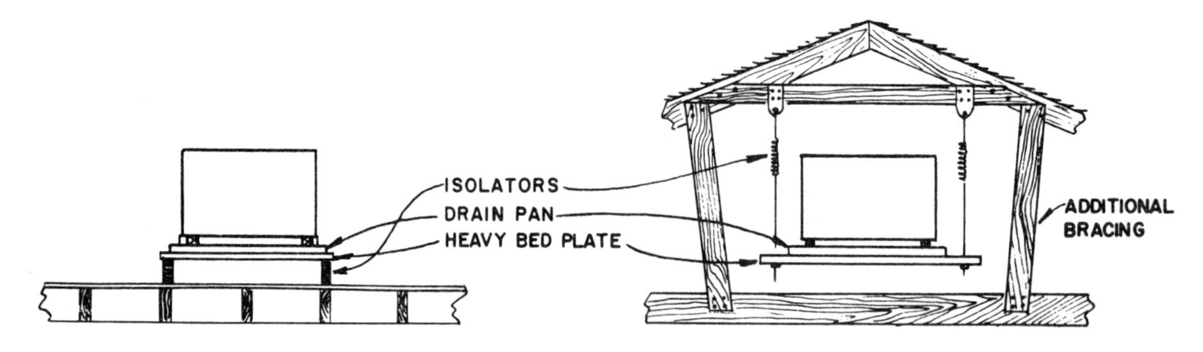

Fig. 21-4 Method of attic installation

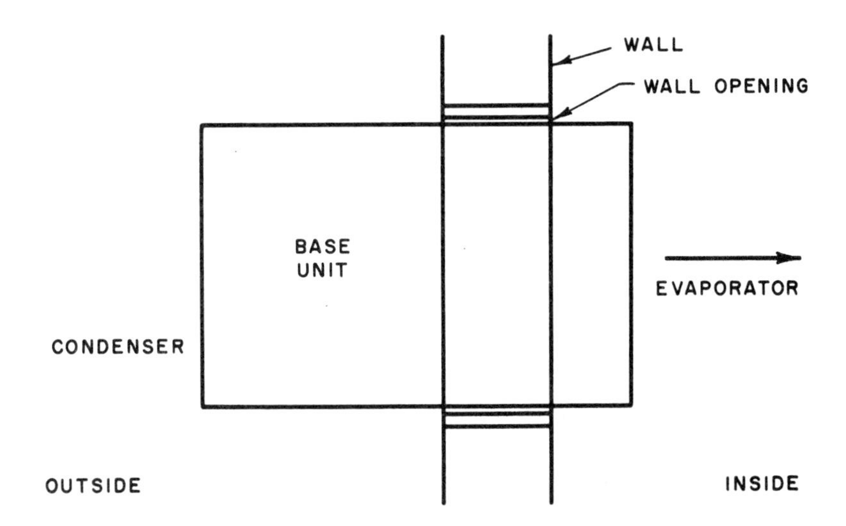

Fig. 21-5 Base unit projecting through wall

maximum size recommended to permit easy weathersealing of the hole. The wall opening construction is the same as that for the inside unit location. The sill of the wall opening must be able to provide substantial and safe support for the weight of the unit.

Construction of the Platform. The unit must be installed on a rigid and well-constructed platform to prevent any dislocation during operation. When determining the dimensions of the constructed platform at the job site, an allowance is to be made for the removal of the side panels for servicing.

Disposal of Condensation. Condensate is removed through 1/2-in. FPT fittings in the base of the unit. In general, a drip pan is not required in through-the-wall installations.

Sound Isolation. The unit must be installed in a manner that prevents the transmission of vibration. An isolation pad or several strips of 1/2-in. rubber can be placed under the entire width of the base unit to deaden the sound and any vibrations.

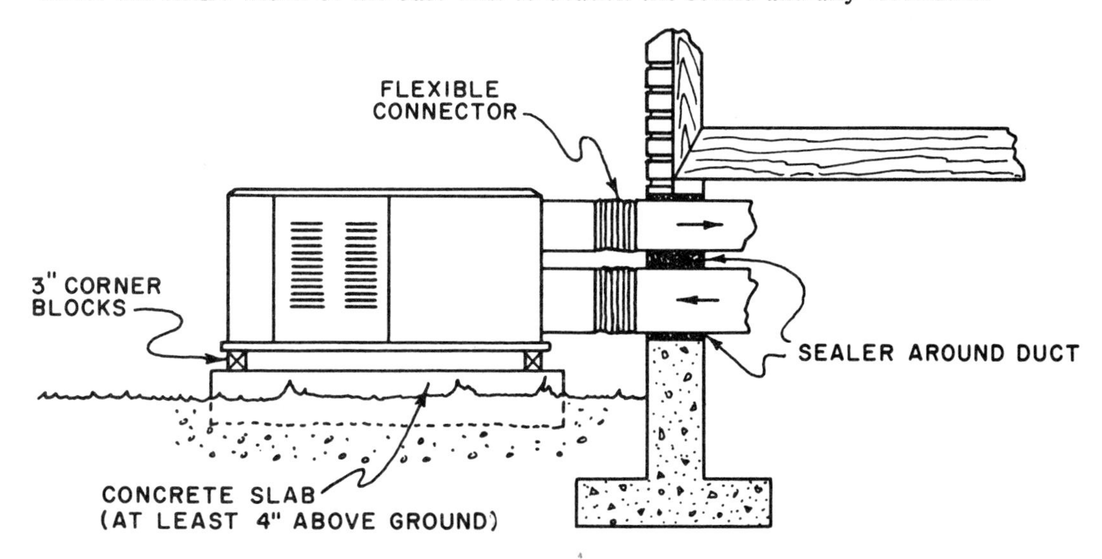

Fig. 21-6 Outdoor installation recommendations

Base Unit Outside the Building (Figure 21-6)

Construction of the Wall Opening. For outside installations, an opening must be made in the wall for the supply and return ducts fabricated at the job site. The size of the hole is determined by the size of the ducts used.

Construction of the Platform. The unit must be placed on a solid and permanent, corrosion- and weather-resistant platform. The location of the platform usually is determined in the survey. The recommended concrete pad construction is as follows: 6 in. thick, 48 in. wide, and 60 in. long with a gravel apron in front of the condenser coils.

The platform must be level and able to support the weight of the unit. Before installing the unit on the platform, the bottom of the unit is to be coated with a liquid or semiliquid waterproofing compound, such as roofing cement or hot tar.

Disposal of Condensation. Condensate is removed through 1/2-in. FPT fittings in the base pan of the unit.

Sound Isolation. If the unit is placed on a concrete foundation on the ground, an isolation pad normally is not required. However, the bottom of the unit should be coated with a liquid or semiliquid waterproofing compound, such as roofing cement or hot tar.

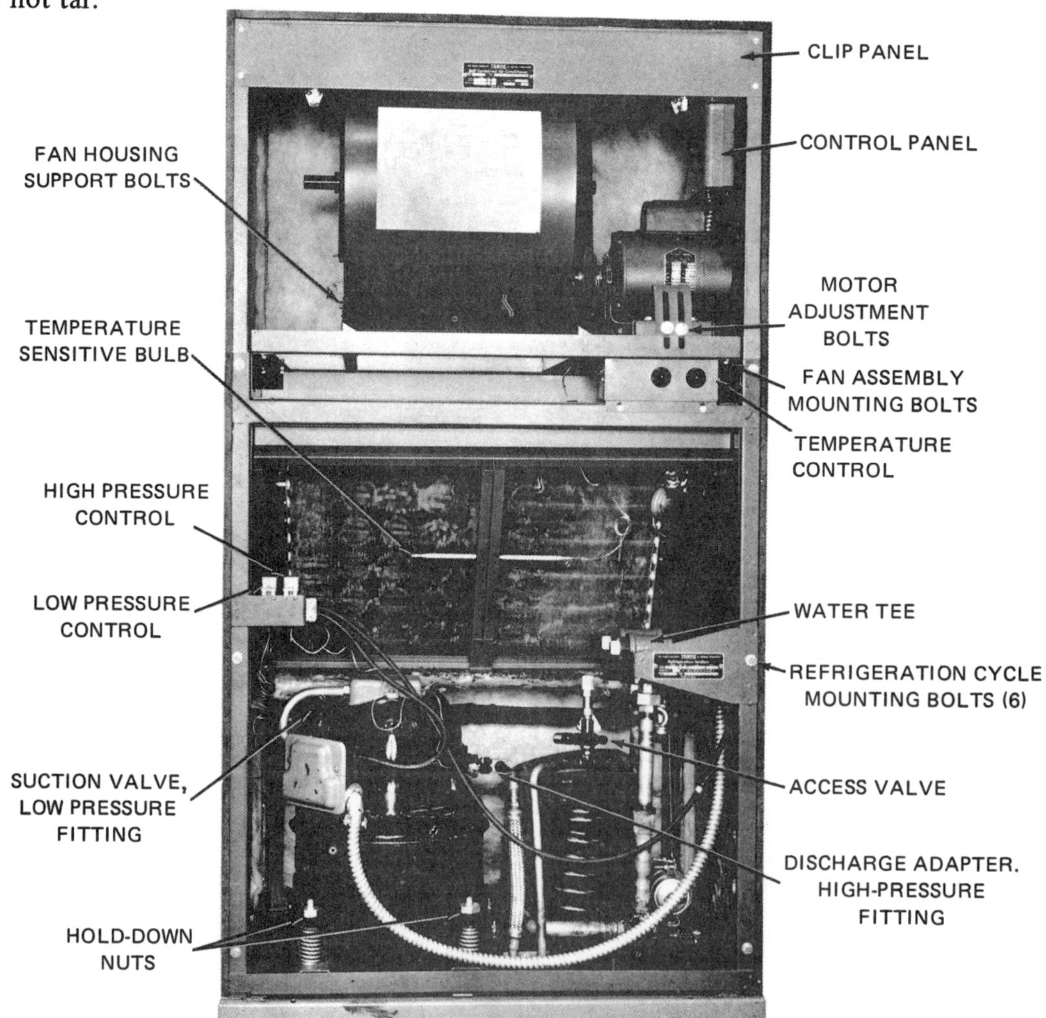

Fig. 21-7 Typical self-contained unit installation

SETTING THE UNIT

Prior to removing the shipping skid, the unit is moved as close as is practical to its final location. Care must be taken to prevent damage to the unit or the building. It may be desirable to build a special handling rig for installations where the unit must be raised more than five feet. For example, a pickup truck carrying an A-frame with a block and tackle is a suitable rig. In addition, local rigging companies can be contracted to do any moving and setting of the equipment.

Unit Inside or Outside the Building

If the unit is to be moved into or through a hole having sleeve dimensions, the shipping skid must be removed; the width of the hole will not accommodate the skid.

A few types of inside installations, such as those in a limited attic or crawl space, may require that the side panels be fastened before the unit is placed in position. If this step is necessary, the instructions for through-the-wall installations are to be followed.

The unit is then moved to its final position. The compressor holddown bolts and the shipping skid are removed. Before the side panels are assembled, the power supply and wiring connections are made (as described in the following paragraph). The installer must insure that the unit is level, fits securely and safely on the platform or in the drip pan, and that adequate sound isolation steps have been taken. The condensate drain line connections are then made.

Power Supply and Wiring

All wiring must comply with local and National Electrical Code requirements. The unit must be grounded.

The units are completely wired at the factory for either 230-volt, single-phase, 60-hertz or 220-volt, three-phase, 60-hertz electrical characteristics. The nameplate characteristics of the unit are to be checked against the available power supply. The power at the unit must be within 10% of the rated voltage. On polyphase units, the phases must be balanced to within 3%. The power company is to be contacted regarding the correction of line voltage that is not within 10% of the standard value or is unbalanced.

A fused disconnect switch is to be provided. The switch must have an adequate size to handle the starting current. The switch is located within sight of the unit. The wiring is completed according to the manufacturer's installation instructions. Figure 21-8, page 262, shows a typical wiring diagram for a self-contained, air-cooled unit.

WEATHERSEALING

The basic unit usually is weatherproofed at the factory. If the unit is installed properly, it will operate satisfactorily in an outside location regardless of weather conditions.

The opening in the wall of the building is necessary in all three types of installation. This opening must be thoroughly weathersealed according to standard carpentry practice, figure 21-9, page 262. In through-the-wall installations, the gap between the unit and the framework must be weathersealed as well. Nonsetting caulking compound is used around the entire unit to complete the weathersealing.

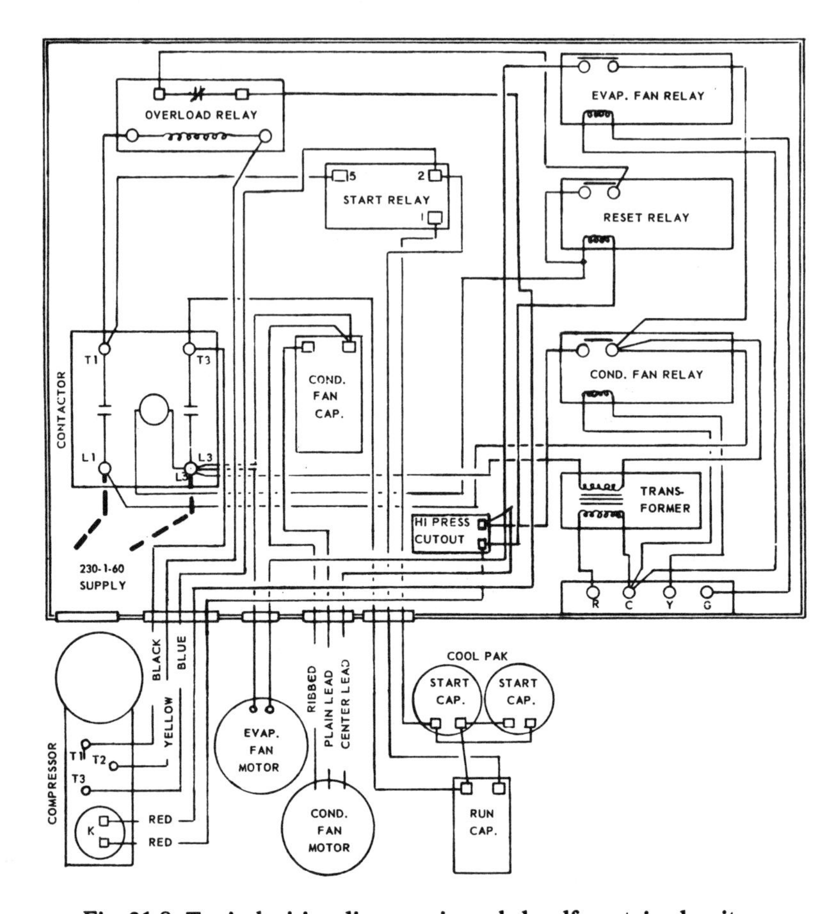

Fig. 21-8 Typical wiring diagram air-cooled, self-contained unit

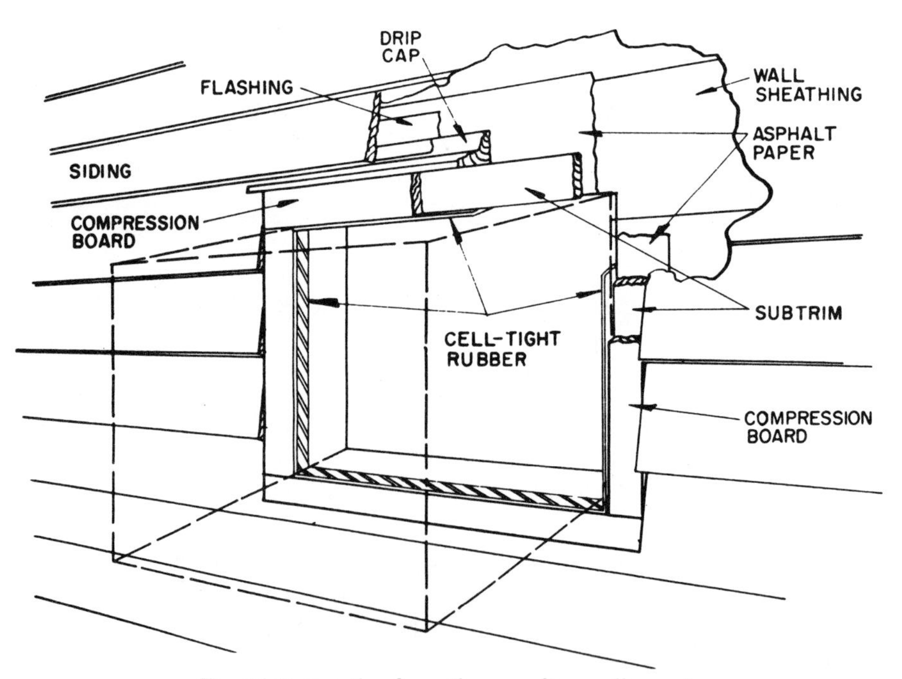

Fig. 21-9 Details of weatherproofing wall opening

The hole or opening is sealed by installing asphalt paper against the wall of the building, flashing under the wall siding, a drip cap, and compression trim boards.

Any additional practical measures are to be taken which will provide a good weatherseal. These additional precautions depend upon local weather conditions and the architectural features of the building.

INSTALLING AIR-COOLED CONDENSERS

Air-cooled condensers can be mounted indoors or outdoors. It is not necessary to place the condenser at the same level as the compressor. Air-cooled condensers must not be located in an area continually exposed to loose dirt and foreign matter which can clog the condenser coil.

Fig. 21-10 Condenser with air deflector

Outdoor Installation

The fan motor is usually located inside the casing of the condenser. As a result, a special hood is not required for weather protection.

The condenser is mounted so that the prevailing wind blows toward the air intake of the unit. If it is not possible to mount the condenser in this manner, an air deflector is used as shown in figure 21-10.

If the air discharge is in the direction of a wall, an air deflector is not required. However, the condenser should be spaced from the wall at a distance of not less than one and one-half times the diameter of the fan, figure 21-11.

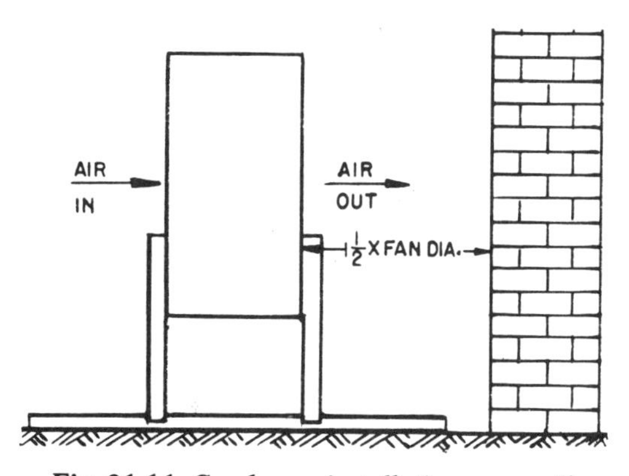

Fig. 21-11 Condenser installation near wall

Indoor Installation

If the condenser is mounted indoors, the air leaving the condenser must be discharged to the outside. This type of discharge prevents the recirculation of heated air through the condenser. When the air is discharged to the outside, a provision must be made for a fresh air intake.

Refrigerant Piping

A hand shutoff valve (L) is installed at the receiver inlet, figure 21-12, page 264. A purge valve or purge fitting is installed in the 1/4-in. pipe tap provided in the condenser inlet header. In systems having a large capacity, an additional purge valve should be installed on the receiver.

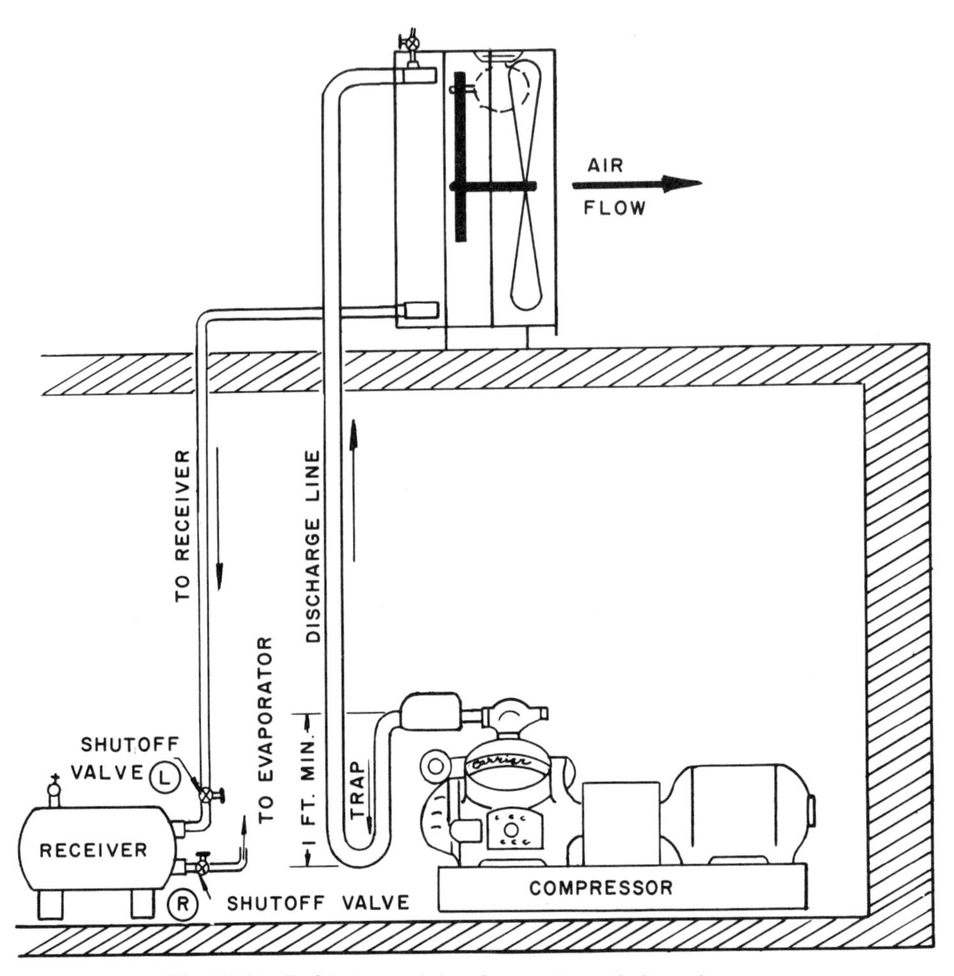

Fig. 21-12 Refrigerant piping for an air-cooled condenser

To prevent condensed liquid refrigerant or oil from damaging the compressor, a trap is installed in the vertical discharge line near the compressor. The height of the trap should be six inches for every 10 feet of vertical discharge line. If the height of the vertical discharge line is such that a single trap is impractical, the loop can be replaced by a check valve or several traps.

When refrigerant piping is being installed, it must be well supported. The piping is installed with sufficient flexibility to insure that vibrations from the compressor, condenser, or piping are not transmitted to the building. A muffler is usually placed in the discharge line to prevent noise due to the pulsation of the refrigerant.

Refrigeration piping should be kept as clean as possible. In many instances, before the piping is connected, it may be necessary to run a swab through the pipe to remove dust and other foreign material. After the tubing is cut to the proper length, all burrs should be removed. Loose filings must not be allowed to remain inside the tubing.

Strainers and strainer-driers are used to filter out most of the foreign material left in the piping. However, some foreign material can still enter the expansion valve and the compressor. Undesirable material, especially moisture and acids, may not be a problem until the unit has been operating for several months. Refrigeration piping systems can never be too clean.

Additional methods of installing the piping for air-cooled condensers located outside the building are shown in figures 21-13 through 21-16.

Electrical Wiring

The electrical installation should be in accordance with local codes and regulations and the National Electrical Code. The fan motor on the condenser operates when the compressor runs and stops when the compressor shuts off. The wiring is to be interlocked so that the compressor cannot run if the fan motor fails to run. The compressor should be equipped with a high-pressure cutout switch and motor overload protection.

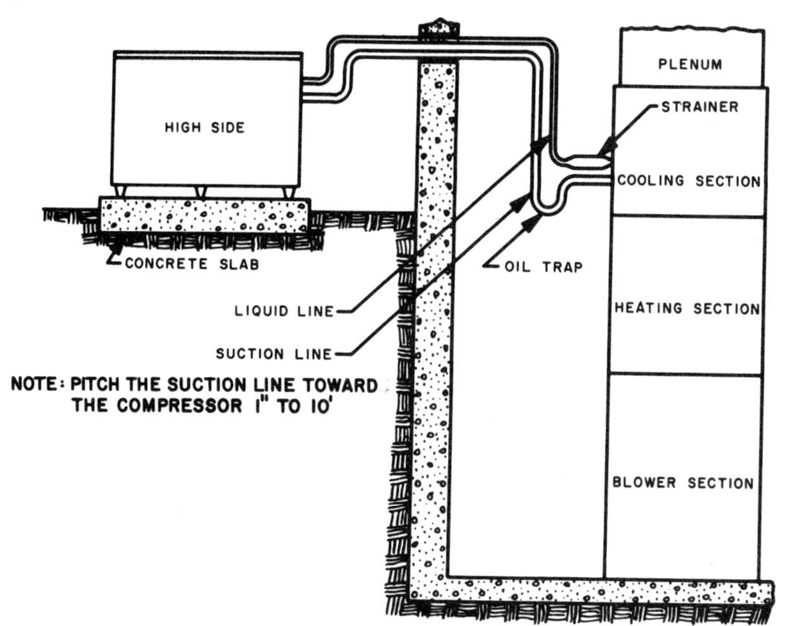

Fig. 21-13 Low side and high side on the same level

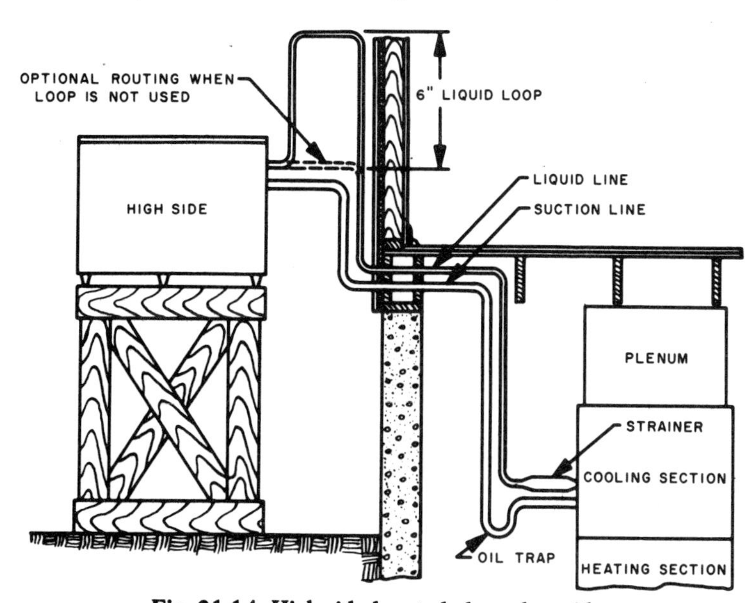

Fig. 21-14 High side located above low side

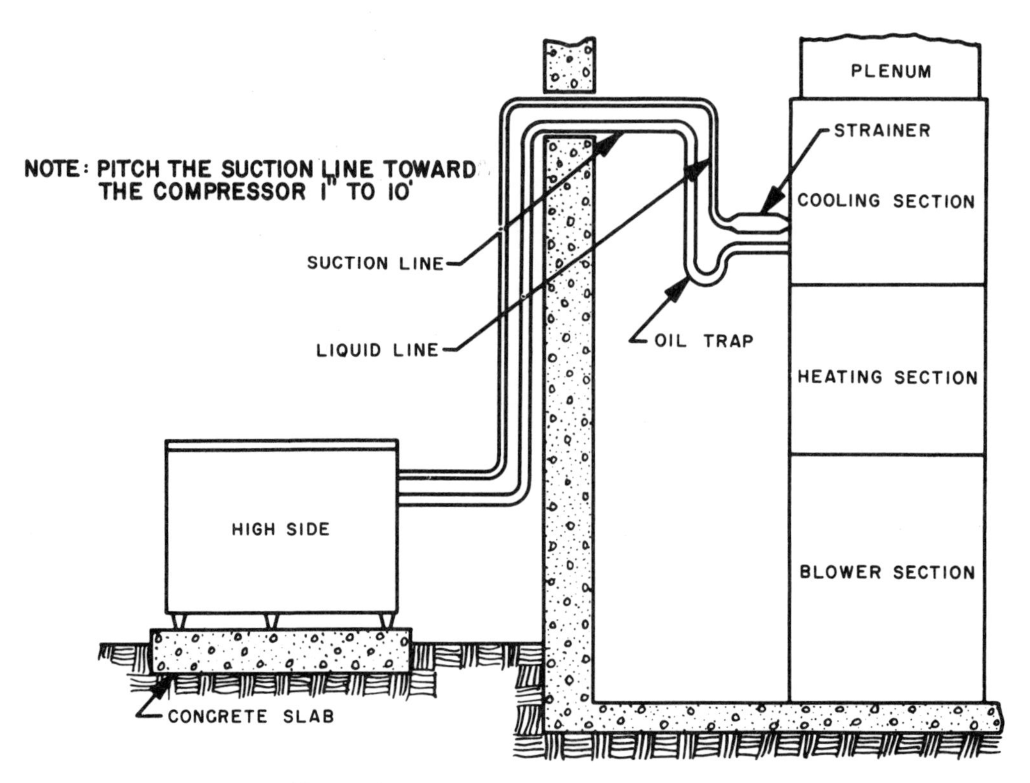

Fig. 21-15 High side located below low side

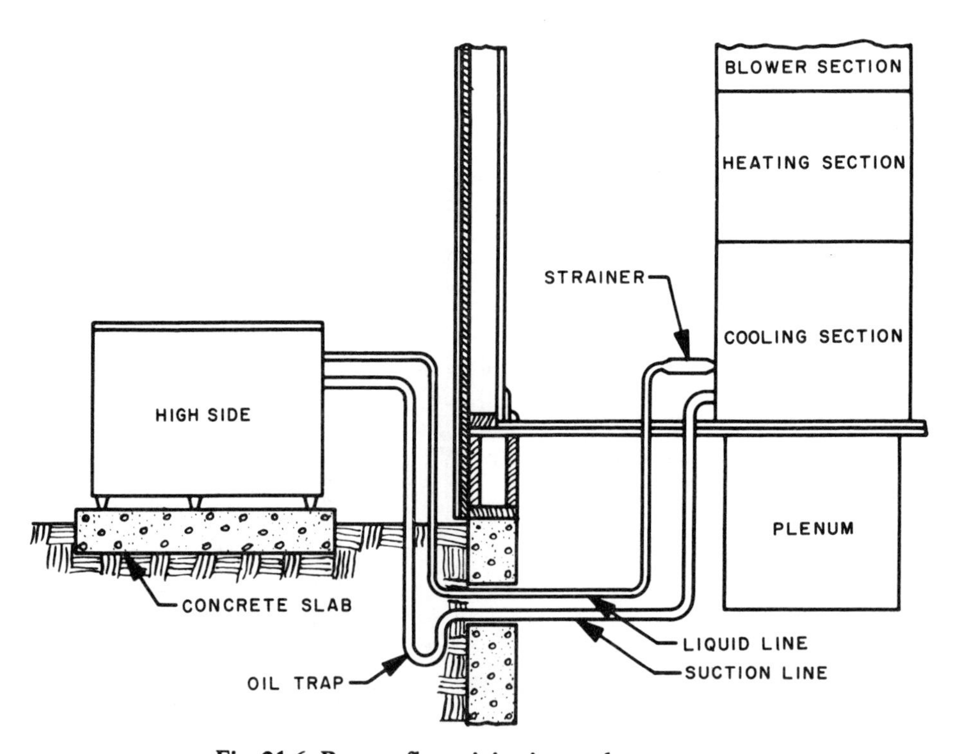

Fig. 21-6 Reverse flow piping in crawl space

SUMMARY

Installing Self-contained, Air-cooled Units

- Prepare the unit by uncrating it, adjusting the fan as required, and removing the shipping holddown bolts from the compressor base.
- Install the unit in one of three possible locations:
  1. Inside the building:
     - Provide a wall opening for condenser air
     - Construct a platform for the unit
     - Provide for the disposal of condensate
     - Install a sound isolation pad
  2. Projecting through the wall of the building:
     - Construct a wall opening
     - Construct a solid base or platform for the unit
     - Provide for the disposal of condensate
     - Install a sound isolation pad
  3. Outside the building:
     - Provide a wall opening for the supply and return ducts.
     - Construct a weatherproof concrete base to serve as a unit platform
     - Provide for condensate disposal
     - Install a sound isolation pad
- Set the unit
- Connect the unit to the power supply
- Seal and weatherproof the wall opening

Installing an Air-cooled Condenser

- Install the condenser indoors or outdoors.
  1. An outdoor installation may require an air deflector.
  2. An indoor installation requires a condenser discharge duct to the outdoors.
- Install the refrigerant piping
- Complete the wiring of the fan motor and condenser.

UNIT 22

COOLING TOWER INSTALLATION AND WATER TREATMENT

OBJECTIVES

After completing the study of this unit, the student will be able to

- list the steps in the installation of –
  1. a forced-draft cooling tower.
  2. a natural-draft cooling tower.
- state three reasons why water treatment is required in systems using water-cooled condensers.
- define the following terms –

hard water	corrosion formation
soft water	corrosion control
scale	algaecide
scale inhibitor	

- list the points at which a scale removing solution can be added to a forced-draft cooling tower and to a natural-draft tower.

There are three basic styles of cooling towers:

- *Atmospheric-* or *natural-draft towers* depend upon natural air currents to cool the circulated water.
- *Draw-through* or *forced-draft towers* pull a constant volume of air through the tower to cool the circulated water. This type of tower is packed with equally spaced slats made of wood or other material. Because these slats provide a greater wet surface for the air to cool, this arrangement yields a greater cooling capacity in an installation having a smaller physical size.
- *Blow-through towers* (also called *forced-draft towers*) differ from the draw-through tower only in that the air supply is blown through the tower rather than drawn through it.

INSTALLING FORCED-DRAFT TOWERS

Setting the Tower

1. Locate the tower so that the prevailing summer wind is in the same direction as the tower discharge air.
2. If possible, locate the tower away from windows or vents to prevent complaints about the noise level.
3. The selected location for the tower must provide adequate clearance.
   a. On the side where the air enters the tower, the clearance should equal the width of the tower.
   b. On the side where the air is discharged, the clearance should equal the length of the tower.

4. A level foundation is to be prepared using wooden beams, concrete pads, or a poured concrete base. This foundation must adequately support the tower when it is filled with water.
5. The fan motor and belt are to be installed as follows:
   a. Bolt the motor to its mounting loosely.
   b. Align the motor sheave and fan sheave grooves.
   c. Place the belt on the sheaves. Adjust the belt by moving the motor mounting. It should be possible to depress the belt about 3/4 in. for each linear foot between the fan and motor sheaves.
   d. Tighten the motor mounting.

Installing the Water Piping and Pump

1. Connect the condenser outlet to the tower intake, figure 22-1.
2. Connect the tower sump to the water pump suction.
3. Connect the pump discharge line to the condenser inlet fitting.
4. Install a water makeup line to the tower sump.
5. Connect the tower drain and overflow connections to a drain, where required by the job specifications.
6. Install a bleedoff connection from the condenser outlet line to the sewer.

   NOTE: *All piping is to be pitched so that it can be drained completely during the winter shutdown.*

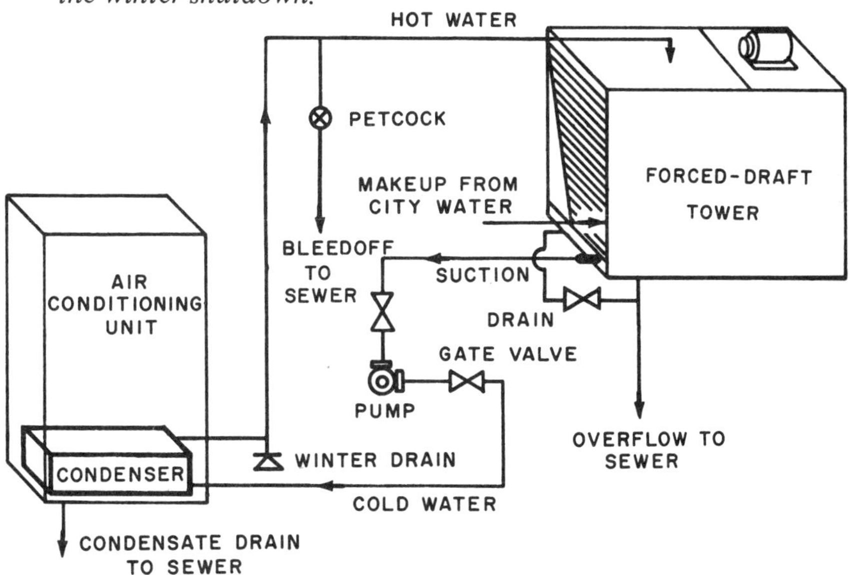

Fig. 22-1 Water piping for forced-draft tower
(Courtesy The Marley Company)

Electrical Connections

NOTE: All wiring must be completed in accordance with the National Electrical Code and applicable local codes.

1. Connect the tower motor. A fused disconnect switch should be installed as close to the tower as possible.
2. Install the water pump wiring.

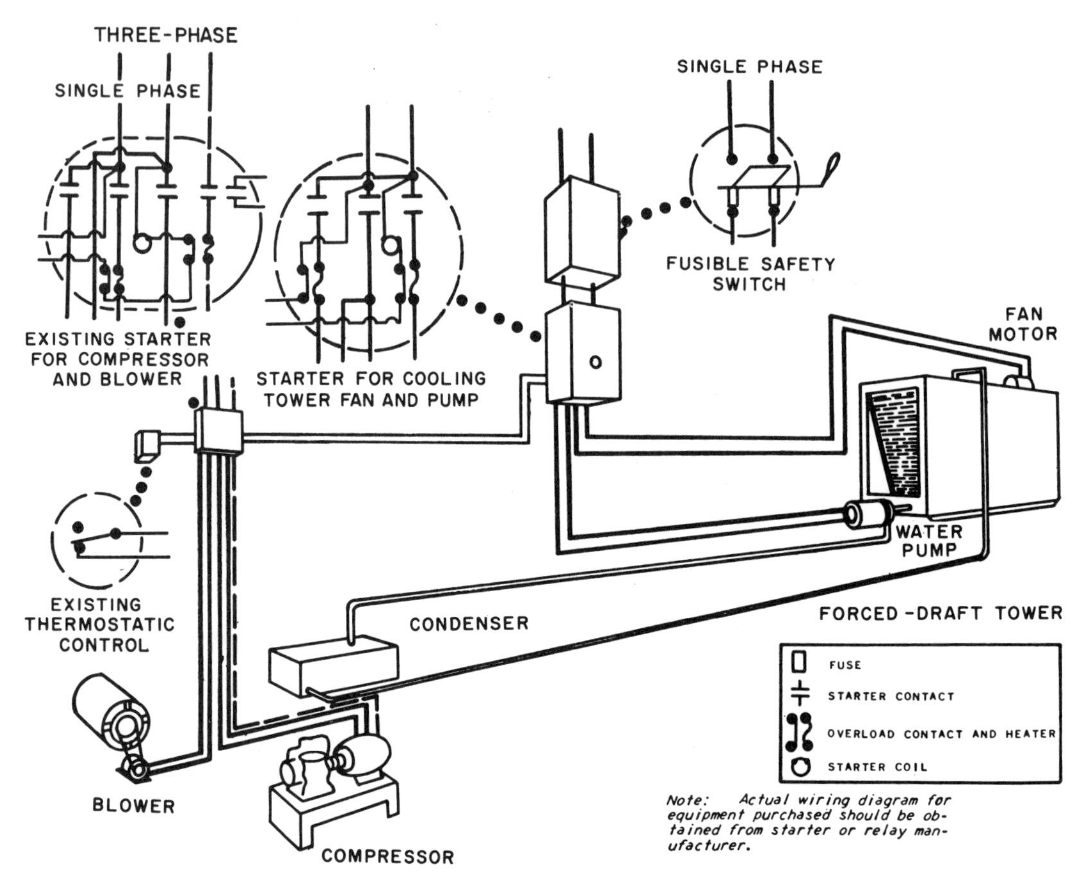

Fig. 22-2 Wiring diagram for forced-draft tower

Figure 22-2 shows a separate starter for 115-V or 230-V single-phase motors driving the cooling tower fan and pump. The starter is energized through an interlock in the compressor motor starter. Either a single-phase or a three-phase motor is provided in the compressor.

INSTALLING ATMOSPHERIC TOWERS

Setting the Tower

1. Locate the tower outside the structure and away from any walls or buildings that can restrict air movement. (Since there is no fan in an atmospheric tower, noise is not a factor.)
2. A level foundation is prepared using wooden beams, concrete pads, or a poured concrete base. This foundation must provide adequate support when the tower is filled with water.

Installing the Water Piping and Pump

Figure 22-3 shows typical piping connections for a natural-draft tower.

Electrical Connections

For an atmospheric or natural-draft cooling tower, only the water pump wiring is to be installed (there is no fan installation). The typical wiring diagram in figure 22-4 shows a separate starter for the single-phase pump motor. The pump motor starter is energized by the compressor motor starter.

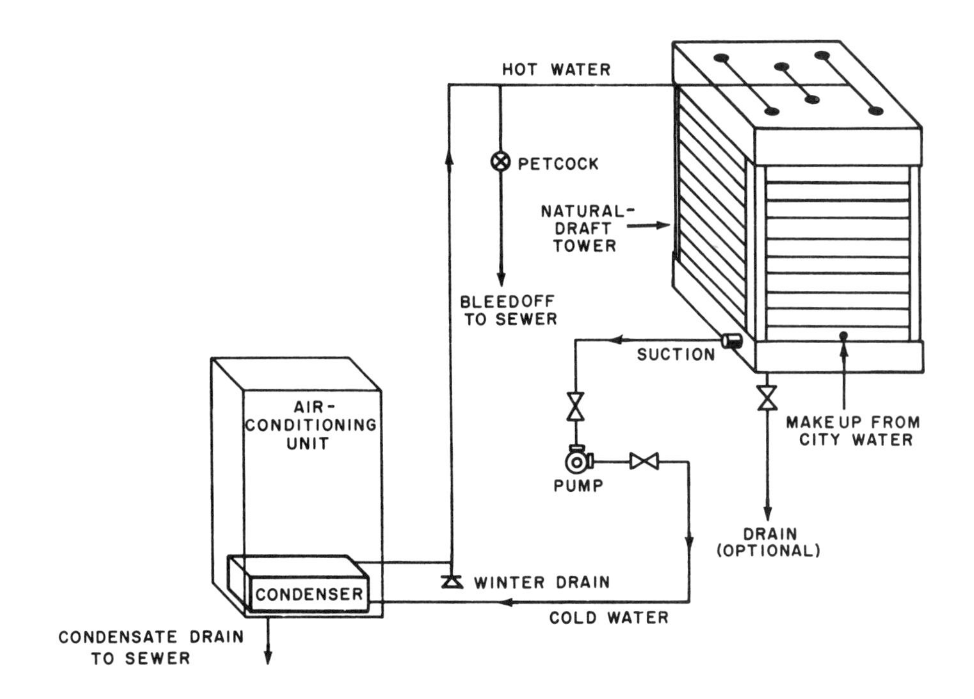

Fig. 22-3 Water piping for natural-draft tower
(Courtesy The Marley Company)

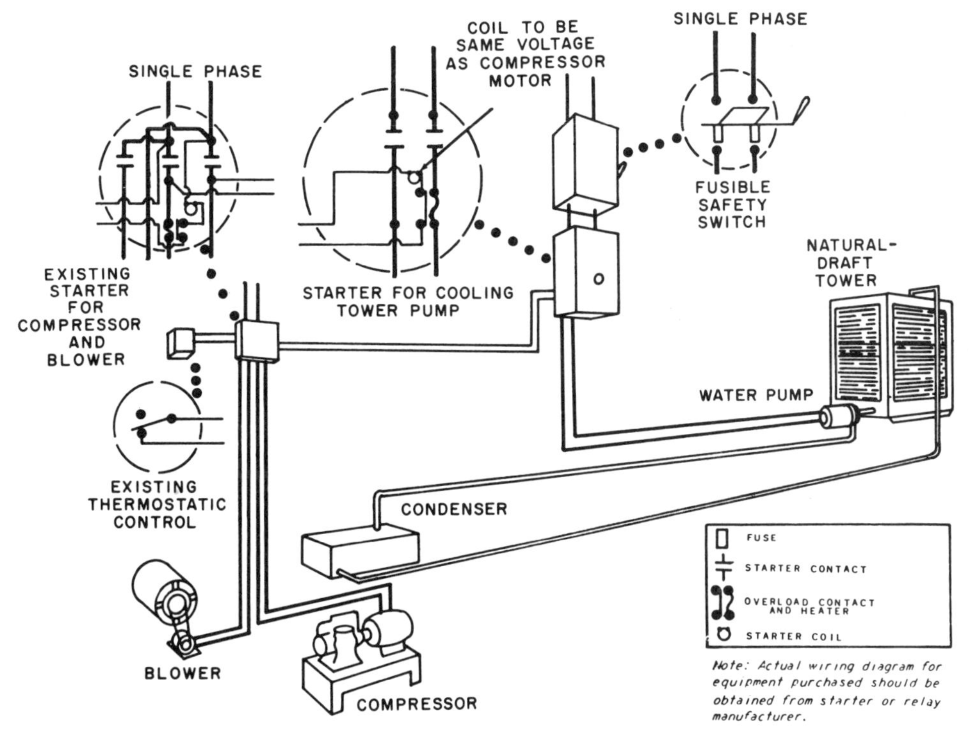

Fig. 22-4 Wiring diagram for natural-draft tower
(Courtesy The Marley Company)

WATER TREATMENT

All air-conditioning systems using water-cooled condensers require some form of water treatment. This treatment is necessary because water (from any source such as wells, lakes, and rivers) contains impurities in the form of minerals, gases, organic matter, algae, and bacterial slime. These impurities act on the condenser water system in the following ways:

- Minerals cause the formation of scale on the metal surfaces.
- Gases in the atmosphere, such as carbon dioxide and sulfur dioxide, form acids when they come into contact with the water. These acids eventually attack and corrode the metal surfaces.
- Algae attach themselves to the surfaces of the system. Since algae are living plants, they grow until the passages of the system are clogged. Bacteria form slime and clog the system in much the same way as algae.

Most of the impurities found in water can be neutralized by chemicals which are prepared to perform a specific function. For example, specific chemical compounds can suspend minerals, dissolve scale, neutralize acids, coat the surface of the metal with a protective film, or attack and kill algae and bacteria.

Some of the chemical compounds used to remove scale or neutralize acids can be harmful to wood and to certain metallic and composition materials used in valves, valve diaphragms, and cooling tower fill. Care must be exercised when these chemical compounds are used in condensers and cooling towers. It is recommended that a water treatment specialist be consulted for each system requiring some form of water treatment.

HARD AND SOFT WATER

Water is classified as hard or soft depending upon the amount of solid impurities contained in the water. In general, water containing less than 200 parts per million (ppm) of solid material is considered to be soft water. Hard water contains solids in amounts greater than 200 parts per million (ppm). In terms of grains of hardness, 100 ppm to 200 ppm is roughly equivalent to 6 to 11 grains of hardness per gallon of water; 200 ppm to 300 ppm is equivalent to 12 to 17 grains of hardness per gallon of water; and 300 ppm to 400 ppm is equivalent to 18 to 24 grains of hardness per gallon of water.

CONTROLLING SCALE

Scale formation is controlled by compounds called scale inhibitors. Some minerals tend to accumulate and become more concentrated as the water in the system evaporates. An inhibitor holds these mineral solids in suspension in the water. If the inhibitor is used in the water and a water bleedoff is provided in the system, the suspended solids flow from the system through the bleed pipe. Thus, the solids do not accumulate and the concentration of solids is maintained at a minimum level. The amount of bleedoff can vary from four gallons per hour for five-ton units to 360 gallons per hour for 100-ton units. For each unit size, the bleedoff also varies according to the hardness of the water.

Based on the specifications determined by a water treatment specialist, one water scale treatment can maintain the amount of solids in the water at a safe level for three months or more.

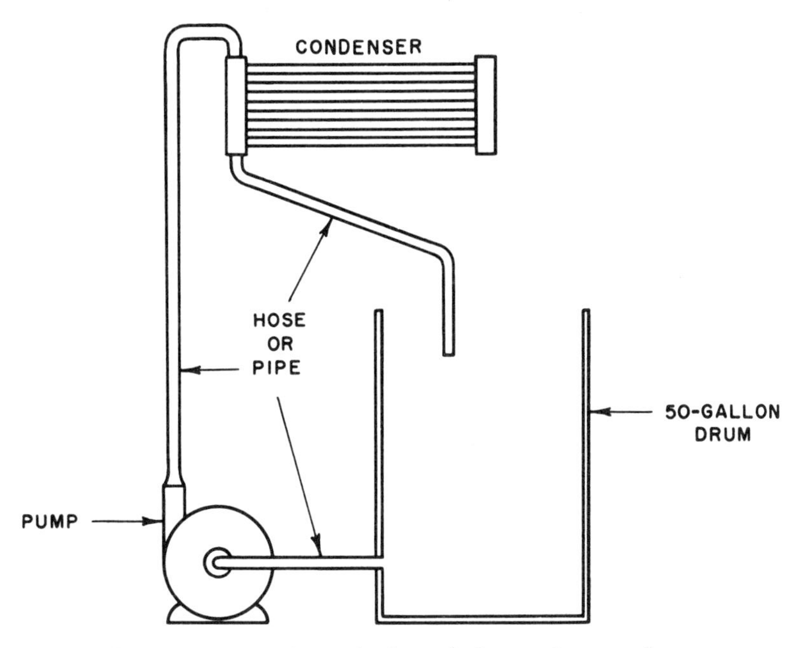

Fig. 22-5 Removing scale through the condenser only

REMOVING SCALE

To remove accumulated scale from a system, the services of a water treatment specialist should be obtained. The specialist can determine the type of scale present. A suitable chemical compound and the quantity of this compound necessary to remove the scale can then be recommended. In most cases, the chemical should not be allowed to pass through the cooling tower. Unless otherwise specified by the water treatment specialist, the chemical should pass through the condenser only, figure 22-5. This precaution is recommended because most scale removers are acid compounds that can damage the materials in the cooling tower. (However, the acid may be mild enough that the water treatment specialist considers it to be harmless to the materials in the cooling tower.)

If the scale removing solution is harmless to the materials in the cooling tower, the solution can be added to the system at the points shown in figure 22-6 for a forced-draft tower. Part (A) in the figure is the tower distributor plate. Part (B) is the tower sump, and Part (C) is the water tank. For natural-draft tower, figure 22-7, page 274, the solution is added at the sump (Part (B)).

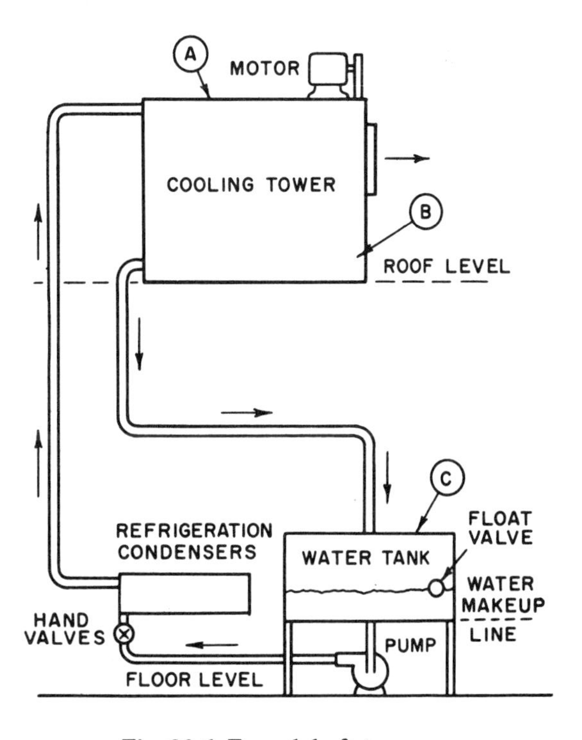

Fig. 22-6 Forced-draft tower

It is a good practice to clean the sump before the scale remover is added. This action is recommended because the sludge in the sump is likely to absorb some of the cleaning strength of the remover.

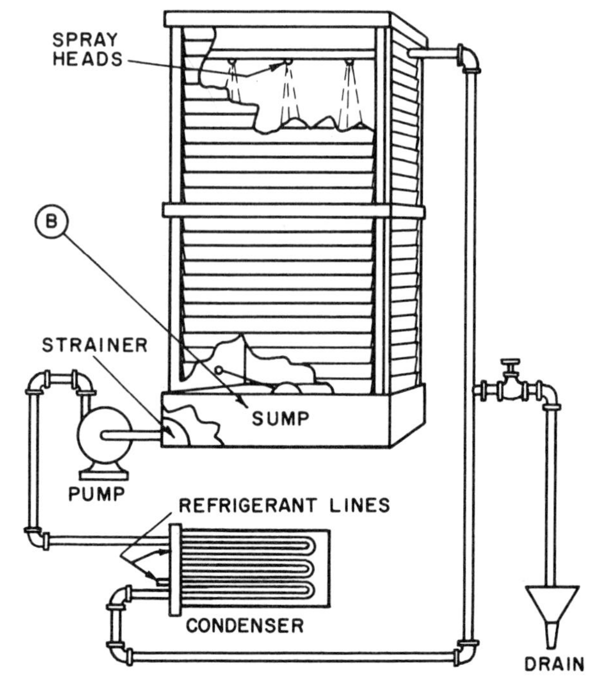

Fig. 22-7 Natural-draft tower

CORROSION

Corrosion in a system can occur in the following ways:

- Acids are caused by gases from the atmosphere dissolving in the water. These acids gradually increase in strength until they begin to attack and etch the metal surfaces in the system.
- One of the metals used in the system is dissolved by the action of electrolysis resulting from the connection of two unlike metals. The two dissimilar metals set up an electric current with the result that one metal dissolves and goes into solution. Some combinations of dissimilar metals react faster than other combinations.
- Oxygen enters the water from the atmosphere resulting in pitting of the metal surfaces in the system.

CORROSION CONTROL

Corrosion can be controlled in the water system by the addition of chemically blended polyphosphates or other similar compounds. The polyphosphate compound deposits a film over the entire surface of the metal. This compound also protects the metal surfaces from some of the milk acids that form from atmospheric gases.

Electrolysis can be prevented by following a few general rules.

- If dissimilar metals are joined in the system, the metals should be electrically insulated from each other.
- If copper and steel are used in the system piping, large surfaces of copper and small surfaces of steel result in rapid corrosion of the steel. Small areas of copper and large areas of steel cause the steel to corrode, but at a rate which is much slower than for the previous case. A similar corrosive action will result when brass and steel are combined.

CONTROLLING ALGAE AND SLIME

Algae and slime are controlled by the use of toxic compounds commonly called algaecides. Chlorine is an algaecide. Since one type of algae grows best in daylight and another flourishes in darkness, one type of algaecide may be suitable and effective for indoor cooling tower installations. Another type of algaecide may be more effective for outdoor installations where the tower is exposed to daylight. In all instances, a water treatment specialist should be consulted to determine the algaecide that is likely to produce the best results. Algae treatment usually lasts one month or more depending upon the conditions of the water and the atmosphere around the tower.

All algaecide compounds are toxic; therefore, the skin and eyes should be protected when these compounds are handled.

The most important rule to be followed regarding water treatment or water treatment problems is to obtain the services of a reputable water treatment specialist.

SUMMARY

- Cooling towers are designed to operate on the principle of –
  1. natural draft where natural air currents cool the circulated water; or,
  2. forced draft in which a fan blows or draws air through the tower to cool the circulated water.
- In forced-draft towers, the discharge air should blow in the same direction as the prevailing wind.
- To install a forced-draft tower:
  1. Locate the tower
  2. Prepare a foundation
  3. Install the fan motor and belts
  4. Install the water piping and pump
  5. Connect the fan motor and water pump wiring
- To install a natural-draft tower:
  1. Locate the tower
  2. Prepare a foundation
  3. Install the water piping and pump
- Water treatment is required in systems using water-cooled condensers because:
  1. Minerals in the water cause scale formation on metal surfaces.
  2. Atmospheric gases form acids in the water; these acids corrode the metal surfaces.
  3. Algae and bacteria clog the system passages.
- Hard water contains relatively large amounts of solid impurities; soft water has relatively small amounts of impurities.
- Scale formation is controlled by inhibitors. Some systems are designed so that the solids suspended by the inhibitor can be removed.

- Scale removers are chemical compounds. In general, these compounds are passed through the condenser only (not the cooling tower), unless the water treatment specialist considers them harmless to the cooling tower.
- Corrosion occurs when atmospheric gases form acids in the water; when electrolysis occurs (the chemical reaction of two dissimilar metals when joined); and through oxidation.
- Electrolysis can be prevented by electrically insulating the dissimilar metals.
- Algaecides are used to control algae and slime. These preparations are toxic and precautions must be taken to protect the skin and eyes when handling them.
- All water treatment problems should be referred to a competent water treatment specialist.

SECTION 7
RESIDENTIAL AND COMMERCIAL CONTROLS

UNIT 23
AIR-CONDITIONING CONTROLS

OBJECTIVES

After completing the study of this unit, the student will be able to

- define the following terms as they relate to devices providing electrical, electronic, or pneumatic control for air-conditioning equipment.

  set point
  control point
  deviation
  corrective action
  cycling
  differential gap
  offset
  lag
  primary element
  bimetallic element
  bellows
  electronic controls
  humidity-sensing primary element
  pressure-sensing primary element

- state six basic functions of a fully automatic control system.
- describe the following types of control action –
  1. simple two-position control
  2. timed two-position control
- list the four configurations of electrical/electronic control circuits.
- describe and state the function of a Wheatstone bridge.
- check the system operation after the installation is complete.
- troubleshoot all of the following system control circuits –
  1. main power
  2. transformer
  3. relay coil
  4. cooling
  5. damper
  6. fan
  7. heating

- list the components of a pneumatic system.
- state the function of the following devices as they apply to a pneumatic system.
  actuator
  damper (normally open type and normally closed type)
  valve (normally open type and normally closed type)
  controller (thermostat)
  relay
  master-submaster control

Air-conditioning equipment can be electrically, electronically, or pneumatically controlled. Numerous combinations of each type of control are possible. Since it is not practical to attempt to describe all of the combinations in this unit, only fundamental terms, functions, operations, and simple service techniques are presented.

DEFINITION OF TERMS

Set point. A value on the control scale (temperature, humidity, or pressure) at which the control indicator is set is the set point. Thus, if a thermostat is set at 75°F, the set point is 75°F.

Control point. The value of the temperature or humidity that is maintained in the conditioned space and recorded by the control is the control point. For example, a thermostat may record a temperature of 76°F (control point) even though it is set at 75°F (set point).

Deviation. The momentary difference between the set point and the control point is the deviation. For example, with a set point of 75°F and a control point of 76°F, the deviation is 76°F - 75°F = 1°F.

Corrective action. The controls will take a corrective action to maintain the control point at a value that is in reasonable agreement with the set point. The following situation is a typical example of a corrective action within a system. A drop in the outside temperature causes a drop in the temperature inside the conditioned space. This temperature drop is recorded by the thermostat in the space, resulting in a lower thermometer or control point reading. The thermostat senses the drop and closes its contacts. In so doing, the thermostat turns on the burner in the furnace. Heat is added to the space and the control point (thermostat reading) rises until it approaches or equals the set point. The thermostat senses the rise in temperature and opens its contacts. This action shuts off the furnace burners and the cycle is completed. In this situation, the thermostat acted to correct and maintain the control point. This same sequence applies in a cooling application, when a rise in temperature occurs in the space.

Cycling. Cycling is a consistent repetition of a change in the control point. For example, if the space temperature (control point) drops several degrees below the set point, the thermostat calls for heat. Heat is added to the space, but the thermostat shuts off the burner before the space temperature reaches the set point. After a very slight drop in temperature, the thermostat calls for more heat. This time the thermostat shuts off the burner only after the space temperature is several degrees above the set point. Cycling is also known as hunting.

Differential gap. This gap is the range through which the conditions of the space must travel to move a two-position control (on-off or open-closed) from one position to the other. If the thermostat contacts are open at 76°F and the thermostat moves to the closed position (contacts are closed) at 74°F, then the differential gap is 76°F - 74°F = 2°F.

Offset. The sustained difference between the set point and the control point is known as the offset.

Lag. Lag is a delay in the effect of a changed condition in one part of the system on a related condition in another part of the system. As an example, consider the case when the condition at the furnace goes from no heat to heat (the burner comes on). Heat passes to the conditioned space and the control temperature increases. In this instance, the lag extends from the time the burner comes on until the time when the temperature condition in the space reaches its former control point.

Primary element. That portion of the control which first uses energy from the controlled medium (such as air or water) is the primary element. For example, the bimetallic element of a thermostat uses the energy resulting from a change in air temperature to open or close a set of contacts. The bimetallic element is a primary element. Such an element senses either temperature, humidity, or pressure changes. Temperature-sensing primary elements can be bimetallic strips, sealed bellows, or sealed bellows attached by capillary tubing to a remote sensing bulb.

Bimetallic elements. This type of element, figure 23-1, consists of thin strips of two different metals securely attached to each other. Since different metals expand and contract at different rates during changes in temperature, the thin strips of metal actually curve toward or away from a given point. In this way, the thermostat makes or breaks its contacts when the temperature changes. Bimetallic elements are usually U-shaped.

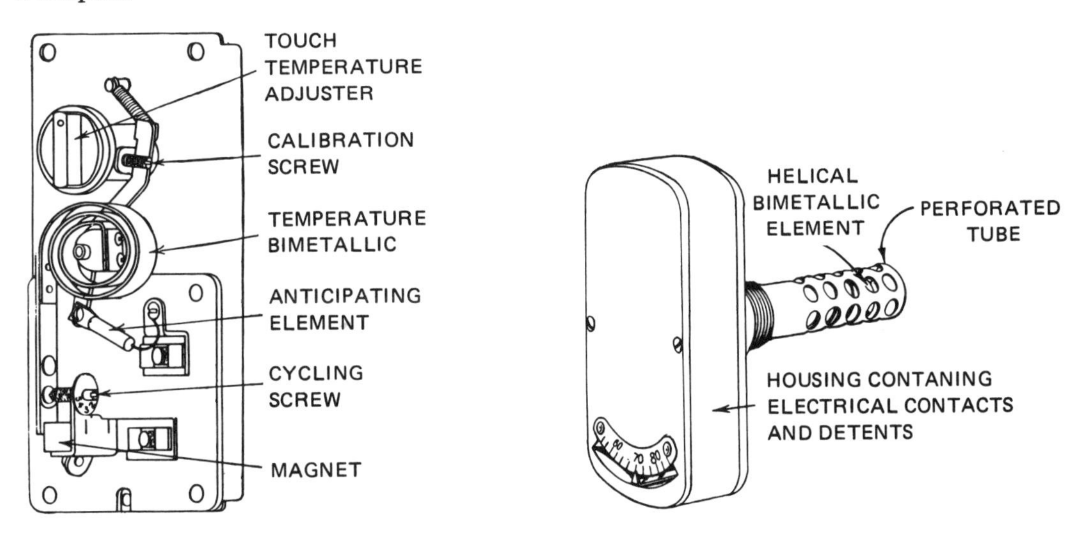

Fig. 23-1 Bimetallic elements in an electric thermostat (left) and in an insertion thermostat for a duct (right)

a temperature-sensing element, a bellows is first put under a vacuum and
lly or completely filled with a liquid and sealed. Changes in temperature
uid in the bellows to expand or contact, resulting in the expansion or
of the bellows. This action of the bellows in turn moves a mechanism
or closes a control valve or similar device.

A remote *bulb bellows* operates in the manner described except that the change in temperature takes place at the bulb location.

Electronic controls. A low-mass primary element is used in an electronic control. This primary element may be as simple a device as a wire whose resistance is altered as a result of changes in temperature. Although a feeble signal occurs due to a resistance change in the wire, the electronic circuit can amplify this signal to a usable strength.

Humidity-sensing primary elements. Any material that responds to a change in the moisture content (such as hair, wood, or leather), can be used as a humidity-sensing primary element. The moisture content of the material changes as it absorbs or gives up moisture. As a result, the substance expands or contracts and moves the control mechanism.

Electronic humidity-sensing primary elements. This type of element consists of two strips of gold fused to a piece of glass. Electrical contact is maintained between the gold strips through a hygroscopic salt which is painted over the glass surface. The resistance of the salt varies according to the amount of moisture that it absorbs or releases. Small changes in resistance cause changes in the flow of current. These changes are detected and used by the electronic circuit.

Pressure-sensing primary elements. These elements can be bellows, diaphragms, and inverted bells or similar devices, figure 23-2. The pressure of the air or water (or other medium) moves the primary element directly. The resulting action operates the pneumatic or electrical control.

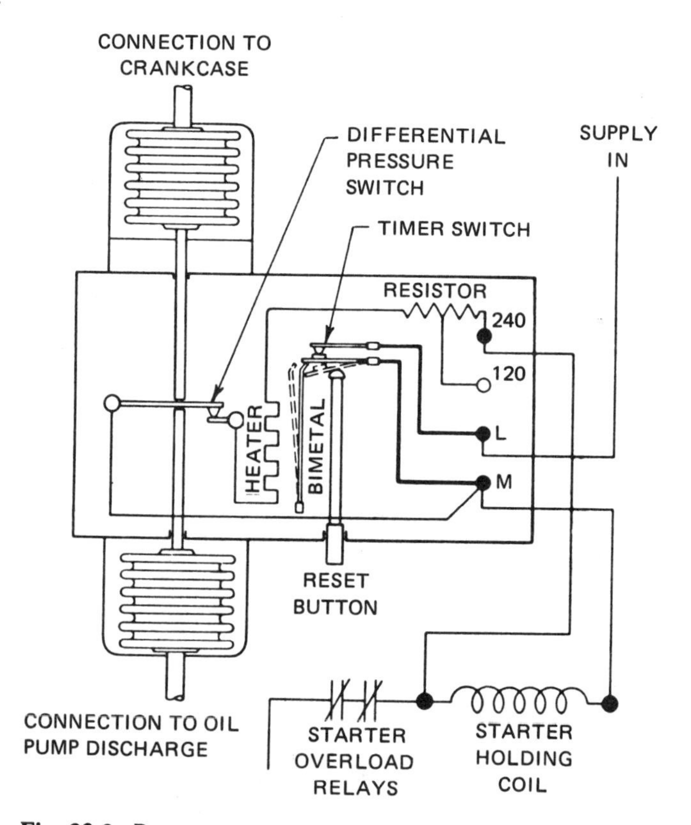

Fig. 23-2 Pressure-sensing elements (the two opposing bellows) operate the oil pressure failure control

BASIC FUNCTIONS OF CONTROL SYSTEMS

A fully automatic control system performs six basic functions as indicated in the following brief descriptions.

- The sensing element measures changes in temperature, pressure, and humidity.

- The control mechanism translates the changes into energy that can be used by devices such as motors and valves.
- The connecting wiring, pneumatic piping, and mechanical linkages transmit the energy to the motor, valve, or other device that acts as the point of corrective action.
- The device then uses the energy in its operation to achieve some corrective action. Motors operate compressors, fans, dampers, and similar devices. Valves control the flow of gas to burners or cooling coils and permit the flow of air in pneumatic systems. Valves also control the flow of liquids such as the water in a water-cooled condenser.
- The sensing elements in the control detect the change in conditions (corrective action) and signal the control mechanism or connecting means.
- The control stops the motor, closes the valve, or terminates the action of the device being used. As a result, the call for corrective action is ended. This action by the control prevents overcorrection.

CONTROL ACTION

Two-position Control

The use of two-position control permits the final control element (motor, valve, or similar device) to occupy one or the other of two positions. Two-position control is used to open and close gas valves for residential heaters and to start and stop motors for fans, compressors, pumps, and similar devices.

Simple Two-position Control

In a system using a simple two-position control, figure 23-3A, the control never catches up to the controlled condition. Rather, the control only corrects a condition that has just passed. The control does not act on a condition that is taking place or is about to take place.

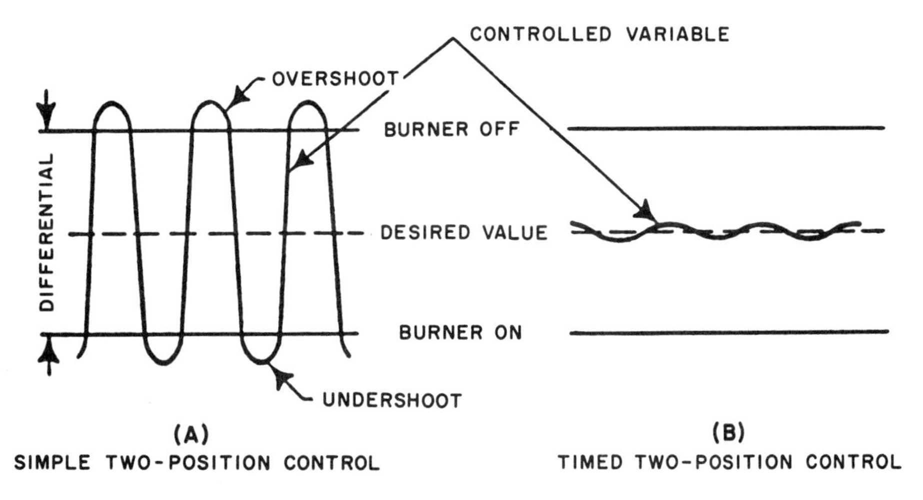

Fig. 23-3 Types of control circuits

Timed Two-Position Control

The ideal method of heating or cooling a space is to replace the heat lost or remove the heat gained in exactly the amounts needed. To do this, the control must respond to a gradual or average change in the air temperature of the conditioned space; this air temperature change is the controlled variable. An average temperature change is produced by adding a tiny heater (timing device) near the thermostat temperature-sensing element, figure 23-3B.

Timer Operation for Heating. While the air temperature at the thermostat is maintained within certain limits, the thermal heater will cool to its *on* point and energize the heating element. The thermal element then heats to its *off* point and de-energizes the heating element. As a result, the thermal heater again cools to the *on* point and the cycle is repeated.

Timer Operation for Cooling. While the air temperature at the thermostat is maintained within certain limits, the thermal heater element will cool to its *off* point and de-energize the heating element. The thermal heater then heats to its *on* point and energizes the heating element. The thermal heater then cools to its *off* point again and the cycle is repeated.

The timing sequence for cooling is the opposite of the sequence for heating. The heater element turns off when the air at the thermostat is cool and turns on when the air is warm. When the element turns on, it adds heat at the thermostat element and causes the thermostat to call for cooling sooner than it would if the heater were not used. This cooling is added to the conditioned space immediately. In this way, the air temperature is prevented from increasing as far as it would if the timer were not provided.

The use of the timer or thermal heater reduces the swing (differential) in the air temperature in the room so that the temperature remains almost constant. The temperature differential is much greater when the timer is not used, as shown in figure 23-3.

ELECTRICAL CONTROL CIRCUITS

Electrical control circuits use electrical energy to transmit signals from the control to the final control device (such as a motor, valve, relay, or other device). Such controls are available with low-voltage and line voltage ratings.

Low-voltage wiring and electrical devices are subjected to a voltage of 25 volts or less. Suitable transformers must be used to decrease the line voltage to a suitable low-voltage value.

The *line voltage* applied to wiring and electrical devices is in the range of 115 to 230 volts. Line voltage may be used at its full value or it may be stepped down by a transformer to supply energy to a low-voltage circuit. The line voltage should not be confused with the term high voltage which is considered to be 500 volts or more.

A *potentiometer*, figure 23-4, consists of a number of turns of resistance wire wound on a cylindrical form. Three connections are provided. The center connection is movable and is wired to complete the circuit at any point along the cylinder.

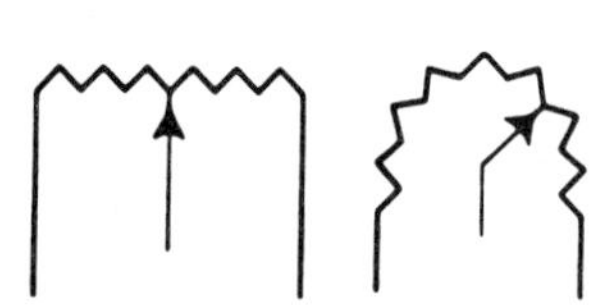

Fig. 23-4 Potentiometer symbols

A *balancing relay*, figure 23-5, contains an armature pivoted at one end which swings between two electromagnetic coils. As the magnetic field in the two coils changes with a changing current, the armature moves toward the stronger magnetic force. The armature is usually provided with contacts so that it can complete circuits as it moves to predetermined positions.

Relay contacts are known as *in contacts* if they complete circuits when the relay is energized. Contacts which complete circuits when the relay is de-energized are known as *out contacts.*

Three combinations of *In* and *Out* contacts are shown in figure 23-6.

In figure 23-6A a single-pole, double-throw relay is shown with one *in* and one *out* contact. Figure 23-6B shows a double-pole, single-throw relay with *In* contacts. Figure 23-6C illustrates a three-pole relay with one single-pole, double-throw switch; one single-pole, single-throw switch with *in* contacts; and one single-pole, single-throw switch with *out* contacts.

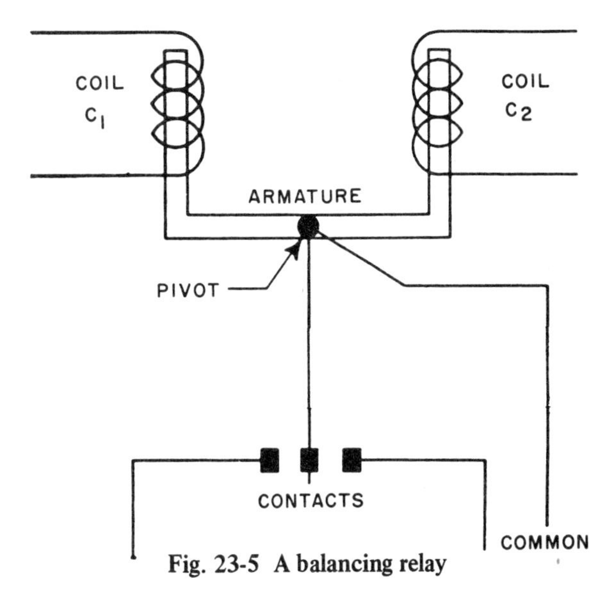

Fig. 23-5 A balancing relay

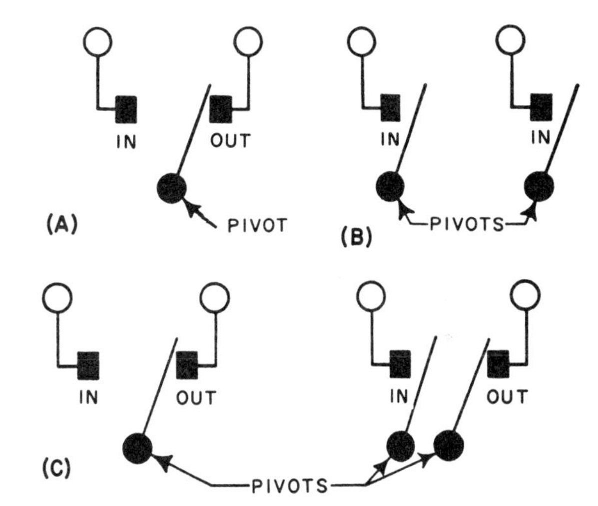

Fig. 23-6 Combinations of In/Out Contacts

TYPES OF CONTROL CIRCUITS

Three-wire, Low-voltage Circuit

One type of three-wire circuit uses a control containing two sets of contacts. The contacts make in sequence to start the control and break in reverse sequence to stop the control.

A second type of three-wire circuit has a control that makes one circuit to start, breaks this circuit, and then makes a second circuit to stop, figure 23-8, page 285.

For the basic control circuit shown in figure 23-8, the following procedures illustrate what corrective action is taken when the temperature in the controlled space drops.

Temperature Drop, Figure 23-8B

- The thermostat blade engages the blue contact.
- The starting circuit is established.
- The motor is energized and starts to rotate clockwise.

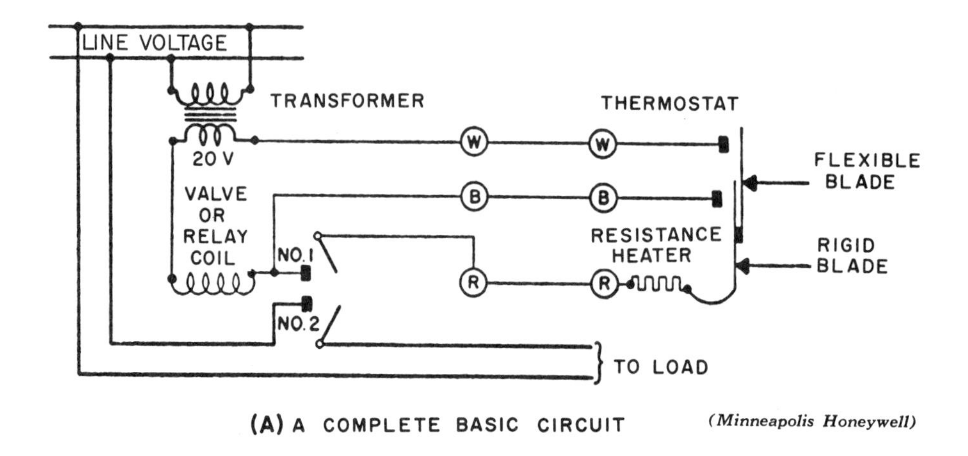

(A) A COMPLETE BASIC CIRCUIT (Minneapolis Honeywell)

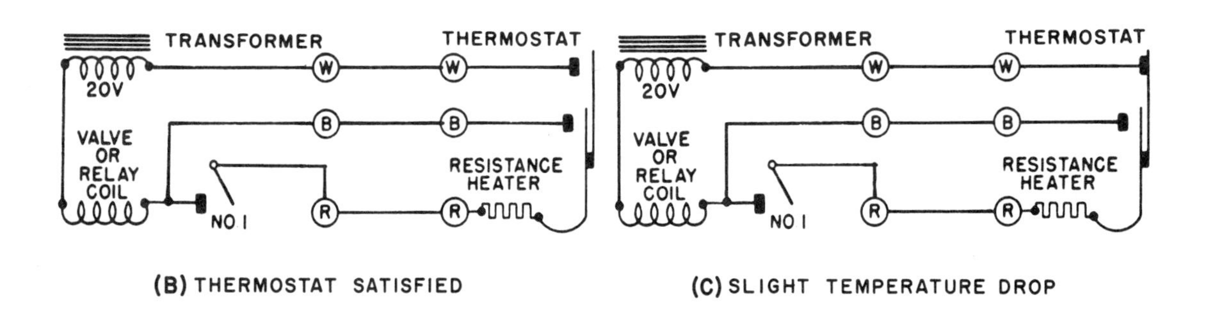

(B) THERMOSTAT SATISFIED

(C) SLIGHT TEMPERATURE DROP

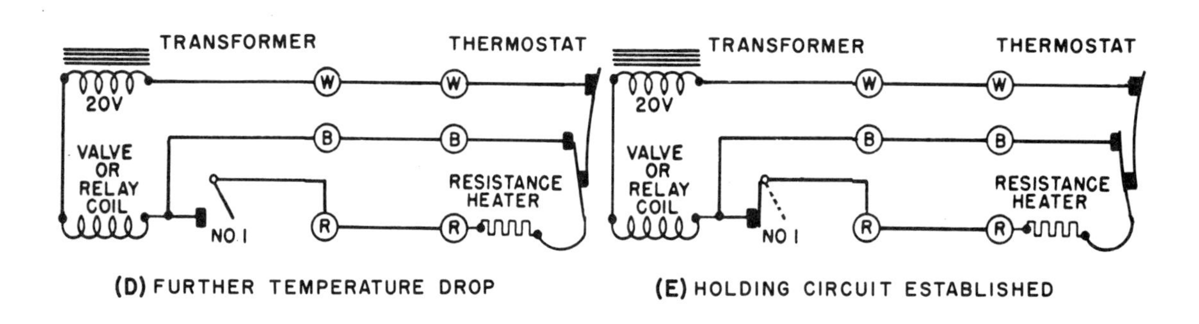

(D) FURTHER TEMPERATURE DROP

(E) HOLDING CIRCUIT ESTABLISHED

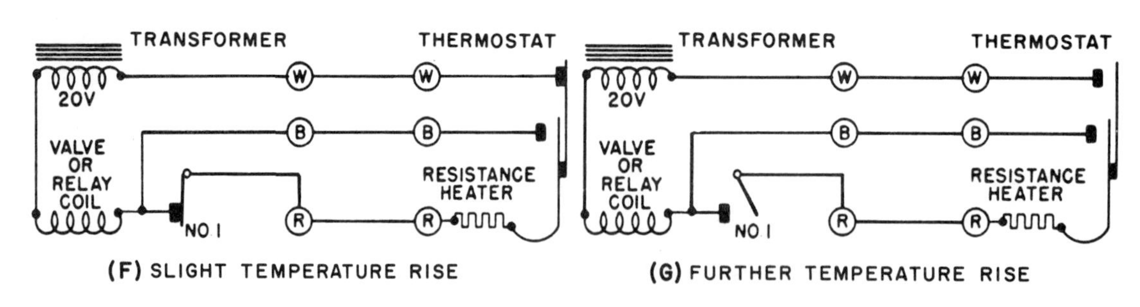

(F) SLIGHT TEMPERATURE RISE

(G) FURTHER TEMPERATURE RISE

Fig. 23-7 A three-wire, low-voltage control circuit

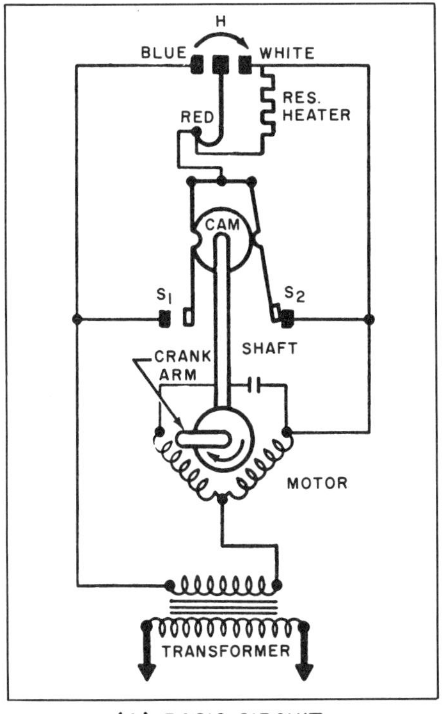

(A) BASIC CIRCUIT

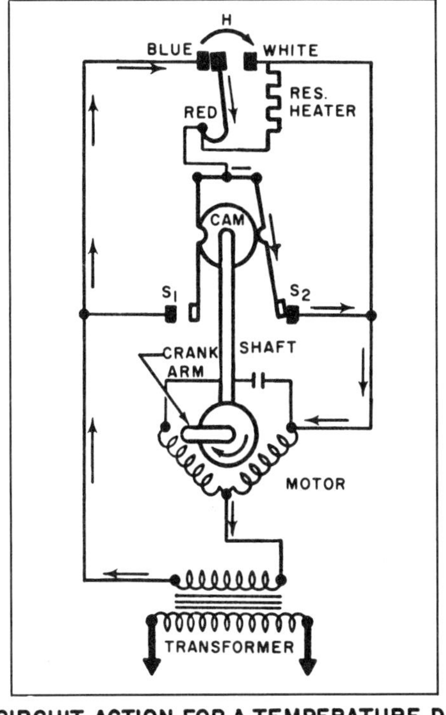

(B) CIRCUIT ACTION FOR A TEMPERATURE DROP

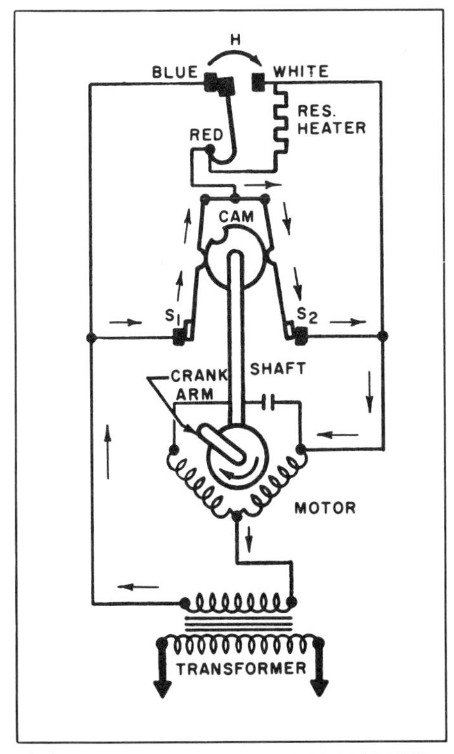

(C) HOLDING CIRCUIT ESTABLISHED

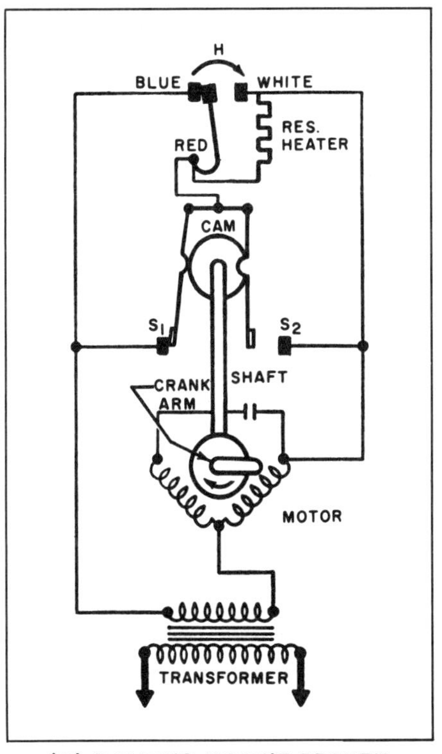

(D) HOLDING CIRCUIT BROKEN

Fig. 23-8 (Cont'd.) (Courtesy Minneapolis Honeywell)

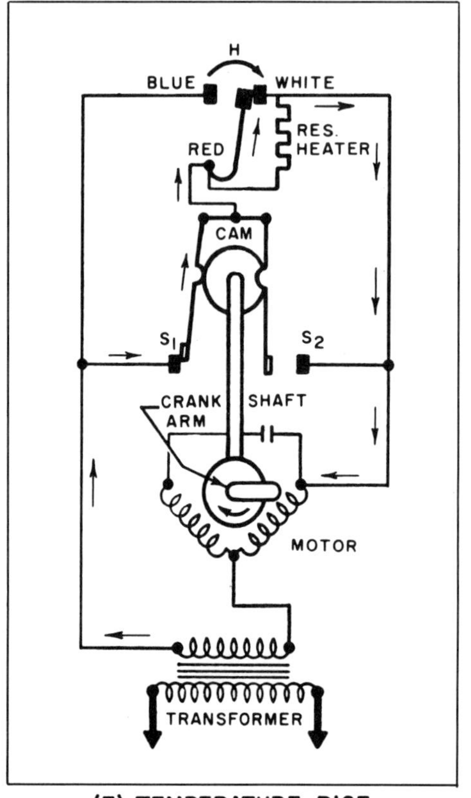

(E) TEMPERATURE RISE

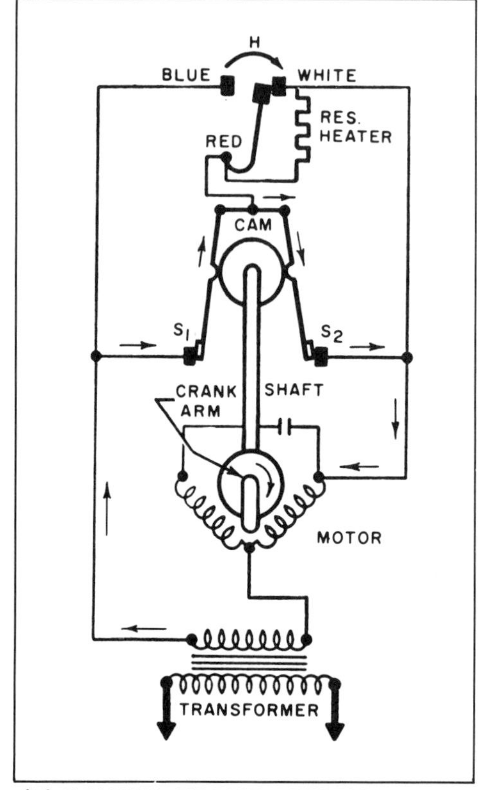

(F) HOLDING CIRCUIT REESTABLISHED

Fig. 23-8

Holding Circuit Established, Figure 23-8C

As the motor and cam rotate:

- The left blade of the maintaining switch makes contact with S1 and the holding circuit is established. This circuit is independent of the starting circuit. Once the holding circuit is made, it furnishes current to the motor regardless of the action of the thermostat.
- If the thermostat holds the blue contact closed, a small amount of current passes through the holding circuit because it offers the path of least resistance.

Holding Circuit Broken, Figure 23-8D

When the motor shaft has rotated 180 degrees:

- The cam breaks contact S2.
- All circuits are incomplete.
- The motor stops.
- Negligible current passes through the heater resistance; the current is insufficient for motor travel.

Temperature Rise, Figure 23-8E

A temperature rise in the controlled space and the heat from the resistance heater produce the following actions:

- The thermostat blade moves to the right and engages the white contact.

- The starting circuit is completed
- The motor starts to rotate clockwise.

As stated previously, the additional heat provided by the resistance heater causes the thermostat to close sooner than it would for a simple two-position control. As a result, the temperature differential is minimized.

Holding Circuit Reestablished

- The right blade of the maintaining switch contacts S2.
- The motor continues to rotate until it makes half a turn.
- At this point, the holding circuit is again broken at contact S1 and the motor stops. One cycle is complete.

Two-Wire Line Voltage Circuit

The two-wire line voltage circuit in figure 23-9 contains a control that makes the circuit when the control switch is closed and breaks the circuit when the control switch is open.

Although the three-wire circuits described in previous sections were low-voltage circuits, they can also be designed as line voltage circuits. Similarly, two-wire, line voltage circuits can be modified to two-wire, low-voltage circuits.

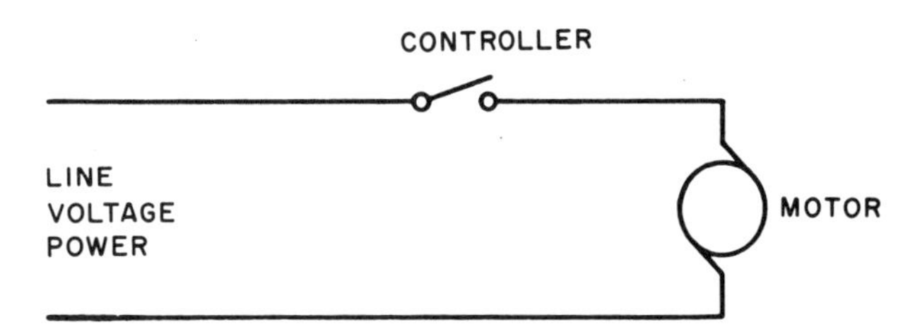

Fig. 23-9 The two-wire, line voltage circuit

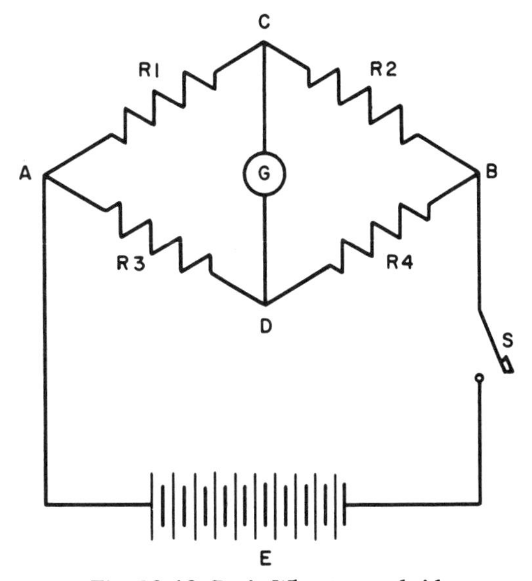

Fig. 23-10 Basic Wheatstone bridge

ELECTRONIC CONTROL CIRCUITS

Electronic circuits are used to control all types of heating and cooling equipment for both commercial and residential applications. Electronic controls feature quick response to temperature changes. As a result, temperature averaging is accomplished readily.

Basic Electronic Control Circuit

The Wheatstone Bridge circuit, figure 23-10, is the basis of many electronic control circuits. The bridge consists of two sets of resistors connected in series. The series resistors, in turn, are connected in parallel across a dc voltage source. One set of series resistors is shown by R1 and R2 in figure 23-10. The second set of resistors is R3 and R4. The voltage source (E) is connected between points A and B. A sensitive current indicator (galvanometer G) is connected across the sets of resistors at points C and D.

When switch S is closed, the voltage from E flows through both sets of resistors. If the potential at point C is the same as that of point D, the galvanometer gives a zero reading. In other words, there is no potential difference between the two points. Thus, the bridge is balanced.

When any one of the resistors has a different voltage, the galvanometer registers a value other than zero. This reading indicates that there is a current between points C and D. Such a current indicates that the bridge is unbalanced. Control manufacturers modify the basic bridge circuit to use it in electronic control circuits. A typical bridge circuit modification is shown in figure 23-11.

Modified Bridge Circuit

In the modified circuit of figure 23-11, the dc voltage is replaced by a 15-volt ac supply. In addition, the switch is eliminated and the galvanometer is replaced by an amplifier-switching relay unit. Resistor R2 is replaced by a temperature-sensitive element, T1.

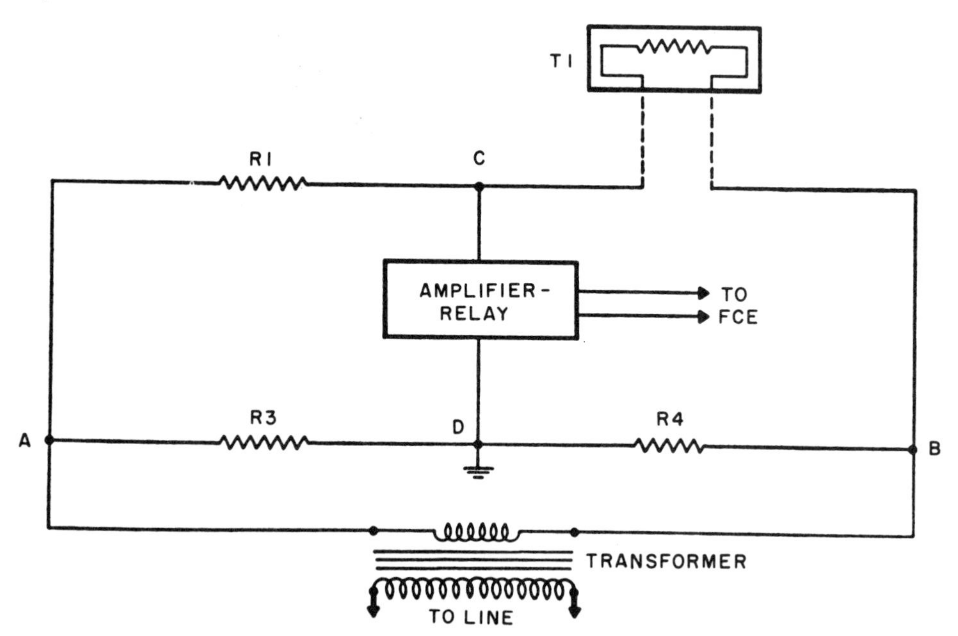

Fig. 23-11 Modified bridge circuit

To simplify the description of the modified bridge, a resistance of 1,000 ohms is assigned to each of the three fixed resistors (R1, R3, and R4) and to the thermostat element T1. When the desired conditions in the space are met, the resistance in T1 does not change. Thus, the voltage across the amplifier-relay is zero. Since there is no voltage across the amplifier-relay, the final control element (a motor, a valve, or other device) cannot be energized. The bridge circuit is balanced.

When the air temperature in the conditioned space changes, the thermostat element senses the change. A corresponding change in the resistance at T1 occurs and the bridge becomes unbalanced. Voltage now flows through the amplifier-relay to the final control element. In a heating application, a drop in temperature in the conditioned space causes a decrease in the resistance of T1. The resulting voltage relationships in the bridge circuit cause the amplifier-relay to increase the voltage to the final control element. As a result, heat is added to the space. As the heat is added, the resistance at T1 increases. The voltage change that occurs in the bridge circuit stops the heating by shutting off the burner.

The modified bridge circuit of figure 23-11 can also be used for cooling applications. Depending upon whether the bridge voltage fed to the amplifier-relay is in phase or out of phase with the supply voltage, the final control element is opened or closed.

SYSTEM CHECKOUT

The control system should be installed according to manufacturers' instructions and applicable codes. After the system is installed, the following checks should be made to insure the proper operation of the system.

- All wiring connections must be examined for tightness. All wires are to be traced to insure that the connections are made according to instructions and code guidelines.
- The thermostat calibration must be checked. If recalibration is necessary, the technique described shortly is to be used.
- The technician must insure that the system starts when the thermostat calls for heating and cooling, and stops when the thermostat is satisfied.

Thermostat Calibration (Heating and Cooling Applications)

The following procedure is to be used if the thermostat requires calibration.

- For a thermostat having a timer heater, a thermostat setting is used that stops the operation of the system. An allowance of ten minutes is to be made for the heater element to settle out.
- The thermostat must be in a level position if it has a mercury switch, figure 23-12.

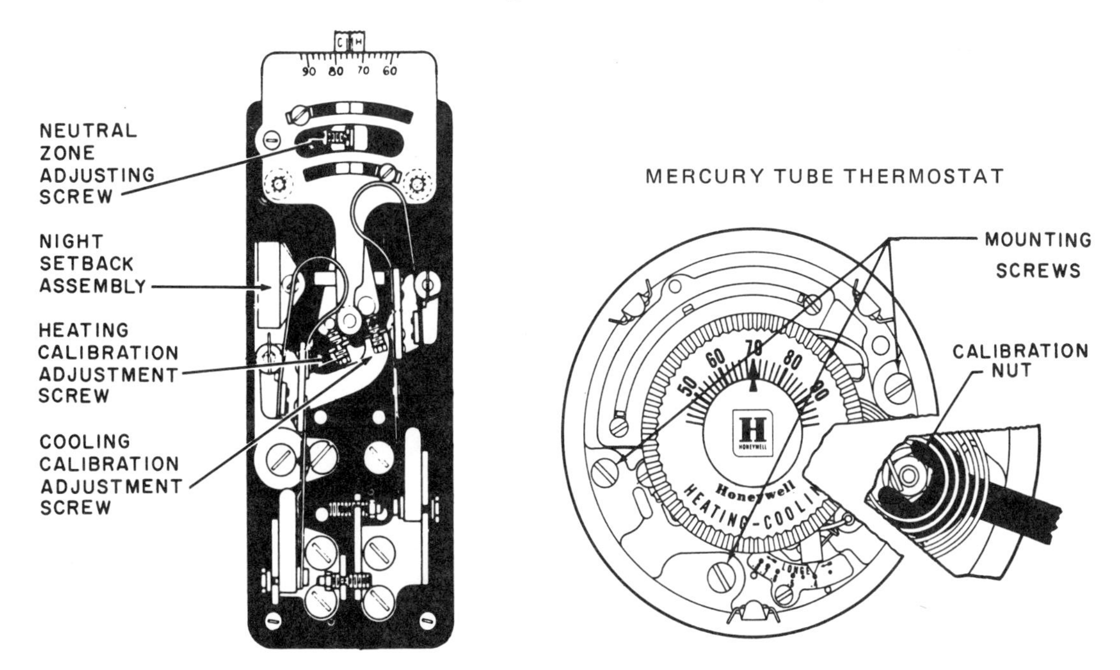

Fig. 23-12
(Courtesy Minneapolis Honeywell)

- Calibration Procedure for Heating:

  1. Place the system switch in the heating position and the fan switch (if used) in the automatic position.
  2. Move the set point indicator to a value above the room temperature. The thermostat contacts should open.
  3. If the thermostat contacts remain closed, turn the heating calibration nut (mercury switch) or screw (snap action switch) counterclockwise until the contacts just open. Do not turn the nut (screw) beyond this point.
  4. Step back and allow the thermostat to dissipate the heat picked up from the hands. (Usually, a few minutes are required.)
  5. Move the set point indicator to the value of the room temperature.
  6. Turn the calibration nut (screw) clockwise only to the point that the contacts close. Do not turn the nut (screw) beyond this point.
  7. Recheck the calibration. Move the set point indicator toward the lower end of the scale. Step back and wait for several minutes until the heat dissipates. Move the set point indicator up-scale until the contacts just close. The contacts should close when the set point indicator has the same value as the room temperature. If the temperature values do not coincide, recalibrate the thermostat.

- Calibration Procedure for Cooling:

  1. Move the system switch to the cooling position and the fan switch to automatic.
  2. Move the set point indicator to the room temperature value. The cooling contacts should just close.
  3. If the cooling contacts remain open, turn the cooling calibration screw or nut clockwise until the contacts just close.
  4. Turn the cooling calibration screw or nut counterclockwise until the contacts just open.
  5. Step away from the thermostat and allow the heat to dissipate.
  6. Recheck the calibration by moving the set point indicator down-scale until the contacts just close. Slowly raise the set point indicator up-scale until the contacts just open. The contacts should open when the indicator reaches the same value as that of the room temperature. If the temperature values do not coincide, recalibrate the thermostat.

TROUBLESHOOTING

Centrally located control panels are commonly used to provide a readily accessible means of checking all of the circuits in the system from one location. A typical troubleshooting procedure for such a central control panel is described in the following paragraphs. If this type of panel is not available, each circuit described can still be checked in the same manner, but at the points in the system where the components (such as the fan motor, damper motor, thermostat, and transformer) are located.

Power Circuit, Figure 23-13

- The troubleshooting procedure always starts with a test for power ahead of the transformer.
- Connect a suitable test light across the power terminals of the panel (Terminals 1 and 2).
- If the light is on, the power source is functioning properly. If the light stays off, check for blown fuses or loose or broken connections. Correct the problem and retest the power terminals.

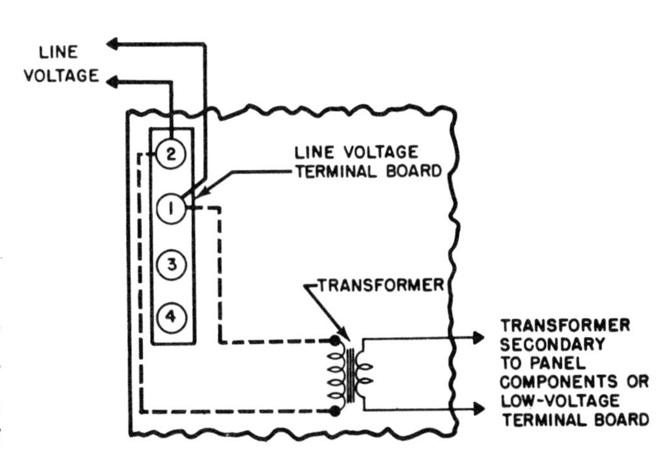

Fig. 23-13 Power circuit

Transformer Circuit, Figure 23-14

- Connect a test lamp of suitable voltage to the power lead on the line voltage (primary) side of the transformer.
- If the light turns on, the power to the transformer is correct. If the light stays off, check the connections and the power circuit. Make any repairs necessary.

Fig. 23-14 Transformer circuit

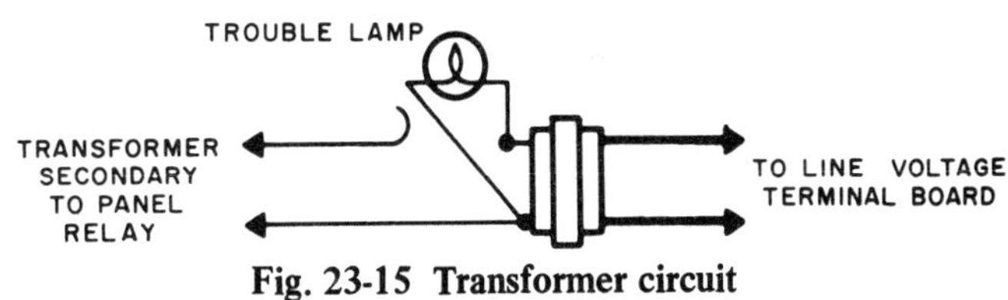

Fig. 23-15 Transformer circuit

- Connect a test light, figure 23-15, across the low-voltage (secondary) terminals of the transformer and disconnect one transformer lead.
- If the light turns on, the transformer is functioning correctly. If the light stays off, check and tighten the transformer line voltage connections. If the light still does not turn on, replace the transformer.

Relay Coil Circuit, Figure 23-16

- Attach a test lamp of suitable voltage to one of the relay leads as shown. If the light turns on, the relay is functioning.
- If the light stays off, test the power to the coil by connecting a test lamp across the circuit from the

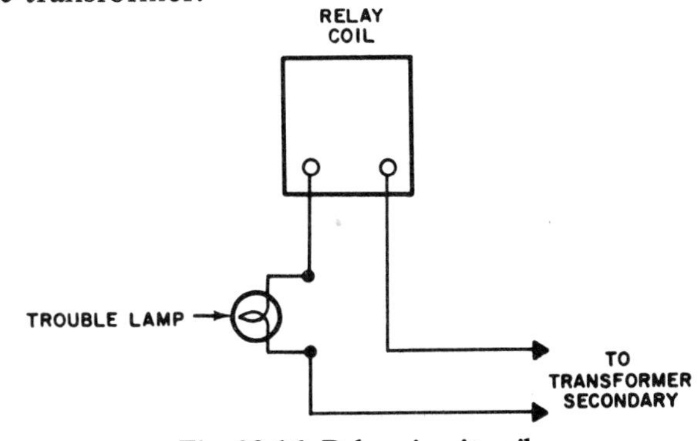

Fig. 23-16 Relay circuit coil

transformer, figure 23-17. If the light turns on, replace the coil. If the light does not turn on, the wiring between the transformer and the relay coil is defective. Repair the wiring.

- Reconnect the test light to one lead of the coil. If the light still does not turn on, replace the coil. If light turns on, the relay coil is functioning correctly.

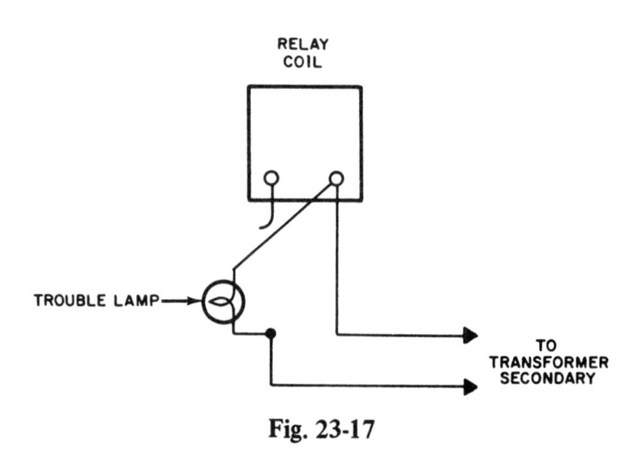

Fig. 23-17

Cooling Circuit, Figure 23-18

- If any cutout devices are used, these devices must be reset before the troubleshooting procedure is started.
- Install a jumper between the thermostat leads (Y to R).
  1. If the starter does not operate, the starter coil, pressure switches, or overloads may be defective.
  2. Jumper the pressure switches and the overload. If the starter still does not operate, the pressure switches or the overload are defective.
  3. Replace all defective parts as necessary. If the starter still does not operate, replace it.

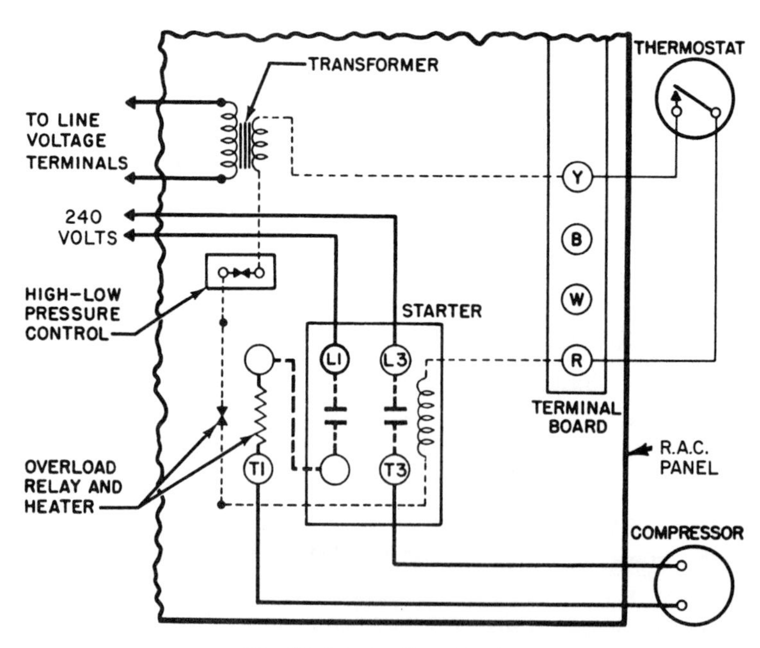

Fig. 23-18 Cooling circuit

- If the starter operates but the compressor does not start up, jumper across the starter (L1 to T1 and L3 to T3).
  1. If the compressor operates, replace the starter contacts.
  2. If the compressor does not operate, the trouble is in the cooling equipment or its wiring. Check out the wiring and each component (such as the motor and relay). Repair or replace any defective devices as necessary.

Damper Circuit, Figure 23-19

- Jumper the thermostat terminals (O to R).
  1. If the damper operates, the problem is in the control (thermostat) circuit. Repair or replace the circuit components as necessary.
  2. If the damper does not operate, the problem is in the wiring to the damper or the damper motor. Repair or replace the wiring as necessary.

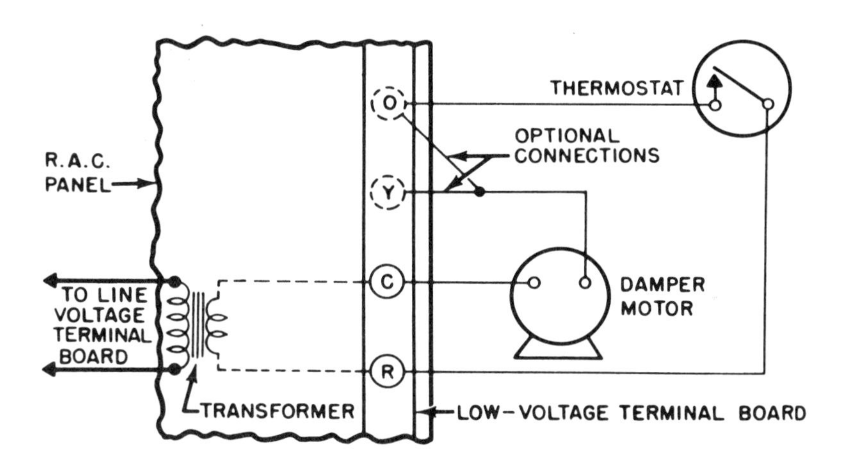

Fig. 23-19 Damper circuit

Fan Circuit, Figure 23-20

- For the fan circuit shown in figure 23-20, page 294, the fan is energized by the thermostat or through the starter.
- Jumper the thermostat terminals (G to R).
- If the fan relay does not operate, replace the fan relay.
- If the fan starts, the control (thermostat) circuit is defective. Repair or replace the circuit components as necessary.
- If the relay operates but the fan does not, jumper the fan relay terminals (F to F).
  1. If the fan operates, replace the relay.
  2. If the fan does not operate, the fan motor wiring or the fan motor is defective. Repair or replace components as necessary.

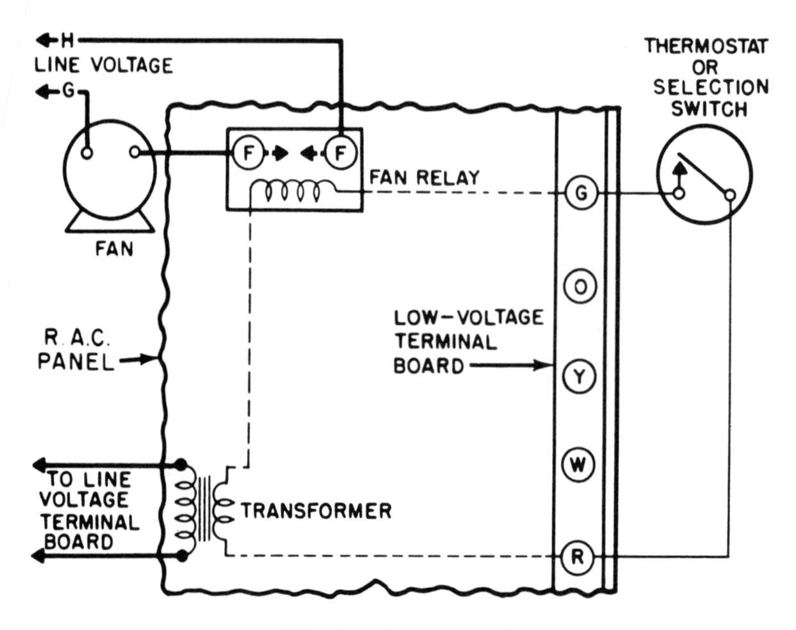

Fig. 23-20 Fan circuit

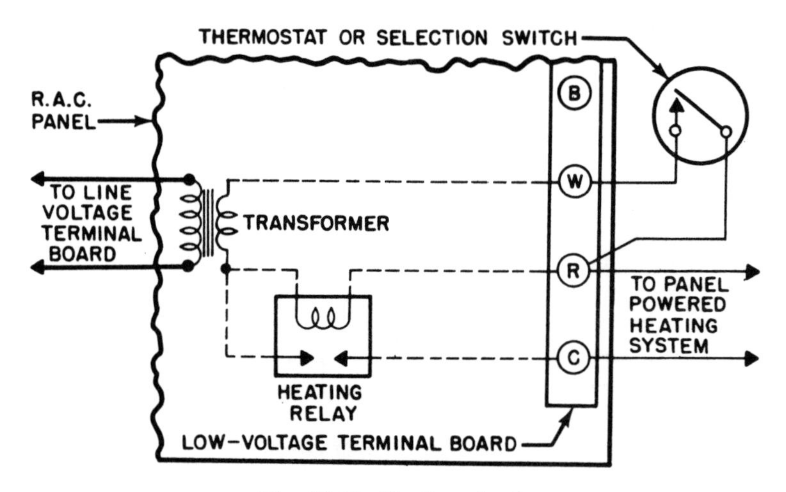

Fig. 23-21 Heating circuit

Heating Circuit, Figure 23-21

- Jumper the thermostat terminals (W to R).
  1. If the heating relay does not operate, replace the relay.
  2. If the burner starts, the control (thermostat) or its wiring is defective. Repair or replace components as necessary.
  3. If the relay operates, but the burner does not start, the heating components or their wiring is defective. Check the wiring and each component; repair or replace as necessary.

PNEUMATIC CONTROLS

Pneumatic controls, figure 23-22, are used principally in large commercial air-conditioning systems. Although they may be used in residential applications. This

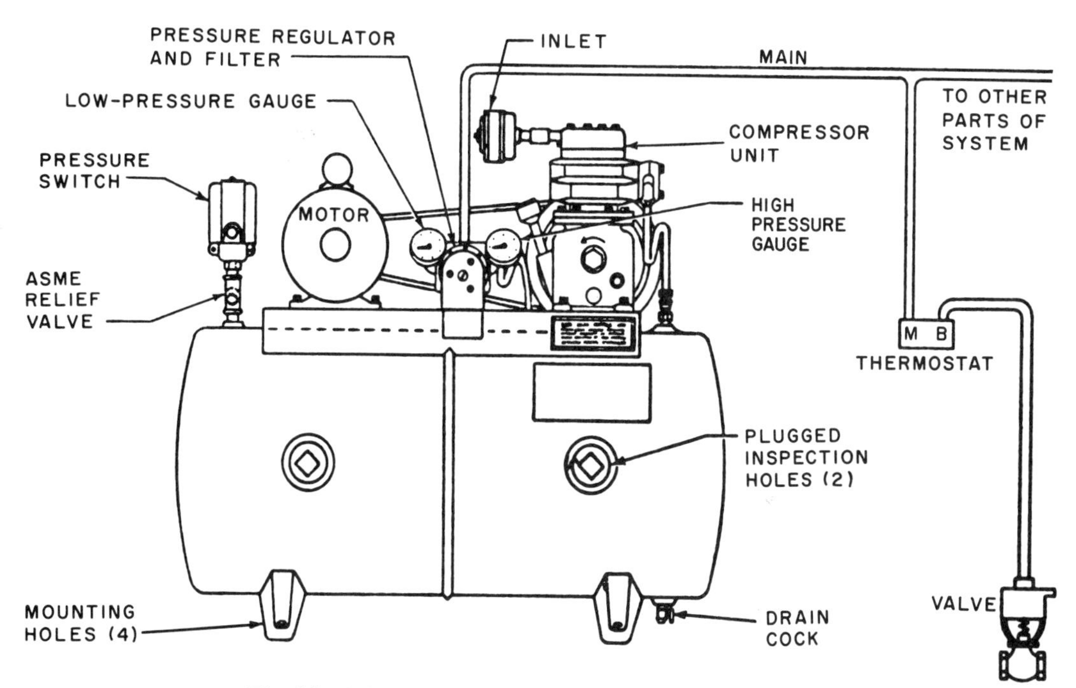

Fig. 23-22 Diagram of a basic pneumatic control system

type of control is operated by air pressure supplied by a compressor. A pneumatic control system consists of the following components:

- Compressor
- Copper tubing to carry the air
- Air temperature, humidity, and pressure controllers
- Relays and switches
- Valves and motors

The compressor supplies air through the copper tubing to the controller (thermostat). The tubing from the compressor to the thermostat is called the *main line* because it carries the main air supply directly from the compressor. The tubing from the thermostat to the operating valve is called the *branch line* because this line does not carry air directly from the compressor, but from a controller.

The air pressure is maintained by the compressor at a value of approximately 20 to 35 psi for low-pressure systems. A reducing valve at the compressor reduces the pressure of the air that enters the main line to 15 psi for systems that supply heating only or cooling only (one-temperature systems). For systems that supply both heating and cooling (two-temperature systems), two reducing valves are used. Thus, two different pressures are available to the main line. These pressures vary according to the type of controller used.

Pneumatic Operation

The air from the compressor flows through the main line to the thermostat (or other controller). The thermostat receives the main line air at a constant pressure and delivers it to the valve (controlled device) through the branch line at pressures that vary according to the temperature recorded by the thermostat. The valve, in turn, moves toward the open or closed position depending upon the variation in air pressure in the branch line. In this manner, a change in room (space) temperature causes the thermostat to change the air pressure in the branch line. The change in pressure in the branch line causes the valve to move toward the open or closed position depending upon whether the room temperature has increased or decreased. When the valve moves toward the open position, more heating is added to the room. When the valve moves toward the closed position, less heat is added. A similar principle applies to the cooling process.

Pneumatic Components

Actuators. An actuator is a motorized valve or damper, figure 23-23. The motor used in each case is usually the same in that it moves (performs work) according to changes in the branch line pressure. A pneumatic motor contains either a bellows, a diaphragm, or a cylinder and piston, and an opposing spring. A damper linkage or valve stem is connected to the bellows (or diaphragm, or piston).

When the branch line pressure to the motor decreases, the bellows expands due to the internal spring pressure. As the branch line pressure increases, the bellows contracts as a result of the air pressure. The expansion and contraction of the bellows moves the linkage or valve stem so that the damper or valve opens and closes.

Normally Open Dampers. A normally open damper is one that is installed so that it moves toward an open position as the air pressure in the damper motor decreases, figure 23-24A.

Normally Closed Dampers. A normally closed damper is one that is installed so that it moves toward the closed position when the air pressure in the damper motor decreases, figure 23-24B.

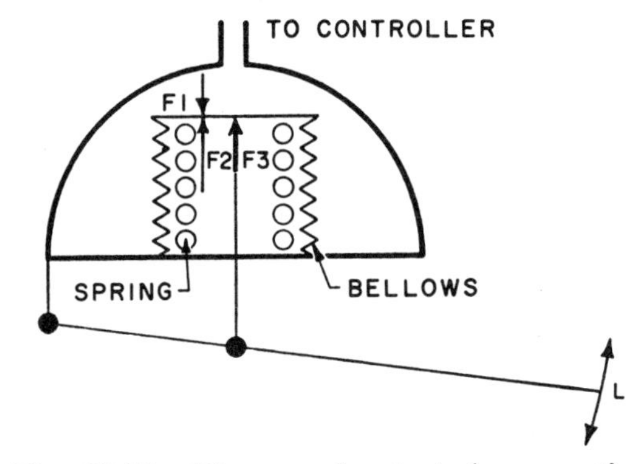

Fig. 23-23 Diagram of a typical pneumatic motor

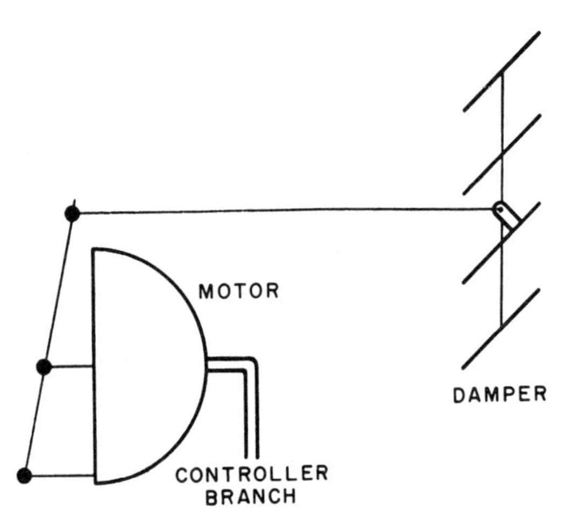

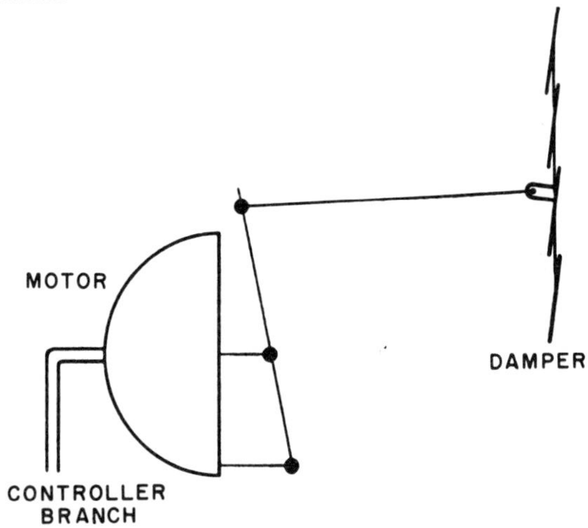

Fig. 23-24

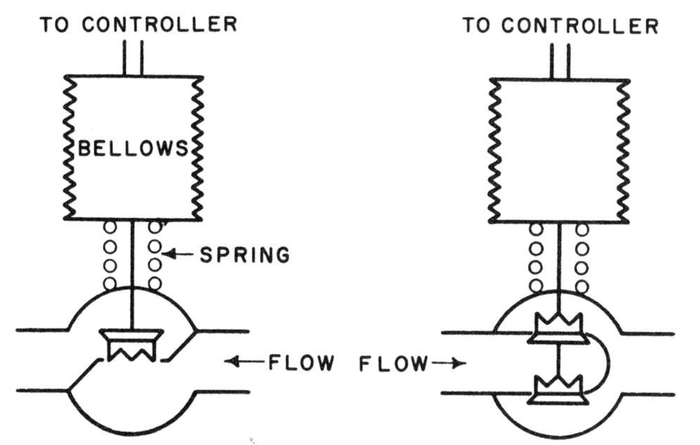

Fig. 23-25 Single-seated normally open valve

Fig. 23-26 Double-seated normally closed valve

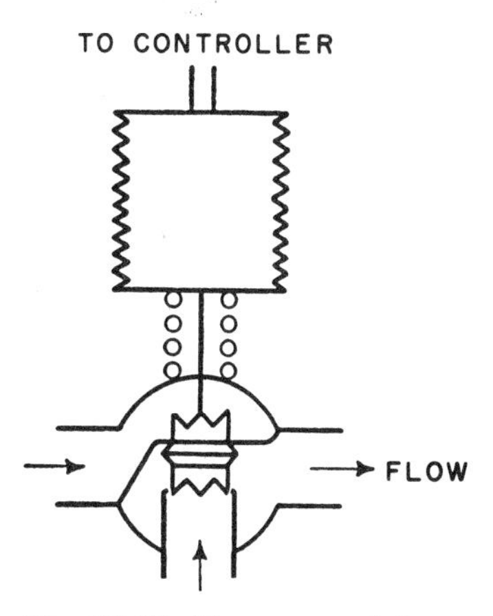

Fig. 23-27 Three-way mixing valve, normally closed to straight through flow

Normally Open Valve. A normally open valve moves toward the open position when the branch line pressure decreases, figure 23-25.

Normally Closed Valve. A normally closed valve moves toward the closed position when the branch line pressure decreases, figures 23-26 and 23-27.

In general, normally open valves are selected for heating applications and normally closed valves are selected for cooling applications.

Controllers (Thermostats). There are four types of controllers: direct acting, reverse acting, graduate acting, and positive acting.

A *direct-acting thermostat* increases the pressure of its branch line when the air temperature increases in the room. A *reverse-acting thermostat* decreases the pressure of its branch line when the air temperature in the room increases. A *graduate-acting thermostat* gradually changes its branch line pressure when the air temperature changes in the room. This type of thermostat can maintain any pressure from 0 psi to 15 psi. A *positive-acting thermostat* abruptly changes its branch line pressure when the room temperature changes. In this case, the branch line pressure is either 0 psi or 15 psi. The thermostat does not maintain pressures between these values. The thermostat has a two-position control action: either full-open or full-closed.

Humidity and pressure controllers operate in the same manner as air temperature controllers (thermostats).

Relays. Relays are installed between a controller and a controlled device (between a thermostat and a valve) to perform a function that cannot be accomplished by the controller. For example, a diverting relay is installed to do the following tasks:

- In one position, the relay can supply branch air; in the second position, it can exhaust the branch air.
- The relay can supply air to either of two branches without exhausting the other branch.
- The relay can shut off the branch air without exhausting the branch.

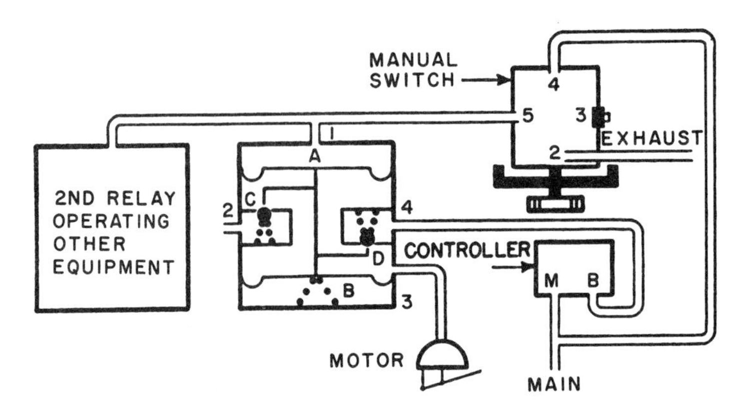

Fig. 23-28 Diagram of a single application for a diverting relay

The operation of a simple diverting relay is shown in figure 23-28. The relay is used to supply and exhaust a branch line.

- With the manual switch in position 2 (the switch makes contacts 2-3 and 4-5), the main line pressure of 15 psi forces diaphragm (A) against spring (B). As a result, valve (C) opens and allows air to exhaust from the pressure chamber and from the relay branch line connected to the motor. Valve (D) remains closed.
- With the manual switch in position 1 (switch makes contacts 2-5 and 3-4), the pressure in the pilot line is 0 psi. Spring (B) forces diaphragm (A) upward. Supply valve (D) opens and exhaust valve (C) closes. The controller (thermostat) now controls the motor through valve (D).

A differential diverting relay can make one or the other of two controllers operate the valve. A *positive diverting relay*, figure 23-29B operates around snap spring (G). The snap spring gives a positive open-closed action, rather than the inherent modulating action of pneumatic controls.

An *averaging graduate relay*, figure 23-30, maintains a branch line pressure to the motor. This pressure is equal to the average of the branch line pressures of the controllers.

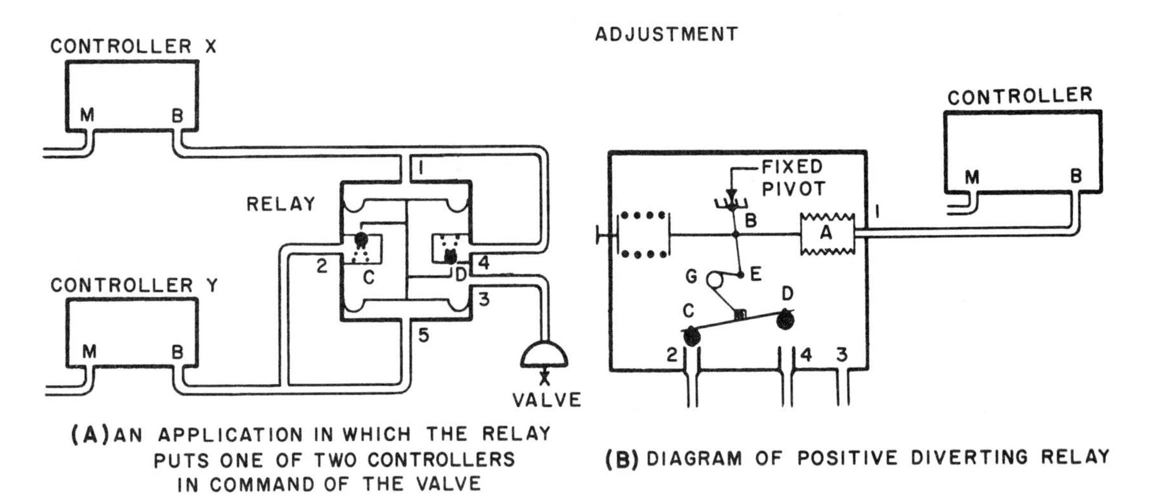

(A) AN APPLICATION IN WHICH THE RELAY PUTS ONE OF TWO CONTROLLERS IN COMMAND OF THE VALVE

(B) DIAGRAM OF POSITIVE DIVERTING RELAY

Fig. 23-29 Positive diverting relay

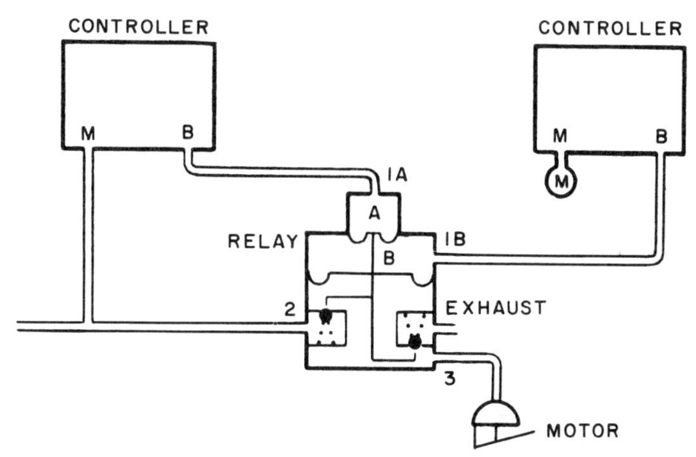

Fig. 23-30 An averaging graduate relay

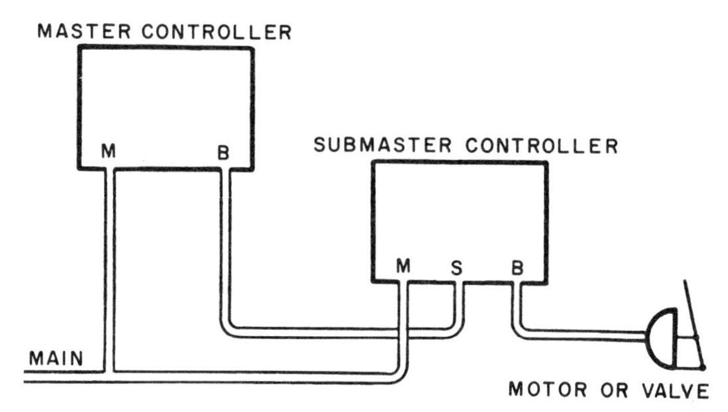

Fig. 23-31 Diagram of a compensated master-submaster control system

Master-Submaster Control. The master control, figure 23-31, regulates the control point of the submaster control. This type of system makes it possible to have compensated control. The master thermostat can be located outside where it will measure the outside temperature; the submaster control can then be located inside to measure the inside temperature. Although the master thermostat changes the control point of the submaster thermostat, the master thermostat never operates the valve or motor directly.

SUMMARY

- Air-conditioning controls can be electrical, pneumatic, or electronic in nature.
- The set point is the temperature, pressure, or humidity at which the control indicator is set.
- The control point is the temperature, pressure, or humidity recorded by the control.
- Deviation is the momentary difference between the control point and the set point.

- Corrective action is action taken to maintain the control point in reasonable agreement with the set point.
- The differential gap is the range through which a controller travels from the point at which its contacts open to the point at which they close.
- Offset is the difference between the set point and the control point.
- The primary element is that part of a control that first uses the energy resulting from a change in the controlled medium.
- Primary elements can be bimetallic strips, bellows, resistance wire, hair, wood, leather, or gold strips.
- A fully automatic control system is one that senses a change in conditions and takes action to correct the change.
- Control action is classified as either simple control or timed two-position control.
- Electrical controls use electrical energy to transmit signals.
- Low voltage is defined as 25 volts or less.
- Line voltage is 115 to 230 volts or less.
- Control circuits are installed in the following configurations:
    1. three-wire low-voltage circuit
    2. line voltage circuit
    3. two-wire low-voltage circuit
    4. two-wire line voltage circuit
- Electrical control circuits can be used on all types of heating or cooling equipment.
- The Wheatstone Bridge circuit is used as the basis for electronic control circuits.
- Check the system operation after the installation is completed.
    1. Examine the wiring for tightness
    2. Check the thermostat calibration
    3. Insure that the system starts and stops on command from the thermostat
- Centrally located control panels simplify troubleshooting because they make it possible to test all of the system circuits at a central point.
- The main circuits that should be tested are:
    1. Main power
    2. Transformer
    3. Relay Coil
    4. Cooling
    5. Damper
    6. Fan
    7. Heating

- Most troubleshooting can be done with suitable test lamps and jumper wires.
- Pneumatic controls are used principally for large commercial buildings; however, they are also suitable for small commercial and residential applications. Pneumatic controls are operated by air pressure.
- A pneumatic system includes:
  1. Compressor
  2. Copper tubing air lines
  3. Temperature, humidity, and pressure controllers
  4. Relays and switches
  5. Valves and motors
- A pneumatic motor or valve is either normally open or normally closed. If normally open, it moves toward the open position when the air pressure decreases; if normally closed, it moves toward the closed position when the air pressure decreases.
- Controllers are either direct, reverse, graduate or positive acting.
- In a master-submaster control combination, the master control changes the control point of the submaster control, but it never operates the valve or motor directly.

REVIEW

1. What are the three major types of controls that can be used for air-conditioning systems?
2. Define the following:

   Set point
   Control point
   Differential
   Offset
   Primary element
   Low voltage
   Line voltage

3. Name five primary elements or materials used in primary elements.
4. Sketch a Wheatstone Bridge circuit and identify the parts of the circuit.
5. What three steps are to be followed to check system operation after the installation is completed?
6. Name seven basic electrical circuits that are part of control systems.
7. What are the two basic tools used to pinpoint troubles in a control circuit?
8. Is it true that pneumatic controls are used only in large commercial buildings? Explain.
9. Why are pneumatic controls called pneumatic?

10. Name five parts of a pneumatic control system and describe four types of pneumatic control action achieved.

    a. Name the parts:

        (1) ______________________________

        (2) ______________________________

        (3) ______________________________

        (4) ______________________________

        (5) ______________________________

    b. Describe the control action:

        (1) ______________________________

        (2) ______________________________

        (3) ______________________________

        (4) ______________________________

SECTION 8
BALANCING THE SYSTEM

UNIT 24
BALANCING THE AIR-CONDITIONING SYSTEM

OBJECTIVES

After completing the study of this unit, the student will be able to

- define the phrase "balancing the air-conditioning system."
- list the steps in the procedure for obtaining a heating system balance.
- perform the procedure for the thermometer method of balancing the system.
- describe the procedures used in balancing the system for the following methods.
  1. velometer or anemometer method
  2. Pitot tube method
- make the necessary adjustments to balance the cooling system.

Once the air-conditioning system is installed, it must be adjusted to insure that the right amount of air is being distributed in the required spaces to accomplish the design objectives. The adjustment of the air-conditioning system is known as *balancing the system.*

The fan must be adjusted to deliver enough air at the proper velocity to provide a satisfactory heating or cooling time across the heating surface or the cooling coil. For this reason, the fan must be properly sized and adjusted to a speed that sends enough heated or cooled air to the rooms.

HEATING SYSTEM BALANCE

The speed of the fan is adjusted to the point that permits a temperature rise of approximately 90°F across the heating surface. For most equipment, this rise should never be more than 100°F. Such a limitation is necessary to prevent damage to the heat exchanger.

Procedure for Balancing

1. Obtain a temperaure reading at the return duct just before the duct enters the unit and another reading at the supply duct just after the duct leaves the unit.

2. Compare the two readings.
   a. If the difference between the readings is greater than 100°F, the fan speed should be increased.
   b. If the temperature difference is less than 85°F to 90°F, the fan speed should be decreased to allow more time for the air to pick up heat from the heating surface. The fan speed is usually adjusted by changing the position of the pulley flanges or by substituting a pulley of another size.
3. Set the fan cut-in switch at approximately 110°F. This setting can be varied depending upon the type of outlets used. If high wall outlets are used, set the cut-in point a few degrees lower than 110°F; if low wall outlets are used, set the cut-in point a few degrees higher than 110°F.
4. Set the fan cutout switch at a value approximately 10°F to 15°F lower than the cut-in temperature. For a 15°F difference between cut-in and cutout, the fan should run for long periods and stop for short periods. Continuous or almost continuous fan operation is desirable for a heating application because it tends to prevent cold drafts.
5. Set the high limit bonnet temperature control according to the manufacturer's instructions; this setting is usually 200°F or lower.
6. Adjust the supply registers and dampers.

The Thermometer Method

The thermometer method of balancing is considered to be reasonably accurate. This method is one of the simplest approaches to balancing and is suitable for residential and small commercial systems.

1. If the system is running, shut it off.
2. Place a thermometer in each room or in each area supplied by the system.
3. Place a thermometer at the thermostat. The thermometers should be located at the same level as the thermostat controlling the system.
4. Open all dampers and registers, figure 24-1.

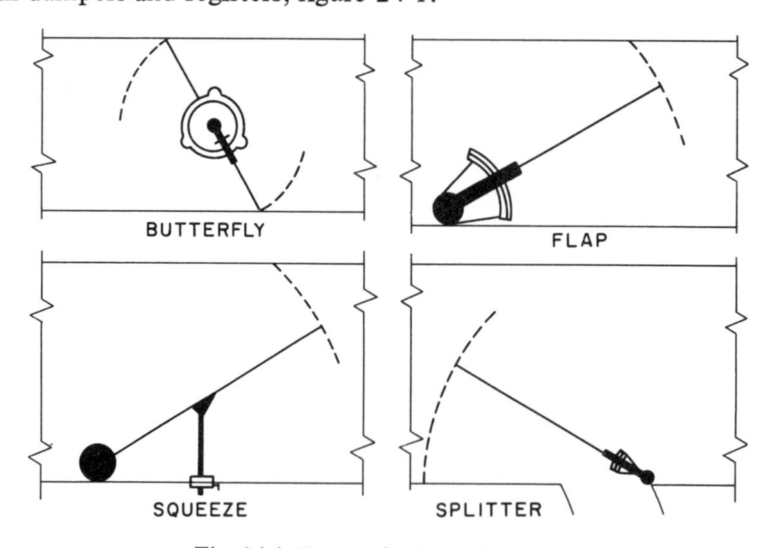

Fig. 24-1 Types of volume dampers

5. Turn the system on.
6. Allow the system to operate until the temperatures stabilize at each thermometer. (Stabilization may take 20 minutes or more.)
7. Read the temperature from each thermometer.
8. If the temperature readings in some rooms or areas are higher than the thermometer reading at the thermostat, slightly close the damper in that room or area. Closing the damper allows less heat to enter one area and permits more heat for other areas.
9. Allow the system to operate until the temperatures stabilize again. Then take a second reading at each thermometer.
10. Repeat this process until the temperature variation from room to room is no more than two or three degrees.

The Short Method

The short method of balancing a system requires the least amount of time of all methods, but should be attempted only by an experienced technician. The procedure for the short method is as follows:

1. Open the dampers in the longest duct runs to the full open position.
2. Open the dampers in the shortest duct runs to the 1/4 or 1/2 damper position, figure 24-2.
3. Open the dampers in the medium length duct runs to a position somewhere between 1/4 and full open, depending upon the number of turns in the duct run.
4. The dampers are then readjusted as necessary, according to the preferences of the owner.

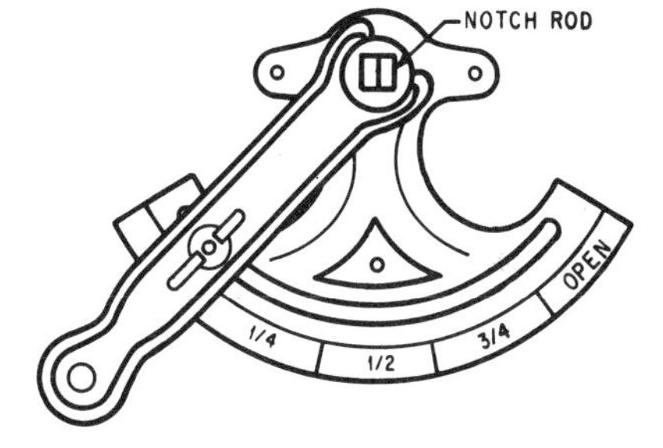

Fig. 24-2 A 3/8-inch damper quadrant

The Velometer or Anemometer Method

The velometer or anemometer method of balancing a system requires instruments. The use of the necessary instruments varies according to the instructions of the manufacturer. In general, accuracy can be achieved by taking readings with the velometer or anemometer at the return grilles. If outside air is used, the readings should be taken at the supply registers.

Procedure

1. Obtain readings with a velometer at several places across the face of the register, figure 24-4, page 306.
2. Add the velocity readings from the velometer. Determine the average reading by dividing the sum by the number of readings taken.

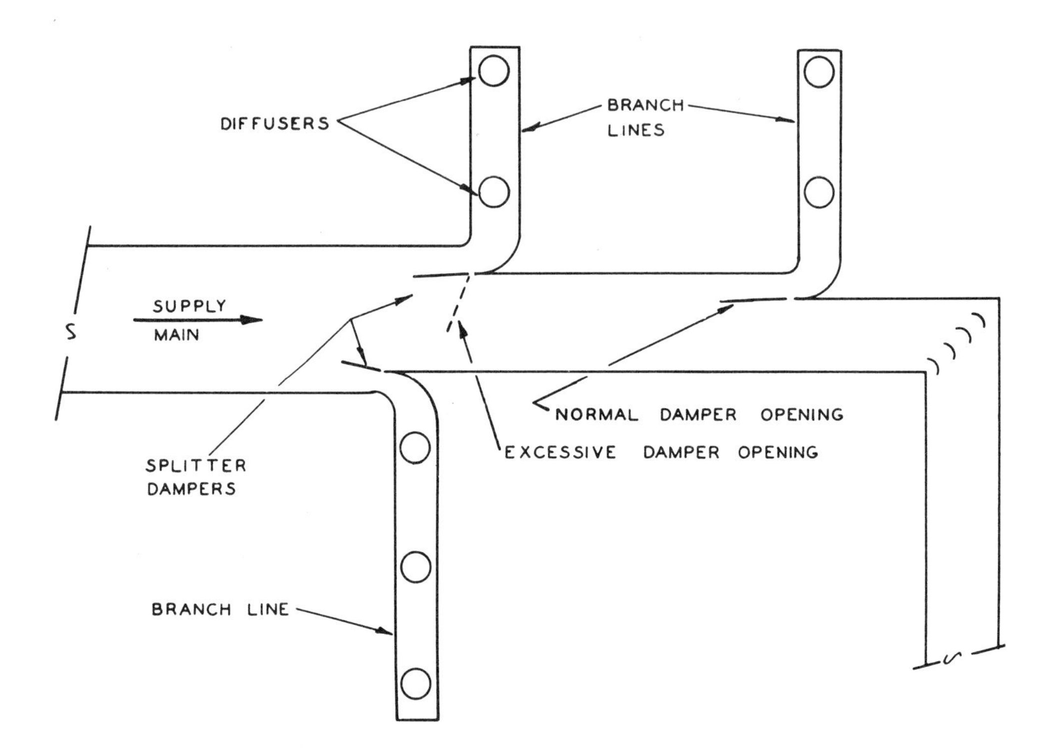

Fig. 24-3 Proper and improper settings of branch line splitter dampers

3. Multiply the average velocity reading by the effective area of the register in square feet. (The effective area is supplied usually by the register manufacturer.) The resulting value is the cubic feet of air per minute (cfm) passing through the register. The cfm value obtained should be compared with the design cfm required for the space. (The design cfm is established when the heating and cooling load is determined.)

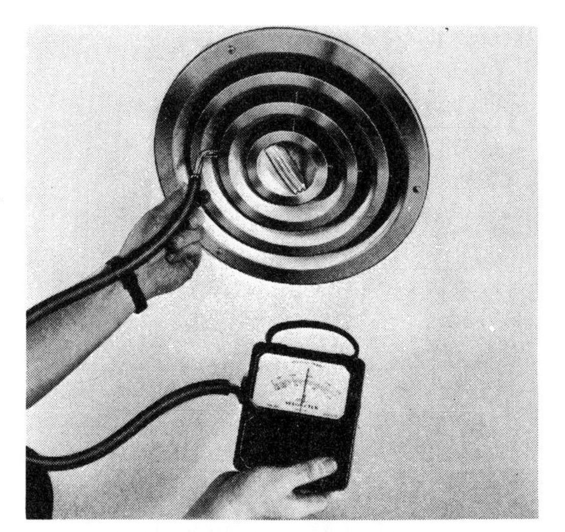

Fig. 24-4 Testing with a velometer

4. If the measured cfm is smaller than the design cfm for the specific register, change the setting toward full open for the damper in the branch run to the register. If the measured cfm is larger than the design cfm, change the damper setting toward the closed position.

5. Take a second velometer reading. Determine the cfm as in steps 2 and 3. Compare this value with the design cfm. Continue the process until the design cfm is achieved at each register.

An anemometer, figure 24-5, is used to take velocity readings across the face of a register. The results, however, do not reflect the actual air velocities since the grille bars in the face of the register deflect the airstream. To determine the cfm accurately, correction factors must be used with the anemometer.

Fig. 24-5 Testing with an anemometer

Pitot Tubes

Pitot tubes are used to balance those duct systems having extensive duct runs. To measure static pressure, velocity pressure, and total pressure in a given duct, figure 24-6, a pitot tube and a manometer are used. Pitot tube measurements are taken before and after the fan section, and at the beginning of each branch duct run. These measurements are used to determine the various fan pressures and capacities and the air quantities being supplied to each branch. The total quantities can be compared to the design quantities required at the registers in the branch. If the total air quantity is satisfactory, the registers are tested with velometers. If the branch air quantity is insufficient, the dampers in the main duct must be adjusted until sufficient air is supplied to each branch.

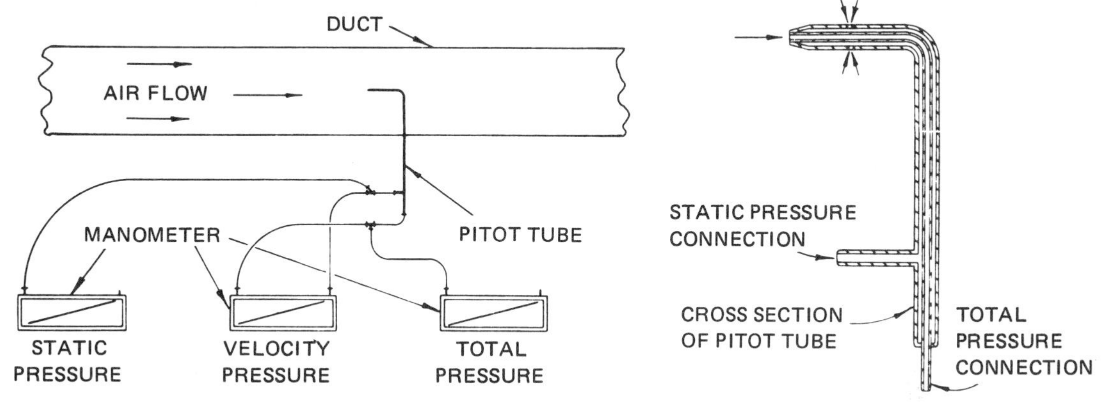

Fig. 24-6 Pitot tube and various measuring arrangements

COOLING SYSTEM BALANCE

The fan speed and damper adjustments made to balance the heating system generally are also suitable for the cooling system. If the cooling conditions are not satisfactory, or if the system is balanced initially during the cooling season, the fan should be set at the lowest speed that gives adequate cooling. The dampers and registers should be set according to any of the three methods given for balancing the heating system.

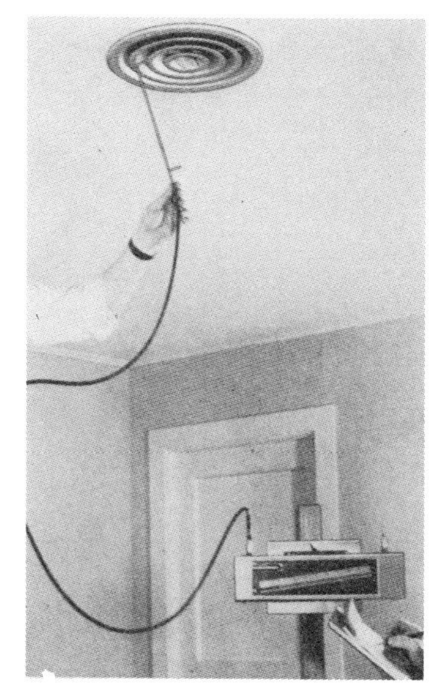

Fig. 24-7 Using Pitot tube and manometer with ceiling diffuser

Approx. Daily Temp. Range	Approx. Drop Across Coil
25 to 30 degrees	15 to 18 degrees
17 to 25 degrees	18 to 22 degrees
12 to 20 degrees	22 to 25 degrees

Table 24-1 Guide for fan speed settings

As a general guide, the fan should be set so that the temperature drop across the coil is in the ranges given in Table 24-1.

These values give approximate but not final fan speed settings. If a greater temperature drop is required across the coil to maintain satisfactory cooling conditions, the fan speed is decreased. If a smaller drop is required, the fan speed is increased.

The thermostat is set at the desired temperature, usually between 75°F and 80°F. The system is operated for approximately 24 hours before the cooling conditions in the space are determined using a sling psychrometer. The dry-bulb temperature should be between 75°F and 80°F at a relative humidity of 50%. If the humidity is higher than 55% to 60%, the volume of air should be reduced. If the humidity is lower than 40% to 45%, the volume of air should be increased. The air volume is increased or decreased by respectively increasing or decreasing the speed of the fan.

SECTION 9
THE METRIC SYSTEM IN AIR CONDITIONING

UNIT 25
METRICS FOR AIR CONDITIONING

OBJECTIVES

After completing the study of this unit, the student will be able to

- use SI metrics as required by the schedule established for the metrication of the air-conditioning industry.

In recognition of worldwide metric activity, the American Society of Heating, Refrigeration, and Air Conditioning Engineers (ASHRAE) has established a metric time schedule. The schedule calls for all ASHRAE handbooks prepared after July 1979 to use metric values only. Thus, a gradual process of metrication is underway in the air-conditioning industry.

THE METRIC SYSTEM

The metric system that has gained popularity in recent years and will eventually be the measurement system in use throughout the world is "Le Système International d' Unités." This system is known simply as the SI Metric system or SI Metrics.

SOME BASIC TERMS AND FAMILIAR EXAMPLES

This general review of common metric units and comparisons of easily recognized measurements will help the student become accustomed to the use of the metric system and will make the transition to metrics easier.

Basic Terms

SI metrics is based on the use of multiples and submultiples of the base unit of ten. Thus, SI metrics is a decimal system. The most common units are described by the prefixes milli-(0.001), centi- (0.01), deci- (0.10), and kilo-(1000). The following table shows the basic relationship between units of the British system (traditionally used in the United States) and the SI metric system.

British System			Metric System		
12 inches	=	1 foot	10 millimeters	=	1 centimeter
3 feet	=	1 yard	100 centimeters	=	1 meter
1,760 yards	=	1 mile	1000 meters	=	1 kilometer
5,280 feet	=	1 mile			

Familiar Examples

The following examples show a number of units of measurement in the British system and their closest comparable metric equivalents.

	Customary English (British) System		Metric System
Length:	1 inch	=	2.5 centimeters
	1 foot	=	30.5 centimeters
Length:	1 yard	=	0.9 meter (a yard is slightly less than a meter)
	1 mile	=	1.6 kilometer (a mile is longer than a kilometer)
Liquid:	1 pint	=	0.47 liter
	1 quart	=	0.95 liter (a quart is slightly less than a liter)
	1 gallon	=	3.76 liters
Area:	1 square foot	=	0.09 square meter
Weight:	1 ounce	=	28.35 grams
	1 pound	=	0.45 kilogram (slightly more than 2 pounds = 1 kilogram)
Volume:	1 cubic foot	=	0.028 cubic meter
Temperature:	32° Fahrenheit	=	0° Celsius

COMPARISON OF CUSTOMARY ENGLISH TERMS WITH CORRESPONDING METRIC TERMS (METRIC CONVERSION)

Several terms currently used in the air-conditioning industry belong to the metric system:

Watt is actually a metric unit of electrical energy commonly used as a customary term. The watt is used to determine the energy efficiency ratio (EER) of an air-conditioning unit: Btuh ÷ Watt = EER.

Kilowatt is a common metric term that is used by the power utilities to measure the electrical energy consumed by equipment such as lights and motors (1,000 watts = 1 kilowatt).

Hertz is a metric term that identifies electric current characteristics (60 hertz is the same as 60 cycles).

It can be seen that a number of metric units have been in use for many years. Table 25-1 is an expanded list of customary units versus metric terms as they apply to the air-conditioning industry. In some instances, several metric units are equivalent to one customary unit (see the bracketed items). In other instances, the same metric unit may apply to several customary units.

	CUSTOMARY ENGLISH		METRIC	
CATEGORY	SYMBOL	UNIT OF MEASUREMENT	SYMBOL	UNIT OF MEASUREMENT
AREA	in.^{2}	square inch	cm^2	square centimeter
			m^2	square meter
	ft.^{2}	square feet	m^2	square meter
	yd.^{2}	square yard	m^2	square meter
ENERGY	Btu	British Thermal Unit	W	watt
			J	joule
			kJ	kilojoule
	H.p.	horsepower	kW	kilowatt
	Tons Refrig.	Tons Refrigeration	kW	kilowatt
ENERGY/TIME	Btuh	British Thermal Units /hr.	kWh	kilowatt-hour
			W	watts
FREQUENCY	Cycle	Cycle	Hz	hertz
LENGTH	in.	inch	mm	millimeter
			cm	centimeter
			m	meter
	ft.	feet	m	meter
	yd.	yard	m	meter
	mile	mile	m	meter
			km	kilometer
MASS	oz.	ounce	g	gram
	lb.	pound	kg	kilogram
	ton	ton	kg	kilogram
	grain	grain	kg	kilogram
PRESSURE	psi	pounds per square inch	kPa	kilopascal
	in. w.g.	inches water gauge	kPa (w.g.)	kilopascal
	in. Hg.	inches mercury	kPa (Hg)	kilopascal
TEMPERATURE	°F	degree Fahrenheit	°C	degree Celsius
TIME	min.	minute	s	second
	hr.	hour	s	second
	day	day	s	second
VELOCITY	fpm	feet per minute	m/s	meters per second
	mph	miles per hour	m/s	meters per second
			km/hr.	kilometer per hour
VOLUME	ft.^{3}	cubic feet	m^3	cubic meter
	yd.^{3}	cubic yard	m^3	cubic meter
	pt.	pint	1	liter
	qt	quart	1	liter
	gal.	gallon	1	liter
VOLUME & TIME				
Air:	cfm	cubic feet per minute	m^3/s	cubic meters per second
			m^3/hr.	cubic meters per hour
	fpm	feet per minute	m/s	meters per second
Liquid:	cfm	cubic feet per minute	1/s	liters per second
	gpm	gallons per minute	1/s	liters per second

Table 25-1 Comparison of customary British system terms to corresponding metric system terms

Table 25-2 lists the conversion factors that are multiplied by the customary unit of measurement to obtain the corresponding metric unit of measure. Many of the English units that appear in Table 25-1 are converted in this table to their equivalent metric values.

CATEGORY	CUSTOMARY ENGLISH UNIT (SYMBOL)	x	CONVERSION FACTOR	=	METRIC UNIT (SYMBOL)
AREA	in.^{2}	x	6.5	=	cm^2
		x	0.0065	=	m^2
	ft.^{2}	x	0.093	=	m^2
	yd.^{2}	x	0.84	=	m^2
ENERGY	Btu	x	1055	=	J
		x	1.055	=	kJ
	H.p.	x	0.746	=	kW
	Tons Refrig.	x	3.51	=	kW
ENERGY/TIME	Btu/hr.	x	0.000293	=	kWh
		x	0.293	=	Wh
LENGTH	in.	x	25.4	=	mm
		x	2.5	=	cm
		x	0.025	=	m
	ft.	x	0.30	=	m
	yd.	x	0.9	=	m
	mi.	x	1.6	=	km
MASS	oz.	x	28.35	=	g
	lb.	x	0.453	=	kg
	ton (2000 lb.)	x	907.	=	kg
	grain	x	0.0000648	=	kg
		x	0.0648	=	g
PRESSURE	psi	x	6.9	=	kPa
	in. w.g.	x	0.25	=	kPa (w.g.)
	in. Hg	x	3.4	=	kPa (Hg)
TEMPERATURE	°F (-32°)	x	5/9 (or 1.8)	=	°C
TIME	min.	x	60	=	s
	hr.	x	3600	=	s
	day	x	86400	=	s
VELOCITY	fpm	x	0.00508	=	m/s
	mph	x	1.6	=	km/hr.
		x	0.45	=	m/s
VOLUME	ft.^{3}	x	0.03	=	m^3
	yd.^{3}	x	0.76	=	m^3
	pt.	x	0.47	=	1
	qt.	x	0.95	=	1
	gal.	x	3.8	=	1
VOLUME & TIME					
Air:	cfm	x	0.0005 (0.000472)	=	m^3/s
		x	1.7	=	m^3/hr.
Liquid:	cfm	x	0.472	=	1/s
	gpm	x	0.06	=	1/s

Table 25-2 Conversion factors

SUMMARY

- A conversion to SI metrics is gradually taking place in the air-conditioning industry.
- SI metrics is the most popular metric system.
- SI metrics is derived from the French "Le Système International d' Unités."
- Several metric units are used in the customary English system of measurement: watt, kilowatt, hertz.
- Comparison of terms and conversion factors are provided for the following categories of measurement:

area	pressure
energy	temperature
energy/time	velocity
frequency	volume
length	volume/time
mass	

REVIEW

1. Name the most popular metric system.
2. What is the numerical expression for:

   a) milli-
   b) centi-
   c) deci-
   d) kilo-

3. Inches, feet, yards, and miles can be converted to what basic unit in the metric system?
4. One inch = _?_ cm.
5. One yard = _?_ m.
6. 32°F = _?_ °C.
7. One inch = _?_ mm.
8. One inch = _?_ m.
9. Twelve inches = _?_ m.
10. Thirty-six inches = _?_ m.

APPENDIX

TABLE A U-FACTORS FOR MASONRY WALLS

[In Btu per (hr.) (ft.^{2}) (°F) difference between the air on the two sides]

	Thickness of masonry, in.	Interior finish and insulation (if indicated)			
		Plain wall, no interior finish	Metal lath and plaster, furred	Gypsum lath (3/8 in.) and plaster, furred	Gypsum lath and plaster plus 1-in. blanket insulation, furred
Solid brick	8	0.50	0.32	0.30	0.14
	12	0.35	0.25	0.24	0.13
	16	0.28	0.21	0.20	0.12
Stone	8	0.70	0.39	0.36	0.16
	12	0.57	0.35	0.33	0.15
	16	0.49	0.32	0.30	0.14
Poured concrete	6	0.79	0.42	0.39	0.16
	8	0.70	0.39	0.36	0.16
	12	0.58	0.35	0.33	0.15
Hollow concrete blocks	Sand and gravel aggregate				
	8	0.56	0.34	0.32	0.15
	12	0.50	0.32	0.30	0.14
	Cinder aggregate				
	8	0.41	0.28	0.27	0.13
	12	0.38	0.26	0.25	0.13

(From ASHRAE Guide, 1965, by permission)

TABLE B U-FACTORS FOR FRAME WALLS

Diagram of wall and exterior finish	Interior finish	Type of sheathing					
		Plywood 5/16 in.		Wood 25/32 in.		Insulating board 25/32 in.	
		No insulation	3-in. insulation	No insulation	3-in. insulation	No insulation	3-in. insulation
Redwood siding	Gypsum lath (3/8 in.) and plaster	0.31	0.081	0.25	0.076	0.19	0.069
	Metal lath and plaster	0.32	0.082	0.26	0.077	0.20	0.07
	Plywood, or wood paneling (3/8 in.)	0.30	0.08	0.24	0.076	0.19	0.069
	Wood lath and plaster	0.31	0.081	0.25	0.076	0.19	0.069
Stucco	Gypsum lath (3/8 in.) and plaster	0.39	0.087	0.30	0.08	0.22	0.073
	Metal lath and plaster	0.42	0.088	0.32	0.082	0.23	0.073
	Plywood or wood paneling (3/8 in.)	0.38	0.087	0.29	0.08	0.22	0.073
	Wood lath and plaster	0.39	0.087	0.30	0.08	0.22	0.073
Brick veneer	Gypsum lath (3/8 in.) and plaster	0.34	0.083	0.27	0.078	0.20	0.07
	Metal lath and plaster	0.36	0.085	0.28	0.078	0.21	0.072
	Plywood or wood paneling (3/8 in.)	0.33	0.083	0.27	0.078	0.20	0.07
	Wood lath and plaster	0.34	0.083	0.27	0.078	0.20	0.07

(From ASHRAE Guide, 1965, by permission)

TABLE C U-FACTORS FOR PARTITIONS AND INTERIOR WALLS

[In Btu per (hr.) (ft.^{2}) (°F) difference between the air on the two sides]

Type of interior finish	Single partition (finish one side only)	Double partition (finish both sides): No insulation between studs	Double partition (finish both sides): 1-in. blanket insulation between studs
Gypsum lath (3/8 in.) and plaster	0.61	0.34	0.15
Metal lath and plaster	0.69	0.39	0.16
Plywood or wood paneling (3/8 in.)	0.59	0.33	0.15
Wood lath and plaster	0.62	0.34	0.15
Gypsum board (3/8 in.) decorated	0.67	0.37	0.16

(From ASHRAE Guide, 1965, by permission)

TABLE D U-FACTORS FOR CEILINGS AND FLOORS

[In Btu per (hr.) (ft.^{2}) (°F) difference between the air on the two sides]

	No insulation	3-in. mineral wool blanket between joists	4-in. blown mineral wool
Ceilings:			
Plaster on 3/8 in. gypsum board, on wood joists	0.61	0.10	0.08
Plaster on metal lath	0.74	0.11	0.07
Wood panel (25/32 in.)	0.34	0.08	0.06
Floors:	Over crawl space or basement		
Hardwood on 25/32-in. subfloor on wood joists	0.37		
Carpet on 1-in. plywood on wood joists	0.30		
Linoleum or asphalt tile on 1-in. plywood on wood joists	0.57		

(From ASHRAE Guide, 1965, by permission)

TABLE E U-FACTORS FOR FLAT ROOFS

[In Btu per (hr.) (sq. ft.) (°F) difference between the air on the two sides]
All numbers in parentheses indicate weight per sq. ft.

TYPE OF DECK	THICKNESS OF DECK (inches) and WEIGHT (lb. per sq. ft.)	CEILING†	No Insulation	1/2 (1)	1 (1)	1 1/2 (2)	2 (3)	2 1/2 (3)	3 (4)
			INSULATION ON TOP OF DECK, INCHES						
Flat Metal	1 (5)	None or Plaster (6)	.67	.35	.23	.18	.15	.12	.10
		Suspended Plaster (5)	.32	.22	.17	.14	.12	.10	.09
		Suspended Acou Tile (2)	.23	.18	.14	.12	.11	.09	.08
Preformed Slabs – Wood Fiber and Cement Binder	2 (4)	None or Plaster (6)	.20	.16	.13	.11	.10	.09	.08
		Suspended Plaster (5)	.15	.12	.11	.09	.08	.08	.07
		Suspended Acou Tile (2)	.13	.10	.09	.08	.08	.07	.06
	3 (7)	None or Plaster (6)	.14	.11	.10	.09	.08	.08	.07
		Suspended Plaster (5)	.12	.10	.09	.07	.07	.06	.05
		Suspended Acou Tile (2)	.10	.09	.08	.07	.07	.06	.05
Concrete (Sand & Gravel Agg)	4, 6, 8 (47), (70), (93)	None or Plaster (6)	.51	.30	.21	.16	.14	.12	.10
		Suspended Plaster (5)	.28	.20	.16	.13	.12	.10	.09
		Suspended Acou Tile (2)	.21	.16	.13	.11	.10	.09	.08
(Lt. Wt. Agg on Gypsum Board)	2 (9)	None or Plaster (6)	.27	.20	.15	.13	.11	.10	.08
		Suspended Plaster (5)	.18	.14	.12	.10	.09	.09	.08
		Suspended Acou Tile (2)	.15	.12	.11	.09	.08	.08	.07
	3 (13)	None or Plaster (6)	.21	.16	.13	.11	.10	.09	.08
		Suspended Plaster (5)	.15	.12	.11	.09	.08	.08	.07
		Suspended Acou Tile (2)	.13	.11	.10	.08	.08	.07	.06
	4 (16)	None or Plaster (6)	.17	.14	.11	.10	.09	.08	.07
		Suspended Plaster (5)	.13	.11	.10	.08	.08	.07	.06
		Suspended Acou Tile (2)	.12	.10	.09	.07	.07	.06	.05
Gypsum Slab on 1/2" Gypsum Board	2 (11)	None or Plaster (6)	.32	.22	.17	.14	.12	.10	.09
		Suspended Plaster (5)	.21	.17	.13	.11	.10	.09	.08
		Suspended Acou Tile (2)	.17	.13	.12	.10	.09	.08	.07
	3 (15)	None or Plaster (6)	.27	.19	.15	.13	.11	.10	.08
		Suspended Plaster (5)	.19	.15	.13	.11	.10	.09	.08
		Suspended Acou Tile (2)	.15	.12	.11	.09	.08	.08	.07
	4 (19)	None or Plaster (6)	.23	.17	.14	.12	.10	.09	.08
		Suspended Plaster (5)	.17	.13	.12	.10	.09	.08	.07
		Suspended Acou Tile (2)	.14	.12	.11	.09	.08	.08	.07
Wood	1 (3)	None or Plaster (6)	.40	.26	.19	.15	.13	.11	.09
		Suspended Plaster (5)	.24	.18	.14	.12	.11	.09	.08
		Suspended Acou Tile (2)	.19	.15	.13	.11	.10	.08	.07
	2 (5)	None or Plaster (6)	.28	.20	.16	.13	.11	.10	.08
		Suspended Plaster (5)	.19	.15	.13	.11	.10	.09	.07
		Suspended Acou Tile (2)	.16	.13	.11	.10	.09	.08	.07
	3 (8)	None or Plaster (6)	.21	.16	.13	.11	.10	.09	.08
		Suspended Plaster (5)	.16	.13	.11	.09	.09	.08	.07
		Suspended Acou Tile (2)	.13	.11	.10	.09	.08	.07	.06

(From ASHRAE Guide, 1965, by permission)

TABLE F U-FACTORS FOR PITCHED ROOFS

[In Btu per (hr.) (sq. ft. projected area) (°F) difference between the air on the two sides]
All numbers in parentheses indicate weight per sq. ft.

PITCHED ROOFS (SHEATHING, EXTERIOR SURFACE, CEILING)		CEILING										
		None		3/8" Gypsum	Metal Lath Plastered		3/8" Gypsum or Wood Lath Plastered		Insulating Board Plain or 1/2" Sand Agg Plastered		Acoustical Tile on Furring or 3/8" Gypsum	
Exterior Surface	Sheathing		3/4" Wood Panel *(2)*	Board (Plaster Board) *(2)*	3/4" Sand Plaster *(7)*	3/4" Lt Wt. Plaster *(3)*	1/2" Sand Plaster *(5)*	1/2" Lt Wt Plaster *(2)*	1/2" Board *(2)*	1" Board *(4)*	1/2" Tile *(2)*	3/4" Tile *(3)*
Asphalt Shingles, *(2)*	Bldg paper on 5/16" plywood *(2)*	.51	.27	.30	.32	.29	.29	.28	.22	.17	.23	.21
	Bldg paper on 25/32" wood sheathing *(3)*	.30	.23	.26	.27	.25	.25	.24	.20	.16	.21	.19
Asbestos-Cement Shingles *(3)* or Asphalt Roll Roofing *(1)*	Bldg paper on 5/16" plywood *(2)*	.59	.28	.34	.37	.33	.33	.31	.25	.18	.25	.22
	Bldg paper on 25/32" wood sheathing *(3)*	.45	.25	.29	.31	.28	.28	.27	.22	.17	.22	.20
Slates *(8)* Tile *(10)* or Sheet Metal *(1)*	Bldg paper on 5/16" plywood *(2)*	.64	.29	.36	.38	.34	.35	.47	.26	.19	.26	.23
	Bldg paper on 25/32" wood sheathing *(3)*	.48	.25	.29	.31	.28	.28	.27	.22	.17	.23	.20
Wood Shingles *(2)*	Bldg paper on 1" x 4" strips *(1)*	.53	.26	.31	.33	.30	.30	.28	.23	.17	.24	.21
	Bldg paper on 5/16" plywood *(2)*	.41	.23	.27	.29	.26	.27	.25	.21	.16	.21	.19
	Bldg paper on 25/32" wood sheathing *(3)*	.34	.21	.24	.25	.23	.23	.22	.19	.15	.19	.17

(From ASHRAE Guide, 1965, by permission)

MECHANICAL SERVICE ANALYSIS

Compressor Will Not Start

CAUSE	SYMPTOMS	REMEDY
Thermostat setting is too high.	Thermostat setting is above the room temperature.	Reset thermostat below room temperature.
High head pressure.	Starter overload cuts out.	Reset starter overload and determine cause of high head pressure.
Defective pressure switch.	Pressure switch contacts remain open regardless of pressure.	Repair or replace pressure switch.
Loss of refrigerant charge.	Pressure switch contacts open.	Check system for leaks; repair leaks and recharge system.
Compressor is frozen.	Electrical system in operating condition with adequate voltage at the compressor.	Repair or replace compressor.

Compressor Short-Cycles

CAUSE	SYMPTOMS	REMEDY
Defective thermostat.	Thermostat differential too close.	Replace thermostat.
Incorrect setting of low-pressure side of pressure switch.	Compressor cycling on low-pressure switch.	Reset low-pressure switch differential.
Low refrigerant charge.	Compressor cycling on low-pressure switch.	Check system for leaks; repair leaks and add refrigerant.
Defective overload.	Compressor cycling on overload.	Replace overload.
Dirty or iced evaporator.	Compressor cycling on low-pressure switch.	Clean or defrost evaporator.
Evaporator blower and motor belts slipping.	Compressor cycling on low-pressure switch.	Tighten or replace belts.
Dirty or plugged air filters.	Compressor cycling on low-pressure switch.	Clean or replace air filters.

Compressor Runs Continuously

CAUSE	SYMPTOMS	REMEDY
Excessive load.	High dry-bulb or wet-bulb temperature in conditioned area.	Check for excessive outside air, infiltration, and excessive source of moisture.
Air or noncondensable gases in the system.	Higher than normal head pressure.	Purge system.
Thermostat setting too low.	Lower than normal temperature in conditioned area.	Reset thermostat.
Dirty condenser.	Higher than normal head pressure.	Clean condenser.
Condenser blower and motor belts slipping.	Higher than normal head pressure.	Tighten or replace belts.
Low refrigerant charge.	Lower than normal suction pressure.	Check system for leaks; repair leaks and add refrigerant.
Overcharge of refrigerant.	Higher than normal head pressure.	Purge and remove excess refrigerant.
Compressor valves leaking.	Pressures equalize rapidly when the system is turned off.	Replace valve plate assembly or the complete hermetic compressor.
Expansion valve or strainer plugged.	Lower than normal suction pressure.	Clean expansion valve or strainer.

System Short of Capacity

CAUSE	SYMPTOMS	REMEDY
Low refrigerant charge.	Lower than normal head and suction pressures.	Check system for leaks; repair leaks and add refrigerant.
Incorrect superheat setting of the expansion valve.	Lower than normal suction pressure.	Adjust superheat setting to 10°F.
Defective expansion valve.	Lower than normal suction pressure.	Repair or replace expansion valve.
Air or noncondensable gases in the system.	Higher than normal head pressure.	Purge system.
Dirty condenser.	Higher than normal head pressure.	Clean condenser.
Condenser blower and motor belts slipping.	Higher than normal head pressure.	Tighten or replace belts.
Overcharge of refrigerant.	Higher than normal head pressure.	Purge and remove excess refrigerant.
Compressor valves leaking.	Pressures equalize rapidly when the system is turned off.	Replace valve plate assembly or the complete hermetic compressor.
Expansion valve or strainer plugged.	Lower than normal suction pressure.	Clean expansion valve or strainer.
Condenser air short circuiting.	Higher than normal head pressure.	Remove obstructions or causes of short circuiting air.

Head Pressure Too High

CAUSE	SYMPTOMS	REMEDY
Overcharge of refrigerant.	Higher than normal head pressure.	Purge or remove excess refrigerant.
Air or noncondensable gases in the system.	Higher than normal head pressure.	Purge system.
Dirty condenser.	Higher than normal head pressure.	Clean condenser.
Condenser blower and motor belts slipping.	Higher than normal head pressure.	Tighten or replace belts.
Condenser air short circuiting.	Higher than normal head pressure.	Remove obstructions or causes of short circuiting air.

Head Pressure Too Low

CAUSE	SYMPTOMS	REMEDY
Low refrigerant charge.	Sight glass indicates bubbles or liquid level valve on receiver indicates shortage of refrigerant.	Check system for leaks; repair leaks and add refrigerant.
Compressor valves leaking.	Lower than normal head pressures and pressures equalize rapidly when system is turned off.	Replace valve plate assembly or the complete hermetic compressor.

Suction Pressure Too High

CAUSE	SYMPTOMS	REMEDY
Excessive load on system.	Compressor runs continuously and capacity is low.	Remove conditions causing excessive load.
Expansion valve is stuck in open position.	Lower than normal head pressure.	Repair or replace expansion valve.
Incorrect superheat setting of the expansion valve.	Lower than normal head pressure.	Adjust superheat setting to 10°F.

Suction Pressure Too Low

CAUSE	SYMPTOMS	REMEDY
Low refrigerant charge.	Sight glass indicates bubbles or liquid level valve on receiver in-indicates shortage of refrigerant.	Check system for leaks; repair leaks and add refrigerant.
Expansion valve or strainer plugged.	Suction line is warm, expansion valve or strainer may be showing frost, and system capacity is low.	Clean expansion valve or strainer.
Incorrect superheat setting of the expansion valve.	Suction line is warm and system capacity is low.	Adjust superheat setting to 10°F.
Evaporator air volume low.	Abnormally cold suction line and low suction pressure.	Increase air over the evaporator.
Stratification of cool air in conditioned area.	System capacity is low and temperature of return air is low.	Increase air velocity through supply grilles.

Compressor is Noisy

CAUSE	SYMPTOMS	REMEDY
Worn or scored compressor bearings.	Noticeable knock in compressor.	Replace the compressor.
Expansion valve is stuck in open position or is defective.	Abnormally cold suction line, high suction pressure, and lower than normal head pressure.	Repair or replace expansion valve.
Overcharge of refrigerant or air and noncondensables in system.	Higher than normal head pressure.	Purge system.
Overcharge of oil.	Oil sight glass in compressor is completely filled during operation.	Remove excess oil.
Liquid refrigerant flooding back to compressor.	Abnormally cold suction line and high suction pressure.	Repair or replace expansion valve.
Shipping or holddown bolts not loosened or removed.	Noticeable transmission of vibration from compressor to rest of unit; compressor held firmly in mounting.	Loosen compressor holddown bolts so compressor is freely floating in mountings.
Lack of oil.	Oil level below midpoint of the oil sight glass in compressor during operation.	Add oil.
Broken compressor valves.	Rapid equalization of the pressures when compressor stops.	Replace valve plate assembly or the complete hermetic compressor.

Compressor Loses Oil

CAUSE	SYMPTOMS	REMEDY
Incorrect superheat setting of the expansion valve.	Visual inspection of suction line indicates long trapped portions of line.	Rerun trapped portion of suction line; locate traps as recommended in installation instructions.
Leaks in system.	Presence of oil at piping joints or connections.	Repair leaks and add refrigerant and oil as required.
Shortage of refrigerant.	Lower than normal suction pressure. Sight glass indicates bubbles or liquid level valve on receiver indicates shortage of refrigerant.	Check system for leaks; repair leaks and add refrigerant.
Expansion valve or strainer plugged.	Lower than normal suction pressure.	Adjust superheat setting to 10°F.

ELECTRICAL SERVICE ANALYSIS

Compressor and Condenser Fan Motor Will Not Start

CAUSE	SYMPTOMS	REMEDY
Power failure.	Test lamp or voltmeter indicates no voltage at disconnect switch.	Call power company.
Fuse blown.	Test lamp or voltmeter shows voltage on line side of disconnect switch but not on unit side.	Replace blown or defective fuse.
Thermostat setting too high.	Thermostat setting above the room temperature.	Reduce temperature setting of the thermostat.
Defective thermostat.	Thermostat contacts do not make when setting is below room temperature.	Repair or replace the thermostat.
Faulty wiring.	Compressor contactor or starter and fan relay do not become energized.	Check wiring and make necessary repairs.
Defective controls.	Compressor contactor or starter and fan relay do not become energized.	Check and replace defective controls.
Low voltage.	Starter overload tripped.	Reset and check for cause of tripping.
Defective dual pressure control.	Dual pressure control contacts remain in open position.	Replace the control.

Compressor Will Not Start, But Condenser Fan Runs

CAUSE	SYMPTOMS	REMEDY
Faulty wiring to compressor.	Test lamp or voltmeter indicates no voltage at compressor.	Check compressor wiring and repair.
Defective compressor motor.	Voltage is available at compressor; but an open winding, ground or stuck compressor prevents operation.	Replace the compressor.
Defective compressor overload (single phase only).	Overload contacts remain in open position.	Replace overload.
Defective starting capacitor. (single phase only).	Starting capacitor does not indicate rated capacitance.	Replace capacitor.

Condenser Fan Motor Will Not Start, But Compressor Runs

CAUSE	SYMPTOMS	REMEDY
Defective fan relay.	Relay contacts do not make when coil is energized.	Repair or replace relay(s).
Faulty wiring to fan motor.	Test lamp or voltmeter indicates no voltage at fan motor.	Check fan motor wiring and repair.
Defective fan motor.	Test lamp or voltmeter indicates voltage available at motor.	Replace fan motor.

Condenser Fan Motor Runs, But Compressor Hums and Will Not Start

CAUSE	SYMPTOMS	REMEDY
Low voltage.	Test lamp or voltmeter indicates inadequate voltage at compressor.	Check line voltage; determine the location of the voltage drop.
Faulty wiring.	Test lamp or voltmeter indicates inadequate voltage at compressor.	Check wiring and make necessary repairs.
Defective compressor.	Test lamp or voltmeter indicates adequate voltage available at compressor.	Replace compressor.
High head pressure.	Higher than normal head pressure for existing conditions.	Check head pressure and complete operation of system to remove the cause of high-pressure condition.
Failure of one phase (three phase only).	Test lamp or voltmeter indicates no voltage from one phase to ground.	Check fuses and wiring.
Defective start capacitor (single phase only).	Starting capacitor does not indicate rated capacitance.	Replace capacitor.
Defective potential relay (single phase only).	Potential relay contacts do not close for starting.	Replace potential relay.

Compressor Starts, But Cycles on Overload

CAUSE	SYMPTOMS	REMEDY
Low voltage.	Test lamp or voltmeter indicates inadequate voltage at compressor.	Check line voltage; determine the location of the voltage drop.
Faulty wiring.	Test lamp or voltmeter indicates inadequate voltage at compressor.	Check wiring and make necessary repairs.
Defective running capacitor (single phase only).	Running capacitor completely dead.	Replace the capacitor.
Defective overload.	Overload is breaking contact under normal operating conditions.	Replace overload (klixon).
Unbalanced line (three phase only).	Test lamp or voltmeter indicates unequal phase voltages.	Check wiring; call power company.

Evaporator Fan Motor Will Not Start

CAUSE	SYMPTOMS	REMEDY
Power failure.	Test lamp or voltmeter indicates no voltage at disconnect switch.	Call power company.
Fuse blown.	Test lamp or voltmeter shows voltage on line side of disconnect switch but not on unit side.	Replace blown or defective fuse.
Faulty wiring.	Evaporator fan relay(s) do not become energized.	Check wiring and make necessary repairs.
Defective fan motor.	Test lamp or voltmeter indicates voltage available at motor.	Replace fan motor.
Defective fan relay.	Relay contacts do not make when coil is energized.	Repair or replace relay(s).

INDEX

P

R

S

THIS PACKET CONTAINS ➡

- Psychrometric Chart for Units 6, 7, 8, 9, 16, and 17.
- A R I Residential Cooling Load Estimate Form for Unit 12
- A R I Commercial Cooling Load Estimate Form for Unit 12
- Residential Plans for Unit 16
- Commercial Estimate (Completed) for Restaurant Unit 17

580(9C1435)